ANGÉNIEUX AND CINEMA

ANGÉNIEUX AND CINEMA

From Light to Image

SilvanaEditoriale

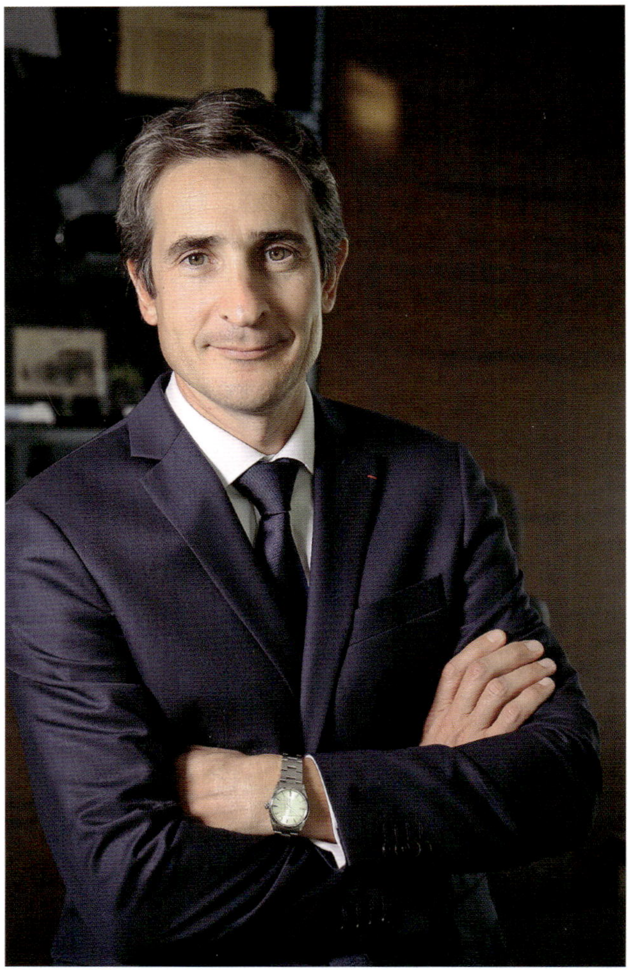

Photo by Pascal Guittet / Usine Nouvelle

We are proud to count Angénieux amongst our jewels.
Taken over by Thomson CSF in 1993, Angénieux is today one of the Group's leading optical centers by virtue of an expertise accumulated in over eighty years at the service of the optical image, a perfect example of pragmatic and inventive French excellence and the starting point of an extraordinary industrial saga.
This excellence is first and foremost the work of an outstanding man, Pierre Angénieux, whose genius goes beyond being a visionary calculator of optical systems. He also succeeded in surrounding himself with the best possible teams and imbuing them not only with a culture of learning and transmitting knowledge and skills but also and above all with his passion for cinema. Angénieux has supplied filmmakers since the 1950s with zooms of exceptional performance, authentic technological gems inseparably bound up with the history of cinema.
Angénieux has played a key part not only in filmmaking but also in photography and television. The Angénieux brand has provided optical systems for the world of medicine, the defense sector and even the conquest of space. The incredible success of the Apollo 11 mission in July 1969 was witnessed in images partly filmed by means of Angénieux optics on all family home television sets.
Since 1993, Angénieux has found new synergies within the Thales group – the plant in Saint-Héand also designs and produces night vision binoculars for elite military personnel and infantry as well as systems for the pilots of helicopters and fighter aircraft like the Rafale – and passion for cinema is still intact.
Angénieux can be proud that this passion is recognised in the highest spheres of the cinema world particularly by those whose opinion really matters its customers and end users, namely directors of photography and other technicians of image production.
The Scientific and Engineering Award received in 2009 was particularly important in that regard, confirming the excellence of our Optimo 15-40 and 28-76 zooms and demonstrating that Oscars can also be won by engineers.
Angénieux can feel pride and face the future with confidence. We have succeeded in making the switch to digital technology and are ready to tackle the great new challenges of the world of images. In a sector of ever-increasing standards, we will continue to assist directors of photography and provide them with the innovative and reliable optical systems required for the exercise of their art.
When high technology and industry, exceptional manual skills and a love of work well done all combine, everything becomes possible. It is this, among so many other things, that makes the company essential to our group. Angénieux stilll has plenty of surprises up its sleeve.

Patrice Caine
President and chief executive of the Thales group

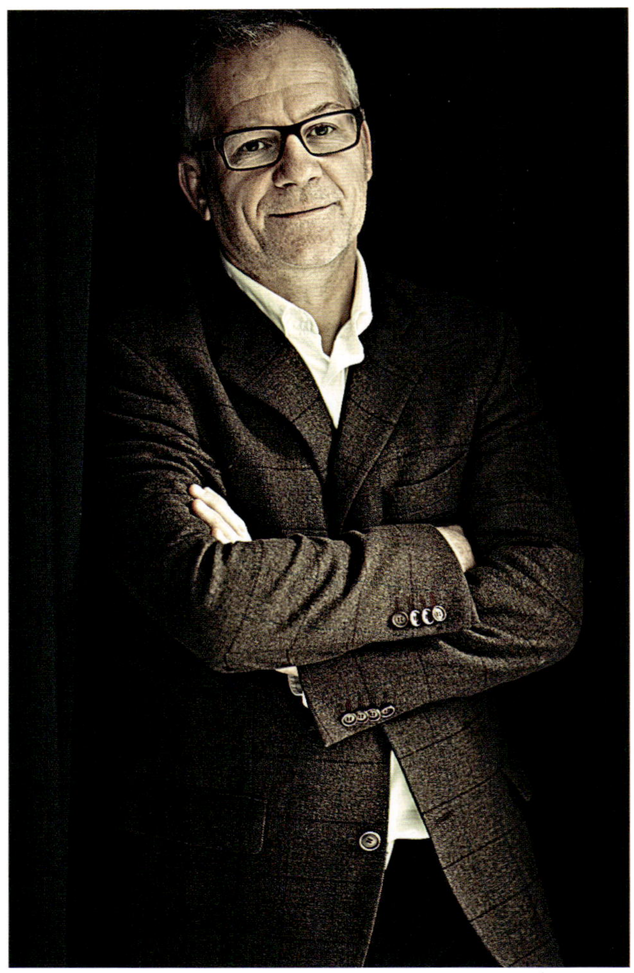

Photo by Marcel Hartmann

It is a great pleasure to write a preface for the splendid book you are now holding. Angénieux is a jewel of French industry as well as a faithful partner of the Cannes Film Festival. The company is also based near Lyon, the home of the Lumière brothers, where one can still feel the same emotions, thinking that this is where Cinema was born in 1895, on the street now known as Rue du Premier Film.

In 2013, when we decided to create the Pierre Angénieux ExcelLens in Cinematography tribute for great directors of photography within the framework of the Cannes Film Festival, we thought: "Who else but Angénieux could pay tribute to the inventors of light, to those without whom films would not be what they are?".

It is this absolute legitimacy that is illustrated here in pages that recount a fantastic saga of more than eighty years: a journey through the past from the early days to the Retrofocus, Pierre Angénieux's greatest invention, and his first zoom, an incredible device without which cinema would not be what it is. Hollywood quite rightly turned to Angénieux for all its filmmaking requirements, offering the company its most splendid rewards and ensuring its place in the history of the world cinema forever through some legendary films. You will discover, as I did, that Angénieux was also involved in the first close-range images of the Moon and color television as well as so many other astonishing adventures. It is French genius that runs through these pages. The Lumière brothers with their many inventions were not the only ones.

The adventure did not end for Angénieux with the advent of digital technology. If anything, the opposite is true, as the company seized this opportunity to demonstrate that the quality of a lens will always make the difference. We celebrate a great director of photography every year in Cannes and I attach such importance to this that I never miss a ceremony. From Edward Lachman to Chris Doyle, from Vilmos Zsigmond to Roger Deakins, from Philippe Rousselot to Peter Suschitzky, we have always been in illustrious company. Nor have these figures ever failed to express their gratitude to all those who work every day at Angénieux, with their exceptional and indeed unique know-how, to provide them with the finest paint brush and enable them to express and sublimate the director's intention in the most accurate and unique way.

I hope that, like me, you will plunge into these pages with passionate curiosity to discover how a French engineer changed the course of world cinema and how his legacy is carefully safeguarded by formidable teams so that we can continue to gaze in wonder at images of an exceptional quality on screens all over the world, and especially at the Cannes Film Festival.

Long live Angénieux and long live cinema.

Thierry Frémaux

Delegate General of the Cannes Film Festival
Director of the Institut Lumière

Contents

« Many testimonies from cinematographers punctuate this book. Find them on pages 94, 210, 216, 258 … »

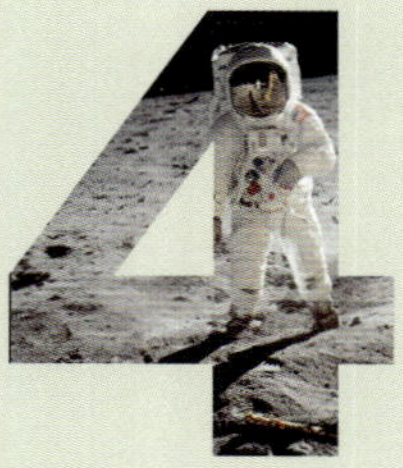

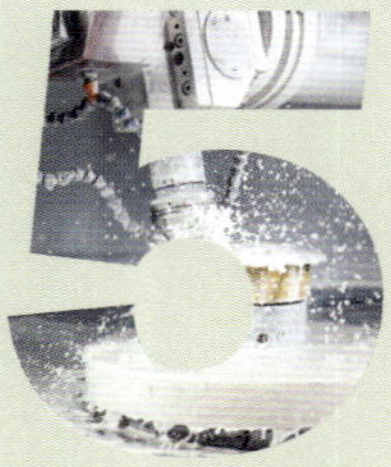

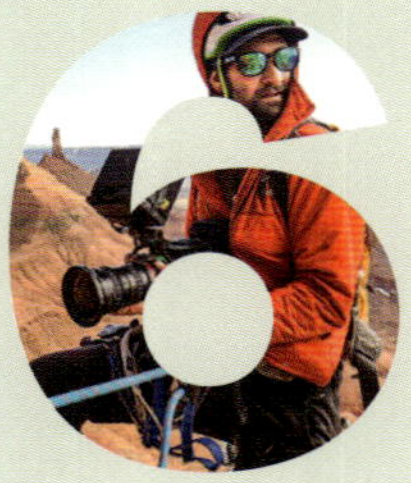

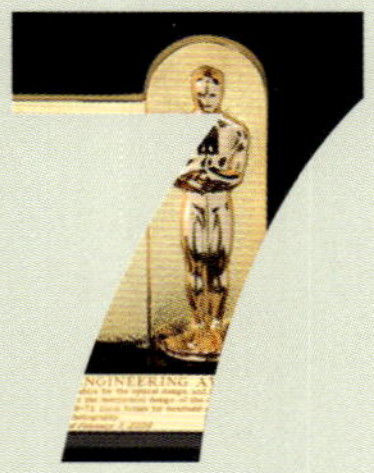

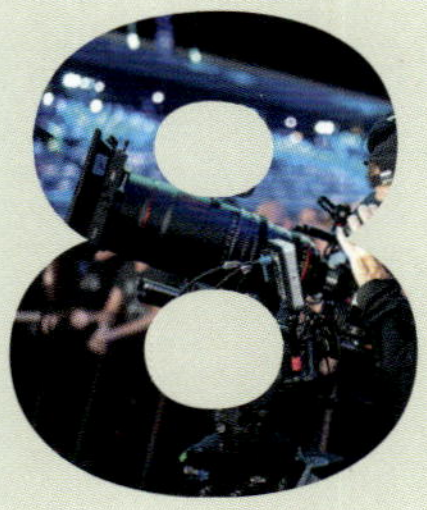

ANGÉNIEUX THROUGHOUT THE YEARS

1947

OBJECTIFS
P. ANGÉNIEUX
PARIS

1958

P. ANGÉNIEUX
PARIS

1965

PARIS
angēnieux

1982

angēnieux

2010

angēnieux®

1930-40

1935
■ Établissements Pierre Angénieux is founded in Paris
■ Start of the partnership with Abel Gance

1938
■ Invention of the Pictograph
■ First lenses under the brand name Angénieux

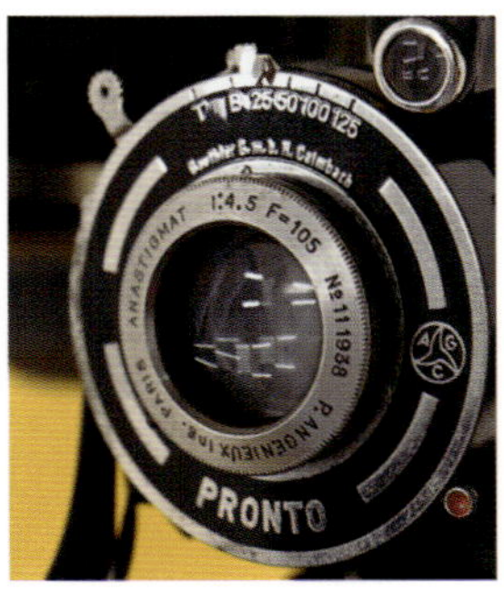

1940
■ Pierre Angénieux moves his Parisian headquarters to Saint-Héand which was still in the free zone

1946
■ A new factory is built.
The company has 80 employees.
Louis Moulard joins the company

1948
■ The company now has 170 employees

1950

1950
■ ■ Invention of the Retrofocus

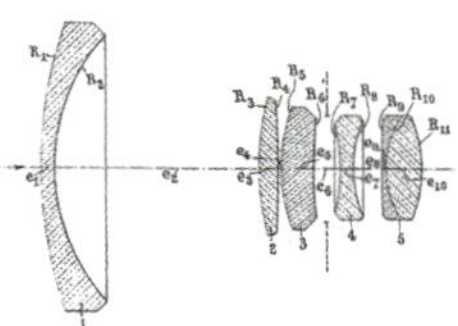

■ Start of the partnership with Kodak
■ First Retrofocus lens (35mm f/2.5)

1951
■ Colour TV first appears in the United States
■ First 16mm and 35mm fixed cine lenses

1953
■ The company has 370 employees
■ Mastery of the f/0.95 aperture

1956
■ Purchase of a Bull Gama 3 electronic calculator
■ All Bell & Howell 70 16mm cameras are now equipped with Angénieux fixed lenses

1957
■ First 4x zoom (17-68mm) for 16mm cinema

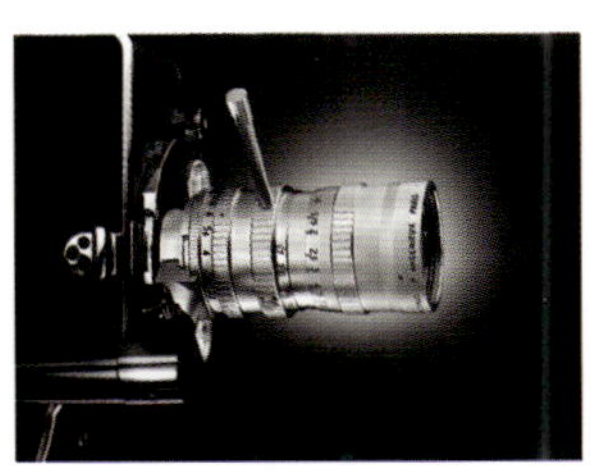

1958
■ The company has 600 employees
■ 40% of sales are from the United States
■ First 4x zoom (35-140mm) for 35mm cinema

1960

1960
■ First 4x zoom (40-160mm) for Orthicon television

1961
■ First 10x zoom (12-120mm) for 16mm cinema

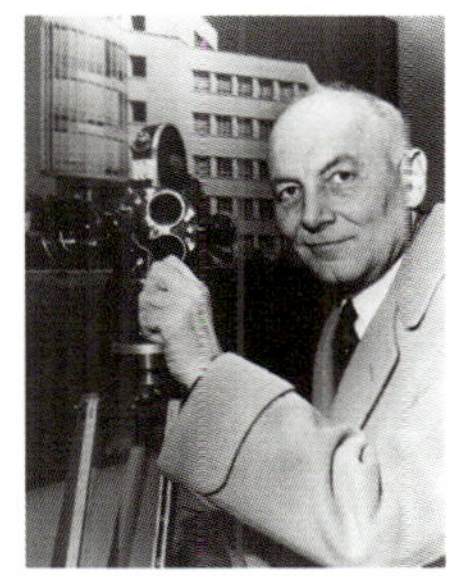

■ First 10x zoom (40-400mm) for Orthicon television
■ Angénieux 40-140 mm Franscope zoom, first anamorphic zoom for 35mm cinema

1962
■ First 10x zoom (25-250mm) for 35mm cinema

1963
■ The new factory is extended
■ First colour TV Zoom for Plumbicon television

1964
■ Pierre Angénieux's son Bernard joins the company
■ ■ The company is assessed for the first time according to American MIL standards with the view to becoming a regular supplier for the NASA programs
■ Scientific and Engineering Award for the Angénieux 10x zoom

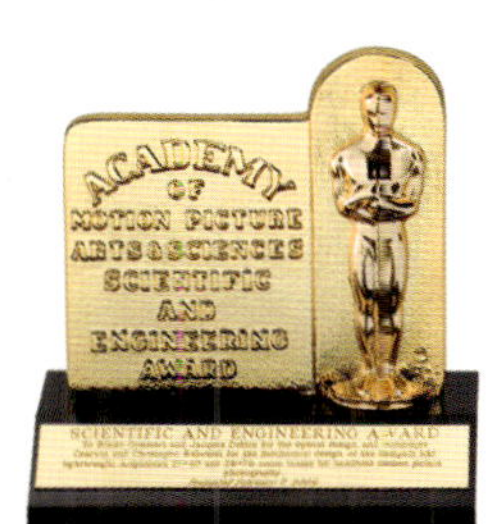

■ *Ranger* 7 mission is a success. The first close-range photos of the moon are taken with a F25mm f/0.95 lens

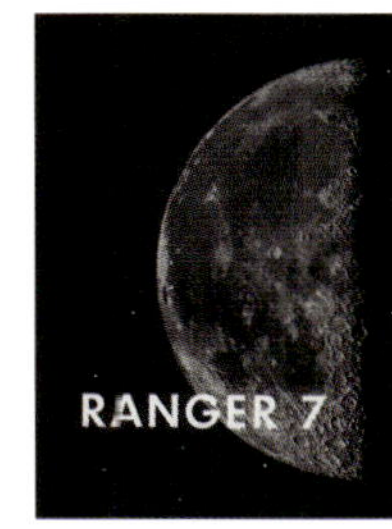

1966
■ Visit of Prime Minister Georges Pompidou.

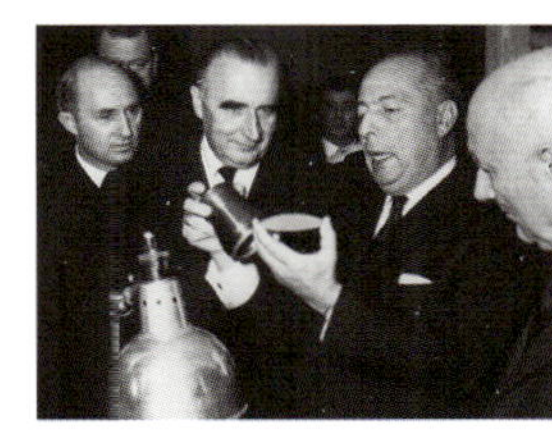

1967
■ 18x tele-zoom (27.5-500) for Plumbicon 1" 1/4 colour TV. The specifications of this zoom (zoom ratio x short focal length in mm: 18x27.5) becomes a standard

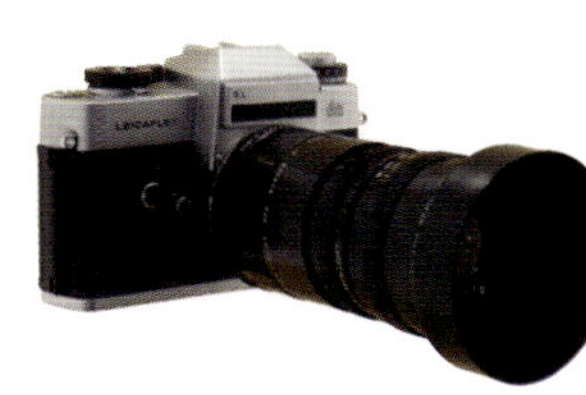

1968
■ First 2x photographic zoom (45-90mm) for Leicaflex

1969
■ The *Apollo 11* mission is a success. On board, 6x25 Angénieux zoom lenses
■ ■ The first steps on the moon are broadcasted in Mondovision

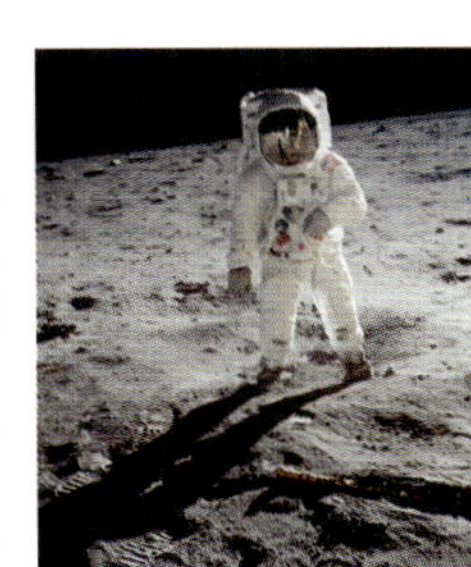

LEGEND

■ Cinema
■ Photography
■ Television
■ Optronics
■ Spatial
■ Medical
■ Company

1970

1970
- Jean Moret, Pierre Angénieux's son-in-law joins the company
- First military lenses and sub-assemblies for Thomson-CSF and TRT
- First heads-up display for civil and military aviation
- Beamsplitter for the Thomson-CSF TTV 1515 Plumbicon 1" 1/4 colour camera

1972
- NASA visits Saint-Héand before the *Skylab* mission

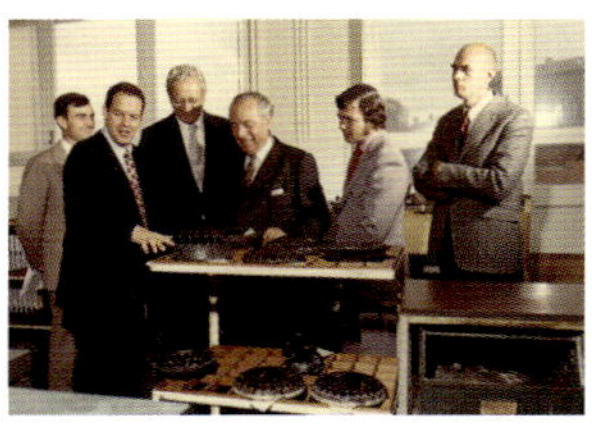

1973
- *Skylab* is launched into orbit on May 14
- Angénieux diversifies into medical optics and decides to reinvent surgery lighting

1974
- Pierre Angénieux retires

1975
- First zoom intended for military use (10x30) is presented at the Salon du Bourget
- *Apollo Soyuz* project. Both the Russians and Americans have 6x25 zooms

1976
- 42x Plumbicon 1" 1/4 camera zoom

1977
- Contract for 100 42x zooms for the Moscow Olympic Games in 1980

1979
- Beamsplitter for the first colour Hawkeye CCD camera by RCA

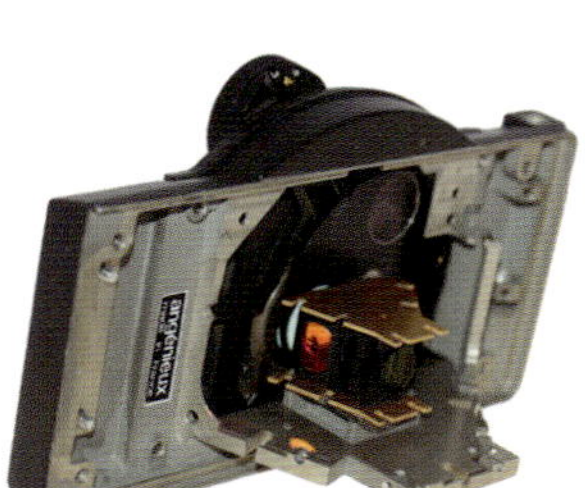

1980

1981
- The American space shuttle Columbia makes its first flight. On board, 3x8.2 and 15x8.5 zooms

1982
- Angénieux goes back to photography with the 35-70mm f/2.5-3.3 zoom
- New Sony Betacam. Arrival of ENG and rollout of 2/3" format

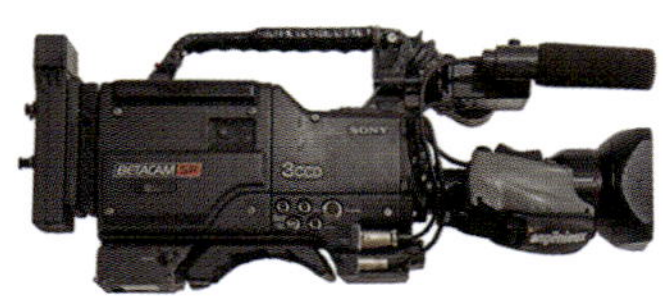

1983
- Jean Moret is awarded the Meilleur Ouvrier de France medal by the French President François Mitterrand

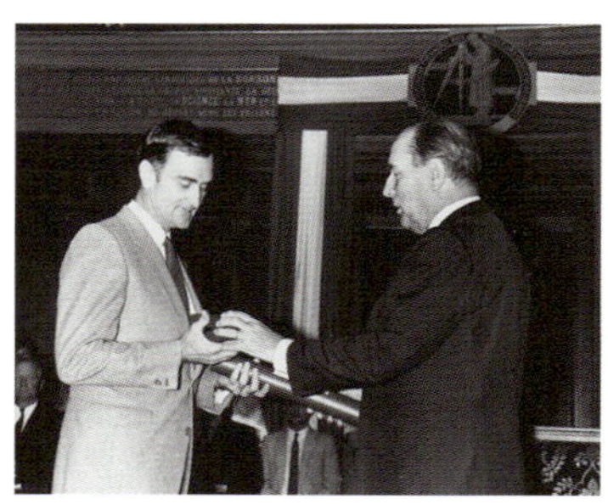

- Angénieux launches a complete range of ENG and OB lenses for 2/3" SD cameras
- 70-120mm zoom f/3.5

1985
- 25-250 zoom f/3.2 HP for 35mm cinema

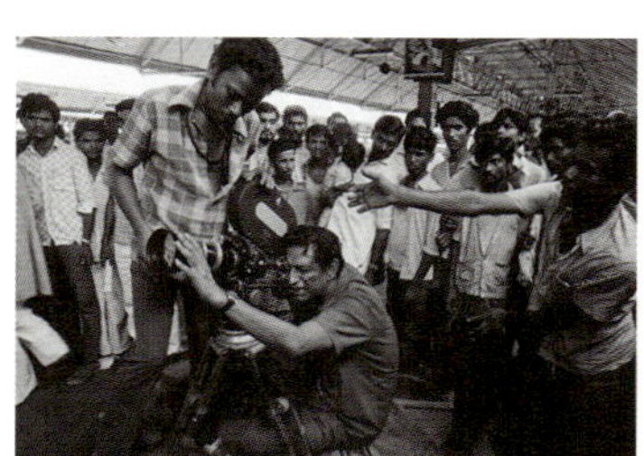

- 100x zoom for Raytheon

1986
- Essilor buys up 49% of the company's capital. The company has 625 employees
- A branch is opened in Japan
- Angénieux is the third biggest manufacturer of lighting for surgery

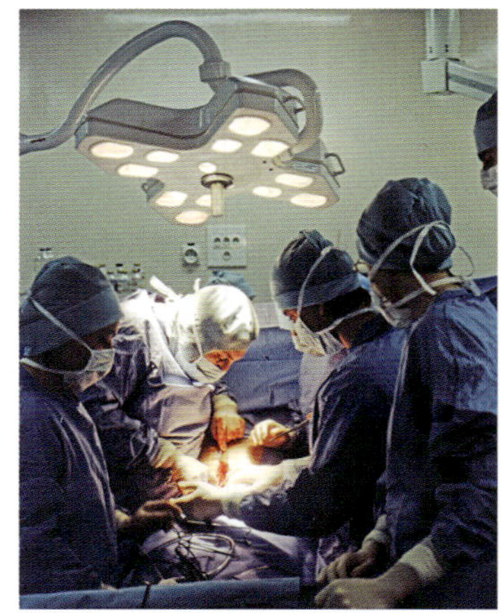

- 20x 8.5 zoom for lasers
- 28-70 f/2.6 zoom, 180mm f/2.3 and 200mm f/2.8 fixed lens
- The observation satellite SPOT1 is put into orbit. Angénieux supplies the optical splitter system
- Beginning of HD TV research and development
- New Sony DXC-3000 2/3", the first professional CCD camera

1988
- Zooms and optical systems for Martin Marietta (EU) and Oerlikon (Canada)

1989
- Essilor increases its share in the company's capital to become the majority shareholder
- Pierre Angénieux receives an Oscar for his lifetime contribution to the film industry

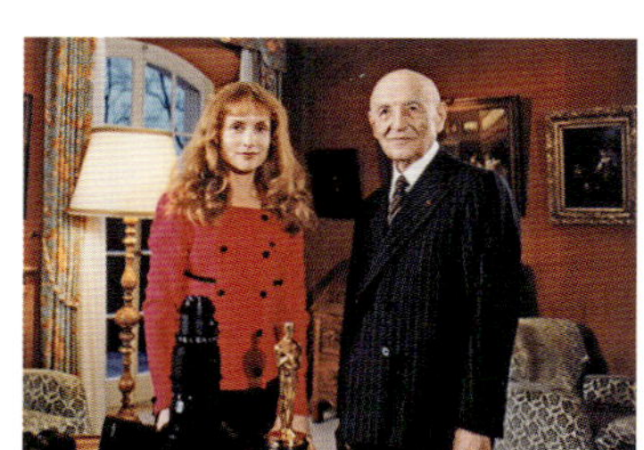

1990

1990
- ISO9001 certification

1992
- Japan branch closes
- HD tests at the Winter Olympics in Albertville. 18x12, 10x12 and 32x15.5 zooms for 1" CCD HD cameras
- First major HD broadcast of the Barcelona Olympic Games

1993
- Ets Angénieux is taken over by Thomson CSF which became Thales in 2000
- Jean-François Pernotte is appointed CEO, Denis Suveran General Manager

1994
- Extension of the Saint-Héand site. A new production area is created

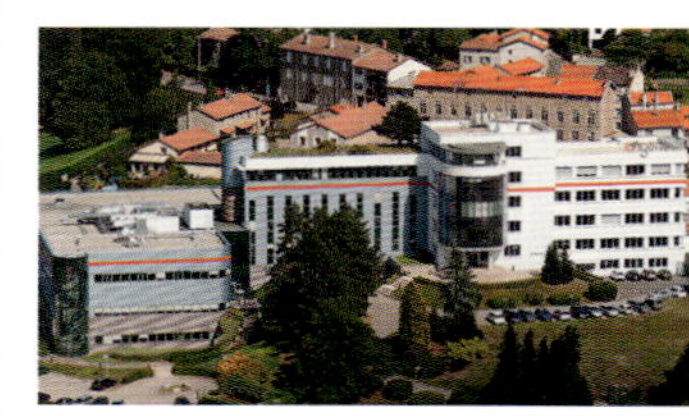

- Transfer of surgical lighting production to Air Liquide Médical but Angénieux remains a subcontractor
- Photography production ends
- 62x9.5zoom for 2/3" SD cameras

1995
- Creation of a digital endoscopic system ATC 3000

1996
- DGA contract for 10,000 pairs of LUCIE night-vision goggles

1998
- Pierre Angénieux dies on June 26, aged 91

1999
- Alain Darius is appointed CEO
- 72x9.5 zoom for 2/3" SD cameras

2000

THALES

2000
- Thomson CSF becomes Thales with the merger of the military activities of Alcatel, Dassault Electronique and Thomson CSF
- First export contract for LUCIE night-vision goggles

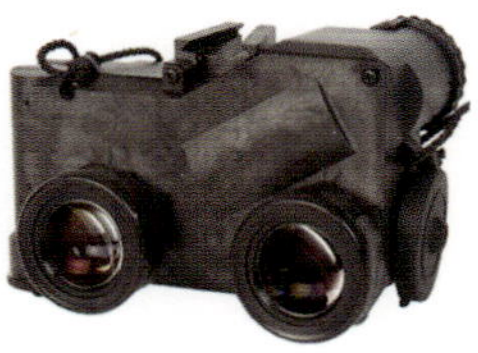

- Angénieux launches a complete range of ENG and OB lenses for 2/3" HD cameras

2001
- Benoît Bazire is appointed CEO and Denis Levaillant General Manager
- Optimo 24-290 zooms for 35mm cinema

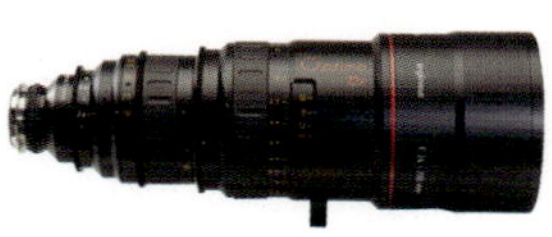

2002
- Cinec Award in Munich for the Optimo 24-290

2004
- Optimo 17-80 zoom for 35mm cinema

2005
- Philippe Parain is appointed CEO in November replacing Benoît Bazire and Denis Levaillant
- Angénieux receives an Emmy Award for its work on HD TV

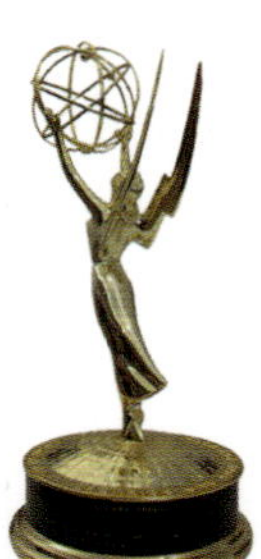

- Delivery of F150mm optics for NASA's *Dawn* mission

2006
- Presentation of the RED One camera. Cinema goes digital
- Optimo 15-40 compact zoom lens for 35mm cinema

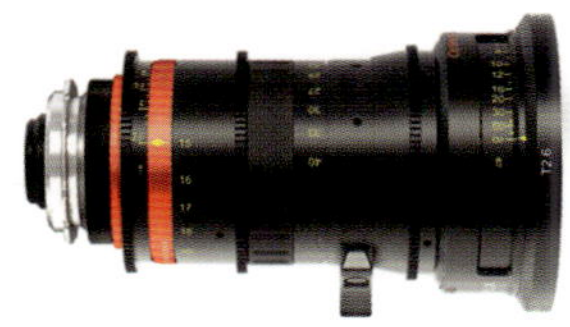

2007
- Optimo 28-76 compact zoom lens for 35mm cinema

2008
- Optimo DP 30-80 compact zoom lens for digital cinema

2009
- Optimo DP 16-42 compact zoom lens for digital cinema
- Scientific and Engineering Award for the Optimo 15-40 and 28-76

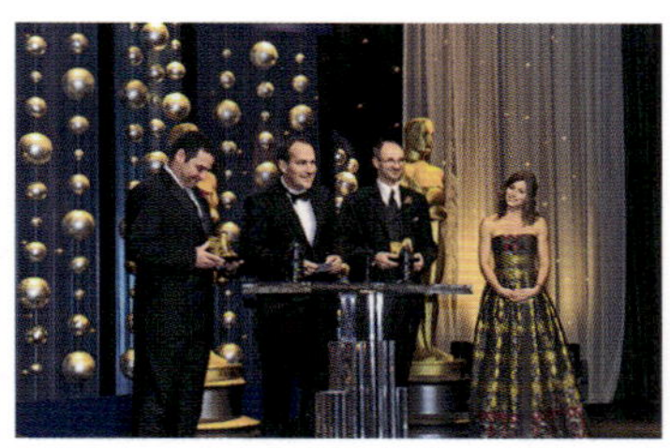

- MINIE-D goggles win the SAGEM contract to equip future French soldiers (FELIN program)

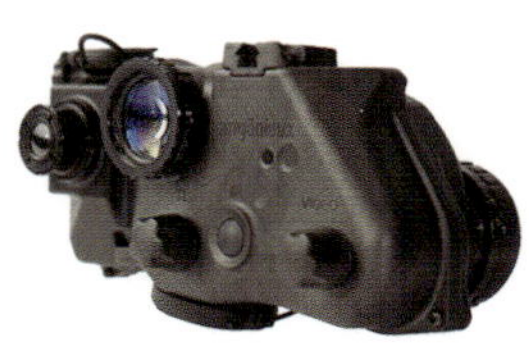

- Visit from the Minister of Defence Hervé Morin
- The French Army adopts HELIE helicopter pilot goggles

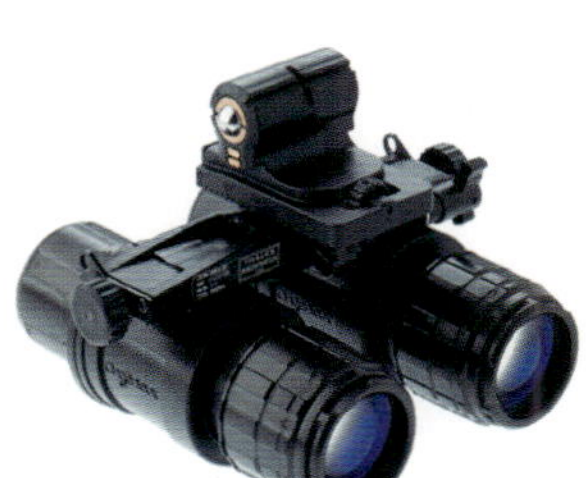

2010...

2010
- Angénieux celebrates its 75th birthday and unveils a new logo
- ISO140001 certification
- Stops production of Broadcast zooms
- Optimo DP pack for 3D cinema production
- Launch of the Optimo 45-120 compact zoom lens for S35mm cinema

2011
- Pierre Andurand is appointed CEO
- Angénieux continues working on 3D HD TV
- French Rafale pilots are equipped with the first DVN

2012
- Angénieux obtains Entreprise du Patrimoine Vivant label
- OHSAS18001 certification
- 19.5-94 and 28-340 zooms (adaptations of the 17-80 and 24-290) for the new-size digital camera sensors
- Cinec Award in Munich for the Optimo 45-120

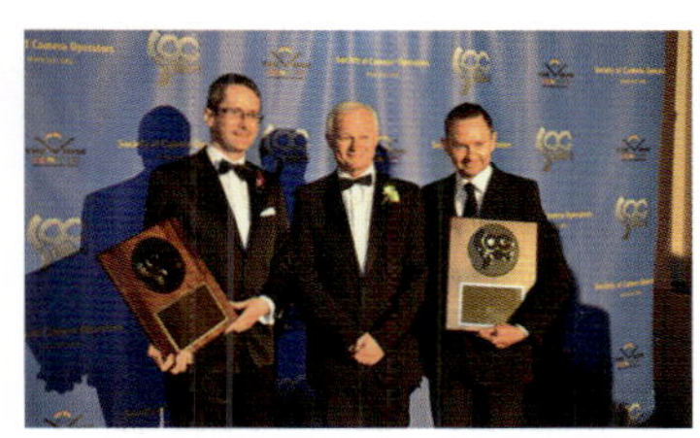

- SOC Technical Achievement Award for Optimo 15-40, 28-76 and 45-120

2013
- First Optimo anamorphic compact zoom lens 56-152 A2S and Optimo zoom DP 25-250
- Angénieux becomes an official partner of the Cannes Festival and creates the "Pierre Angénieux ExcelLens in Cinematography" tribute to honour the Director of Photography Philippe Rousselot

2014
- Cinema is digital. The Optimo DP range is renamed Optimo Style. Optimo 30-72 A2S anamorphic compact zoom
- BSC Bert Easey Technical Award for Angénieux zoom lenses
- Angénieux pays tribute to Vilmos Zsigmond in Cannes

2015
- Angénieux pays tribute to Roger Deakins in Cannes

2016
- Launch of IRO technology, a system offering greater flexibility for Angénieux zooms
- Optimo 44-440 A2S anamorphic zoom, new EZ Type range for S35mm and Full Format
- Angénieux pays tribute to Peter Suschitzky in Cannes

2017
- Optimo Ultra zoom 12x for S35mm, U35mm and Full Format
- Angénieux pays tribute to Christopher Doyle in Cannes
- Thales chooses the Saint-Héand teams to assemble Sophie cameras

2018
- Thales LAS France creates a Cinema board (having merged with Thales Angénieux): Emmanuel Sprauel (President of the Angénieux brand), Pierre Créniault (Director of Cinema Optics), Severine Serrano (International Sales and Marketing Director), Christophe Remontet (Product Policy Director)
- Angénieux pays tribute to Edward Lachman in Cannes
- Design and production of complex optical sub-assemblies for Thales HUD and HMD systems for fighter pilots (mainly Rafale) and for helicopter pilots (Topowl helmets)
- Just 50kms from Ceres, the *Dawn* probe fitted with Angénieux optics transmits incredible images of the asteroid which are of great interest to NASA
- Cinec Award in Munich for the Optimo Ultra 12x

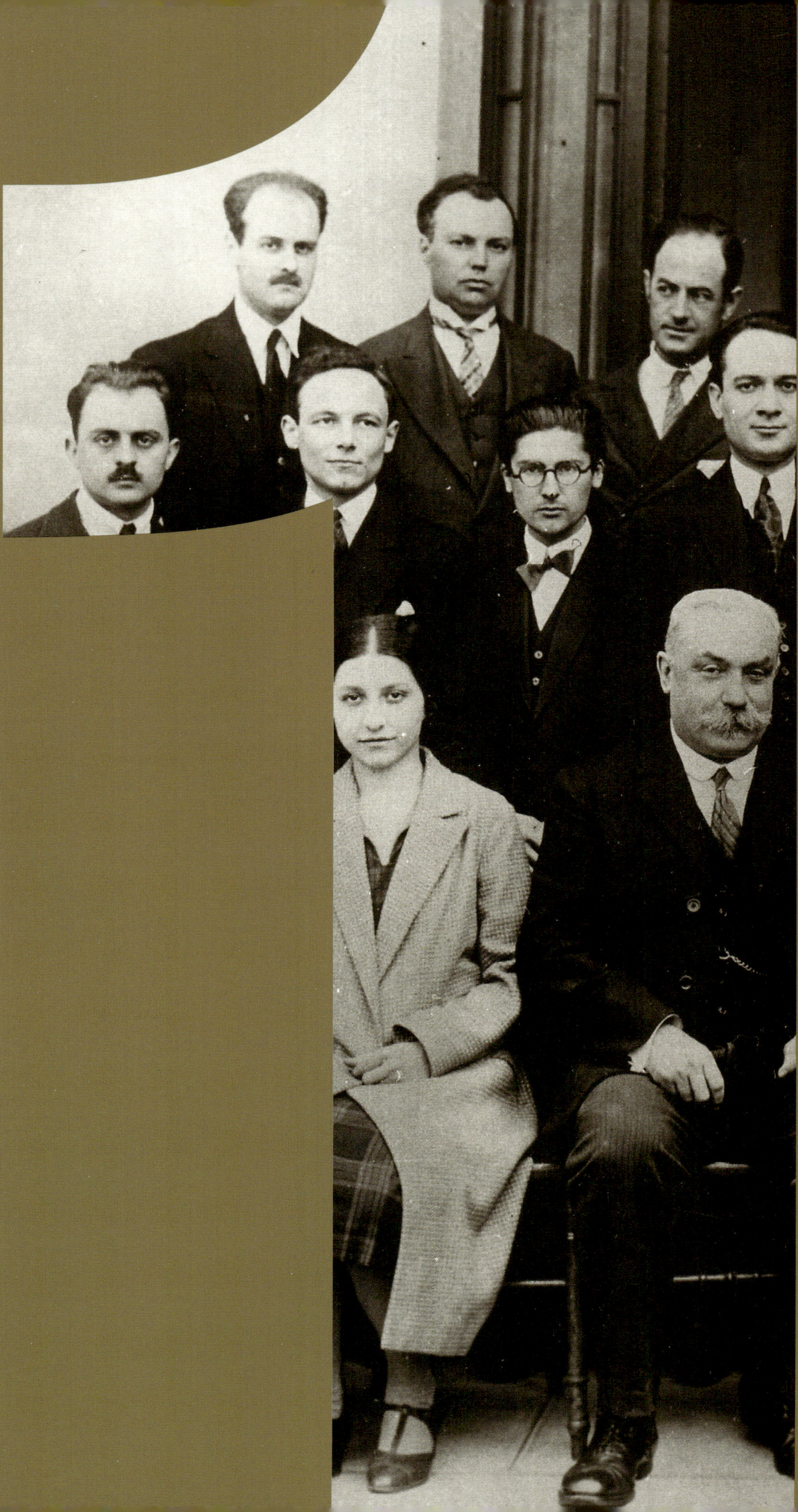

THE PORTRAIT OF A MAN
PIERRE ANGÉNIEUX

The Angénieux saga starts with a man, Pierre Angénieux, who displayed exceptional **inventiveness** from his company foundation in 1935 to his end of life, making the Angénieux brand a worldwide point of reference for optics. The employees of the firm take pride today in carrying on the legacy of over **80 years of excellence**.

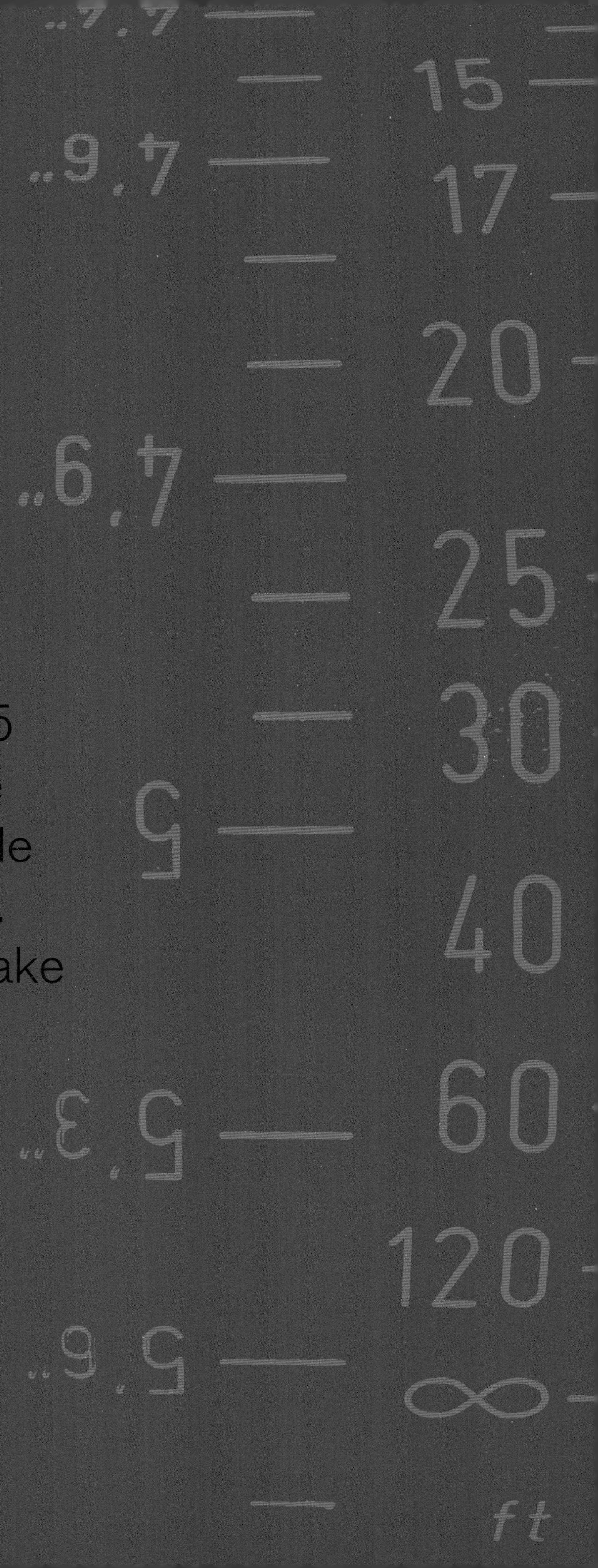

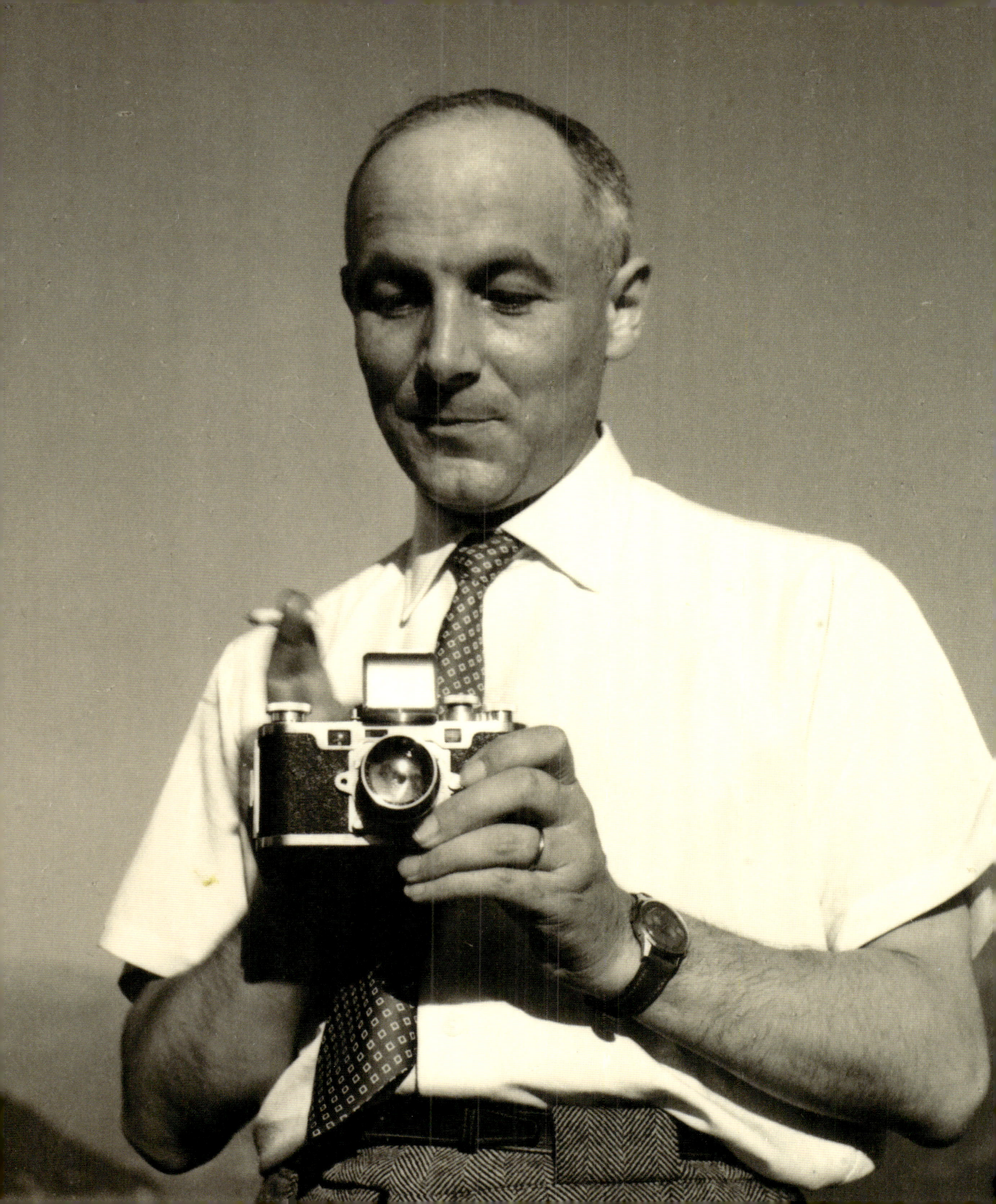

Pierre Angénieux was born on July 14, 1907 in Saint-Héand, a village in the central-eastern part of France near Saint-Étienne.

The Angénieux were a family of weavers working from their home in Saint-Héand.
His father was a market gardener.
Already noticed by his teacher at the village school as a particularly bright and attentive pupil, Pierre Angénieux went on to attend the Notre-Dame de Valbenoîte secondary and high school where he studied mathematics in Saint-Étienne.
Angénieux's evident gifts suggested a career in engineering.
He enrolled in the École des Arts et Métiers in Cluny in 1925 at the age of 18.
Fascinated by optics and attracted by the teaching of the physicist Charles Fabry (1867–1945), he chose to continue his studies in 1928 at the École Supérieure d'Optique, where he attended the courses of Henri Chrétien, inventor of the **Hypergonar**[1].
Called up with the class of 1927, Pierre Angénieux was granted a deferment of two years for his studies. Assigned to the air force as suitable for employment as a meteorologist, he joined the 32nd regiment of observers in the fort of Saint-Cyr in Montigny-le-Bretonneux in the Yvelines near Paris on October 22, 1929.
Declared unfit for military service on December 4, 1929, Angénieux quickly returned to civilian life.

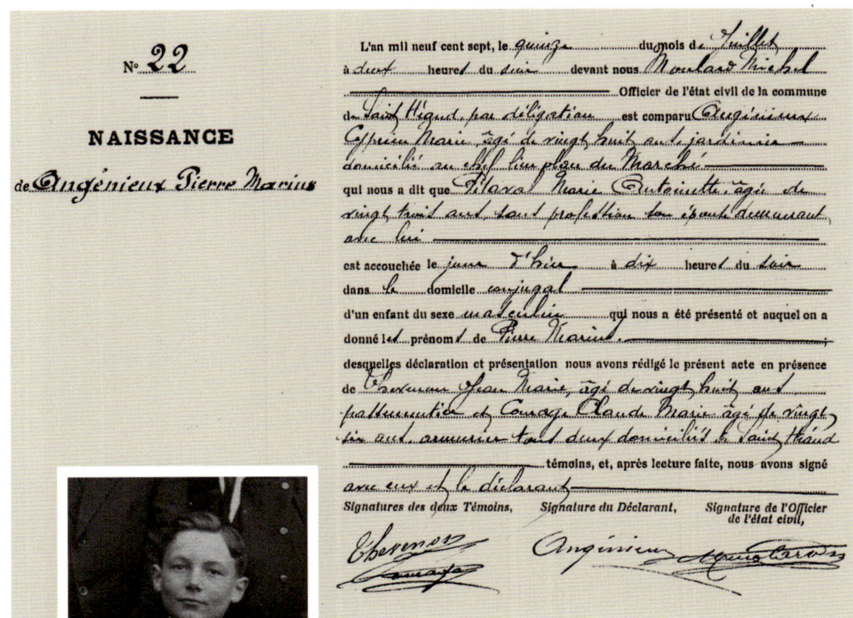

N° 22

NAISSANCE

de Angénieux Pierre Marius

L'an mil neuf cent sept, le quinze du mois de Juillet à deux heures du soir devant nous Moulard Michel Officier de l'état civil de la commune de Saint Héand, par délégation est comparu Angénieux Cyprien Marie, âgé de vingt huit ans, jardinier domicilié au chef lieu place du Marché qui nous a dit que Vilaval Marie Antoinette, âgée de vingt trois ans, sans profession, son épouse demeurant avec lui est accouchée le jour d'hier à dix heures du soir dans le domicile conjugal d'un enfant du sexe masculin qui nous a été présenté et auquel on a donné les prénoms de Pierre Marius. desquelles déclaration et présentation nous avons rédigé le présent acte en présence de Chevenon Jean Marie, âgé de vingt huit ans passementier et Courage Claude Marie âgé de vingt six ans, armurier tous deux domiciliés à Saint Héand témoins, et, après lecture faite, nous avons signé avec eux et le déclarant.

Signatures des deux Témoins, Signature du Déclarant, Signature de l'Officier de l'état civil,

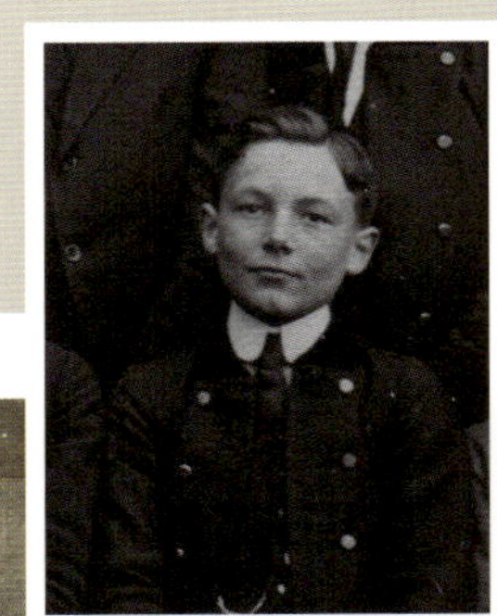

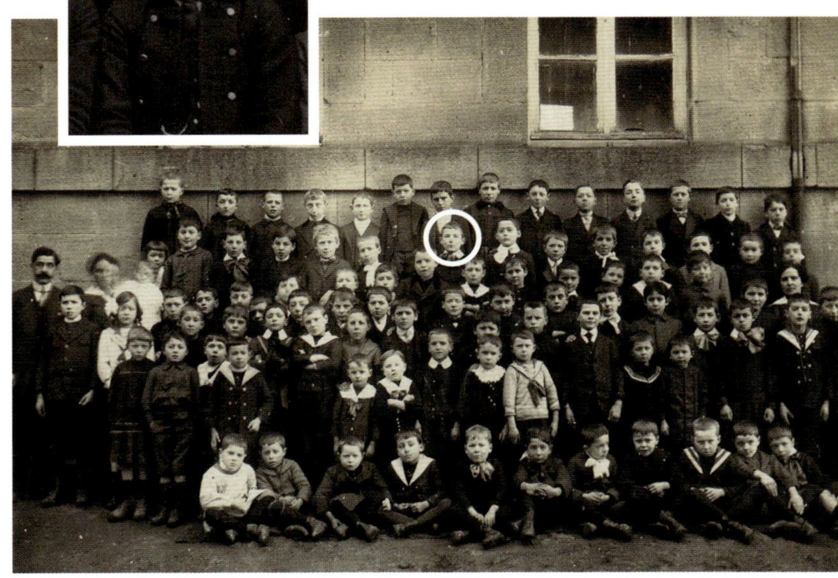

above
Certificate of the birth of Pierre Angénieux in Saint-Héand on July 14, 1907.

Pierre Angénieux at the nursery school in front of what was to become his factory (collection of Lucie Thomas) and, in foreground, during his schooldays in Saint-Étienne (collection of the Collège Notre-Dame de Valbenoîte).

opposite
Class of 1928–29 at the École Supérieure d'Optique (collection of the École Supérieure d'Optique) and class at the Arts et Métiers in Cluny (collection of the Granet family).

[1]Invented by Henri Chrétien for cine cameras in 1926, the Hypergonar is an anamorphic optical system that compresses the image on the film and makes it possible to shoot a much wider scene than with a conventional lens. When the compressed images are projected, the lens mounted on the projector expands them and returns them to their original shape.
One of the first directors to use the Hypergonar was Claude Autant-Lara for *Construire un feu* in 1928. Its use remained limited, however.
At the beginning of the 1950s, plunged into crisis by the development of television, the film industry offered spectacular productions to fill the cinemas again. On January 12, 1953, the patents were already in the public domain. 20th Century Fox bought up the Hypergonars produced before the war. This system was then marketed under the name CinemaScope.

The École Supérieure d'Optique
The extent to which France was lagging behind Germany in the field of optics, having been pitilessly revealed during World War I, the Institut d'Optique appliquée and École Supérieure d'Optique were founded in 1917. Operations did not commence until 1920, however, with the enrolment of the first year at the latter.
Founded by the Duc de Gramont, a rich industrialist and member of the Académie des Sciences, the astronomer Henri Chrétien and the scientist Charles Fabry, the school trained optical engineers.
It was there that Pierre Angénieux studied and from there that he was to recruit many of his assistants.

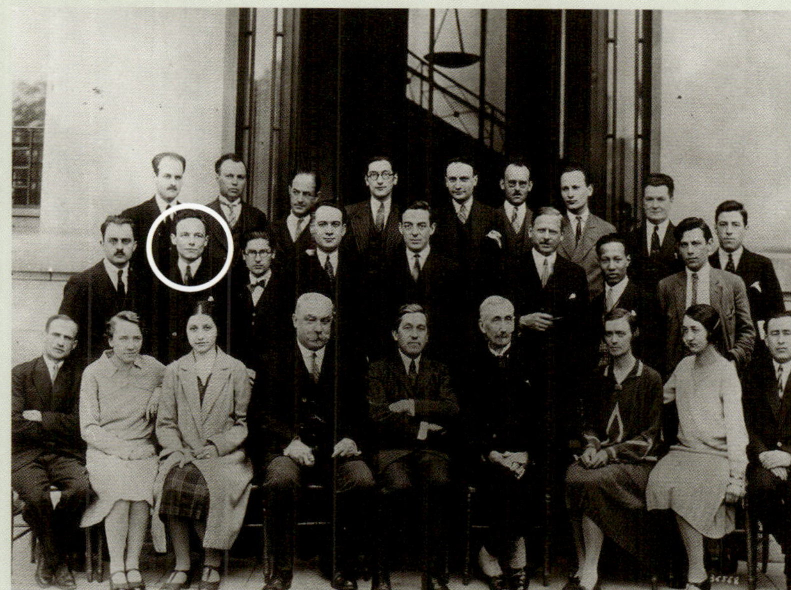

« From the École Supérieure d'Optique, Pierre Angénieux entered the industry.
An independent man, he foresaw the advent of the market and civilization of the photographic image, which truly came about in that period. At the personal level, it was cinema with its artistic results that fascinated him. »

André Masson

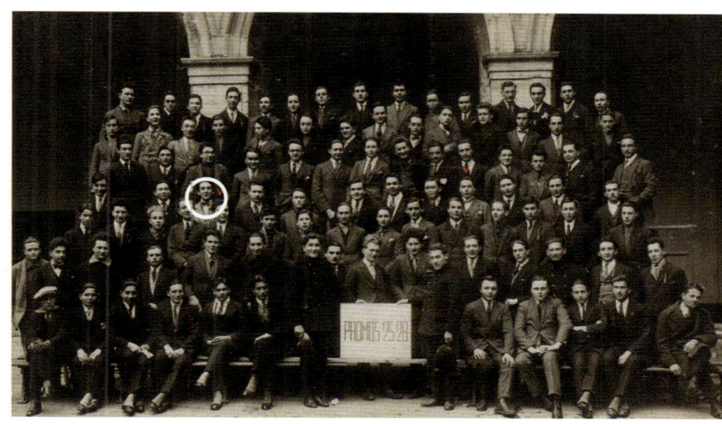

André Masson, a graduate of the École Supérieure d'Optique, joined Angénieux in 1952 as head of the optical workshop before becoming scientific director from 1954 to 1973 and associate chief executive until 1988.
Before entering the company, Masson carried out a great deal of research, especially on cathodic deposition (thin layers).
The holder of various patents, he was made a knight of the French Order of Merit in 1984 and then an officer in 2005.
Retired but still active, he was prompted to found the Club Nanotechnologies in 1989 by Hubert Curien, Minister of Scientific Research, and served as its president until 2000. He died in January 2011.

Pierre Angénieux embarked on his professional career in 1930 with a solid background in optics and mechanical engineering.

right
Pathé-Baby projector (collection of Sylvain Tomasini, photo by Eric Perrin).

below
Poster of the Pathé-Baby Motocaméra marketed in 1928 (collection of the Fondation Jérôme Seydoux-Pathé ©).

opposite
Pathé factory in Joinville-le-Pont, Val-de-Marne, René Denilauler archives, 1928 (collection of the Fondation Jérôme Seydoux-Pathé ©) and *posographe* (calculator of aperture and exposure time) of the Pathé-Baby kit (collection of Sylvain Tomasini, photo by Eric Perrin).

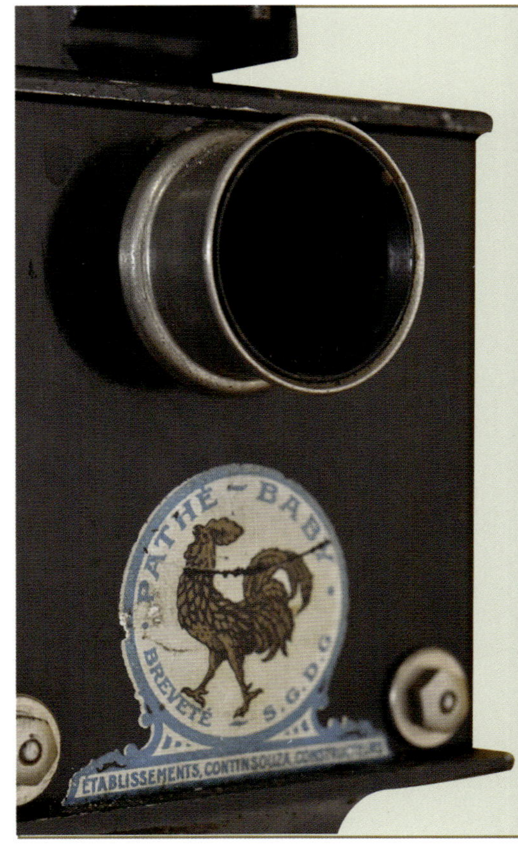

Cinema was well established as the seventh art between the two World Wars. The period from the first public showing of the Lumière brothers, at the Grand Café in Paris on December 28, 1895, to the end of World War I saw the rise of the American film industry. Germany invented expressionist cinema with films released in 1922, like Murnau's *Nosferatu, eine Symphonie des Grauens* (*Nosferatu*) and Fritz Lang's *Dr. Mabuse the Gambler*. In France, directors like Jean Renoir, Paul Fejos and Abel Gance took an interest in the more technical and aesthetic aspects of the image.

The 1930s saw the advent of talkies, a new phenomenon that attracted many eminent intellects. Pierre Angénieux decided to take part. The young engineer started out with **Pathé**, first cinema empire, where he appears on the payroll of the plant in Joinville-le-Pont as "Baby technician" from March 1930 to November 1931, the date of his resignation.

This plant also produced color film and included shooting studios that attracted many directors in search of technical innovations. It was there that Pierre Angénieux met Jean Renoir, Abel Gance and others directors then making films in the Pathé studios. Passionately involved in this universe and particularly attentive to their needs, he was now himself in search of new optical equipment.

This period at Pathé marked his entrance into the world of cinema, which he was never to leave.

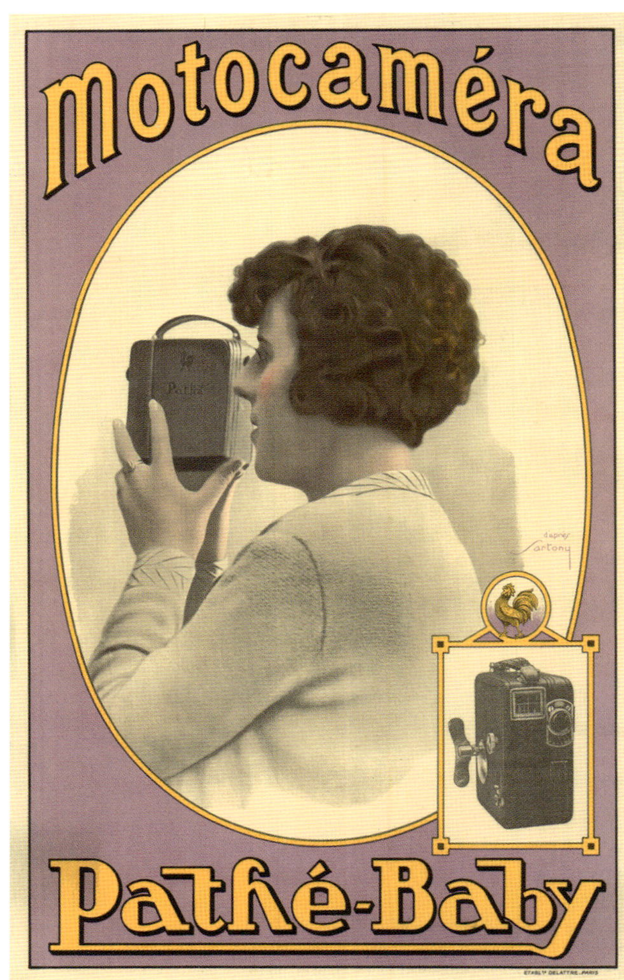

Pathé and Pathé-Baby

Charles and Émile Pathé founded Pathé Frères in 1896. Foreseeing the possibilities opened up by nascent industry, they decided to extend their activities to the production of equipment for cinema.
The Compagnie générale des Cinématographes, Phonographes et Pellicules was founded on December 11, 1897.
The two brothers gradually built up an empire. In addition to the factory producing phonographic material in Chatou in the Yvelines and the movie studios in Joinville-le-Pont and Montreuil, they were to produce films under the Pathé label with worldwide distribution.
A factory producing film in Vincennes was to challenge the monopoly of the American George Eastman.
Together with the Continsouza company, Pathé was also a pioneer in the production of professional movie cameras and projectors. Its activities reached their peak at the beginning of the 20th century.
The Pathé brothers were, however, convinced of the supremacy of the American film industry and no longer believed in their own success. They gradually dismantled their empire. The cinema sector was taken over by Pathé Cinema with Charles as sole executive. In 1922, for the sale and distribution of the films shot in his studios, he asked Continsouza to design an economical and user-friendly home projector to show highly abridged versions. Thousands of Pathé-Baby projectors were put on the market accompanied by a catalogue of no fewer than 192 titles from the Pathé film library.
The success of this venture led to the marketing of a home movie camera in 1923.
Pathé-Baby was founded as a French company distinct from Pathé Cinema in 1924, and a whole range of cameras and projectors were to be produced and sold under this general name until 1951.

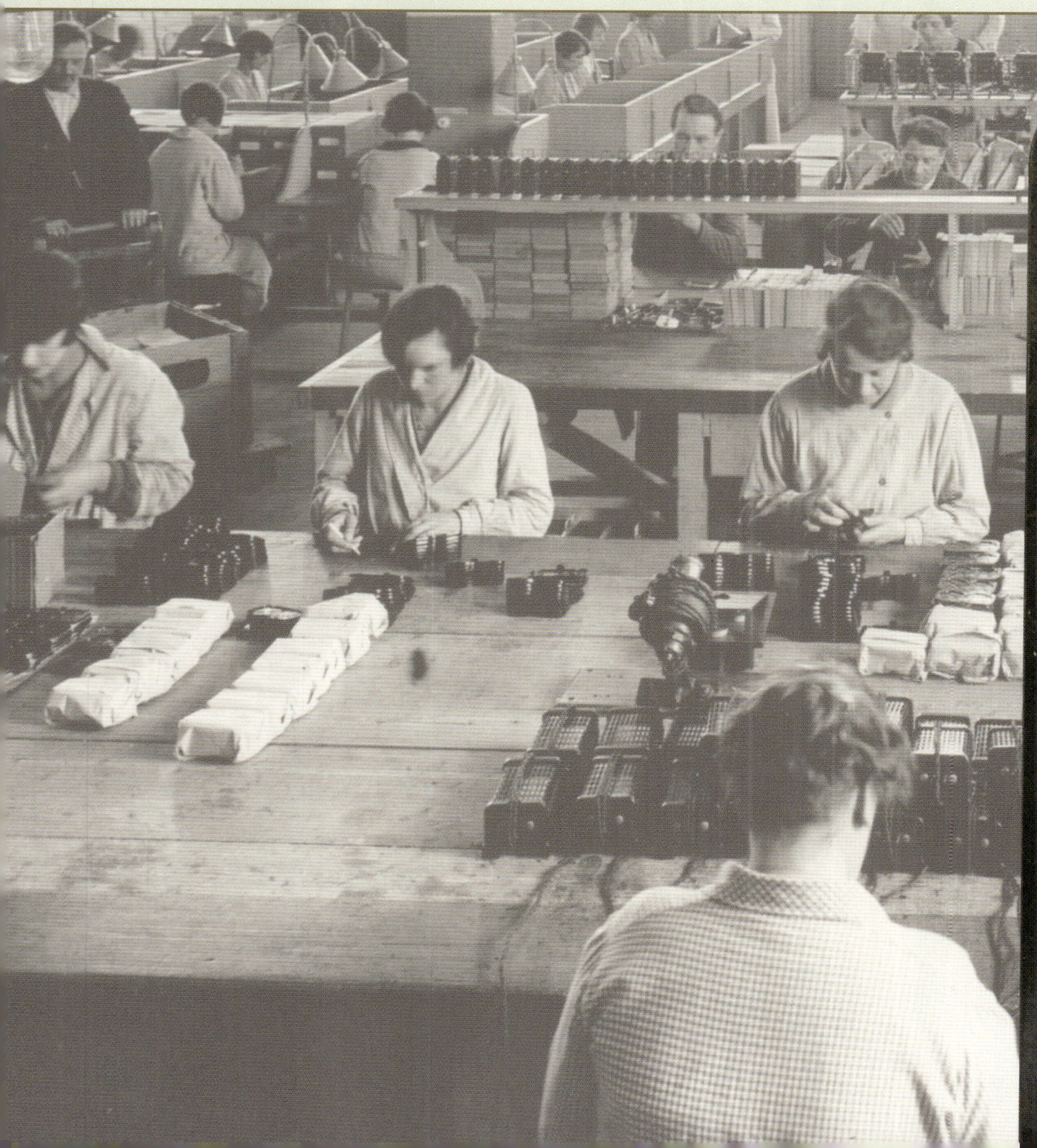

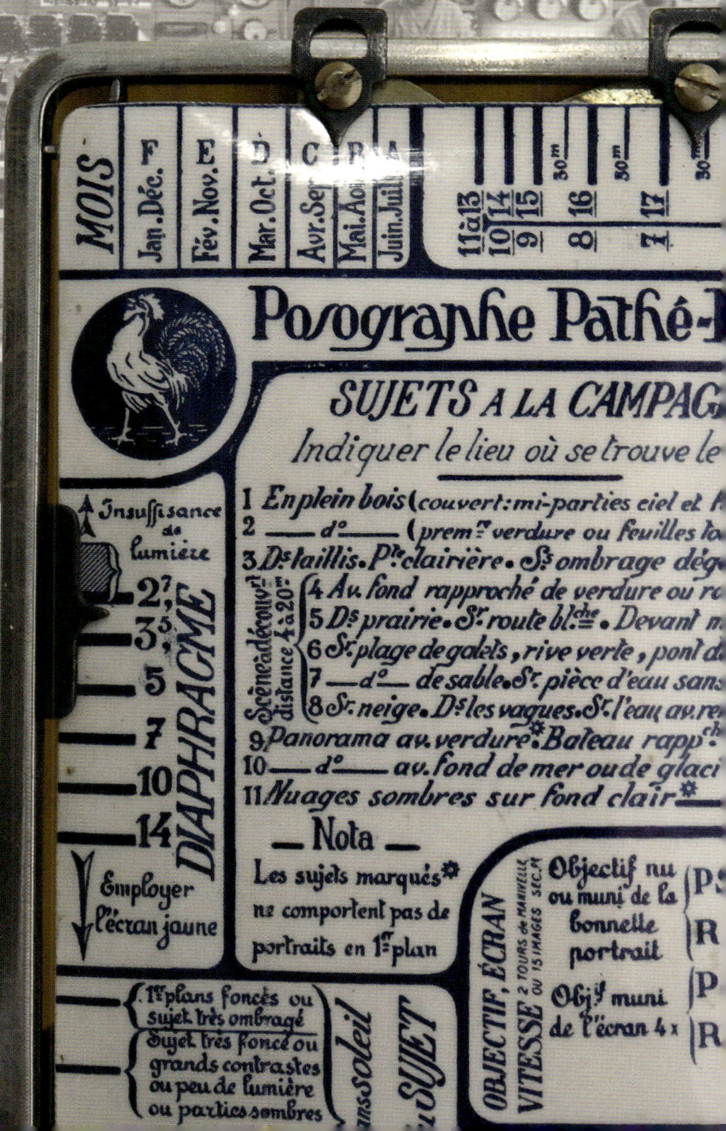

In 1932, Pierre Angénieux was working with André Debrie, a manufacturer of cameras and projectors second only to Pathé in the production of professional equipment for the French cinema industry.
In 1933, together with two other optical engineers, Angénieux founded the ASIOM (Applications Scientifiques et Industrielles de l'Optique et de la Mécanique). While the head office was in Paris at 7 rue Henri Murger, the first optical engineering workshop was set up in Saint-Héand in the building of the old school. ASIOM produced elements for OPTIS.
He married Marguerite Ducastel on August 9, 1934 and they had four children: Bernard, Martine, Daniel and Jacques. Bernard and Jacques were later to join their father's company together with Martine's husband Jean Moret.
The name of Pierre Angénieux was then associated with those of Jean-Marie Gutmann, Georges Laborie, and Edmond Laurent for a number of patents connected with cinematography in natural colors.
The ASIOM additive color process aroused the interest of Realita, which bought the company in the spring of 1935. The practice of photography was then developing in France. While a great deal of material was imported from Germany, French manufacturers like Lumière, Demaria-Lapierre, Cornu and Gallus offered fierce competition. Fostered by Pathé, home cinema displayed great success. The demand for optical material was enormous.
Pierre Angénieux decided that the time was right to set up on his own and started manufacturing lenses.

The company that bears the name of Pierre Angénieux was founded in Paris on September 20, 1935.

This was the start of a great adventure.
Throughout his professional life, Pierre Angénieux never ceased to invent new equipment for the world of images.
In 1950, his work on the retrofocus contributed to the development of 24x36 reflex cameras with interchangeable lenses. He introduced radically new methods of optical calculation and his design for a zoom with mechanical compensation was universally adopted. These innovations revolutionized the practice of photography and cinema. The renown of Angénieux zooms spread far beyond the French borders and they came to form part of the equipment of motion picture and television studios all over the world.
Their fame was such in the United States that NASA chose Angénieux for its space programs. The *Ranger7* space probe used Angénieux 25mm f/0.95 M1 lenses to take the first close-range photographs of the moon in 1964. Angénieux zooms were used by the *Apollo* missions and played their part in transmitting the most outstanding televised event of all time, namely the landing of *Apollo 11* and man's first steps on the moon on July 21, 1969.
A visionary as well as an untiring calculator, Pierre Angénieux laid the foundations of lasting success in the 1970s by diversifying the company's activities. He retired in 1974 but continued to provide his teams with a wealth of advice.
Pierre Angénieux died on June 26, 1998.
Above and beyond his achievements, his incredible legacy is the lasting nature of his work. All of the present-day zooms are still inspired by the principle developed by Pierre Angénieux.

Cinema Pays Tribute to Pierre Angénieux

Pierre Angénieux received the prestigious Gordon E. Sawyer Award for his lifetime contribution to the film industry from the Academy of Motion Picture Arts and Sciences in 1989.
Then 82 years old, he did not make the trip to Hollywood. It was the French actress Isabelle Huppert who brought the precious statuette to him on February 28, 1990 at his home in Geneva.
The sober recorded message of acceptance played at the Oscar presentation ceremony said simply: "I thank the Academy for this great honor. Thank you very much. Merci".

right
© Academy of Motion Picture Arts and Sciences (photo by Marc Salomon).

below
Angénieux archives.

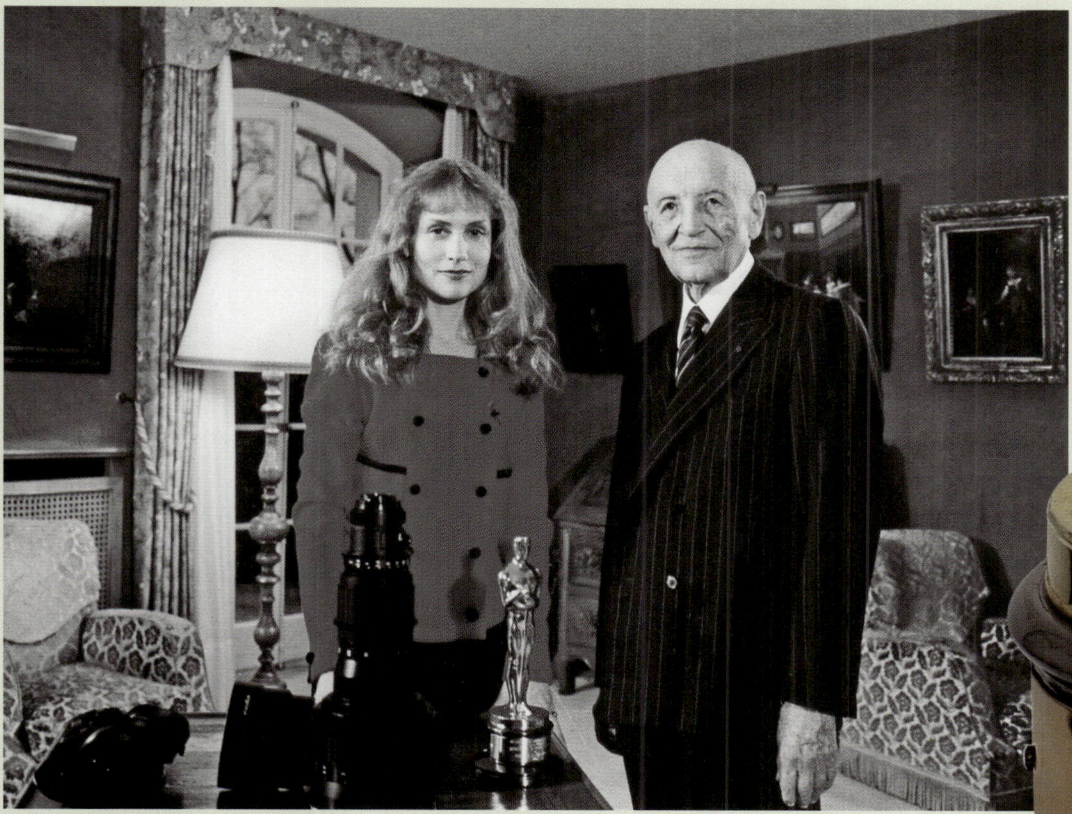

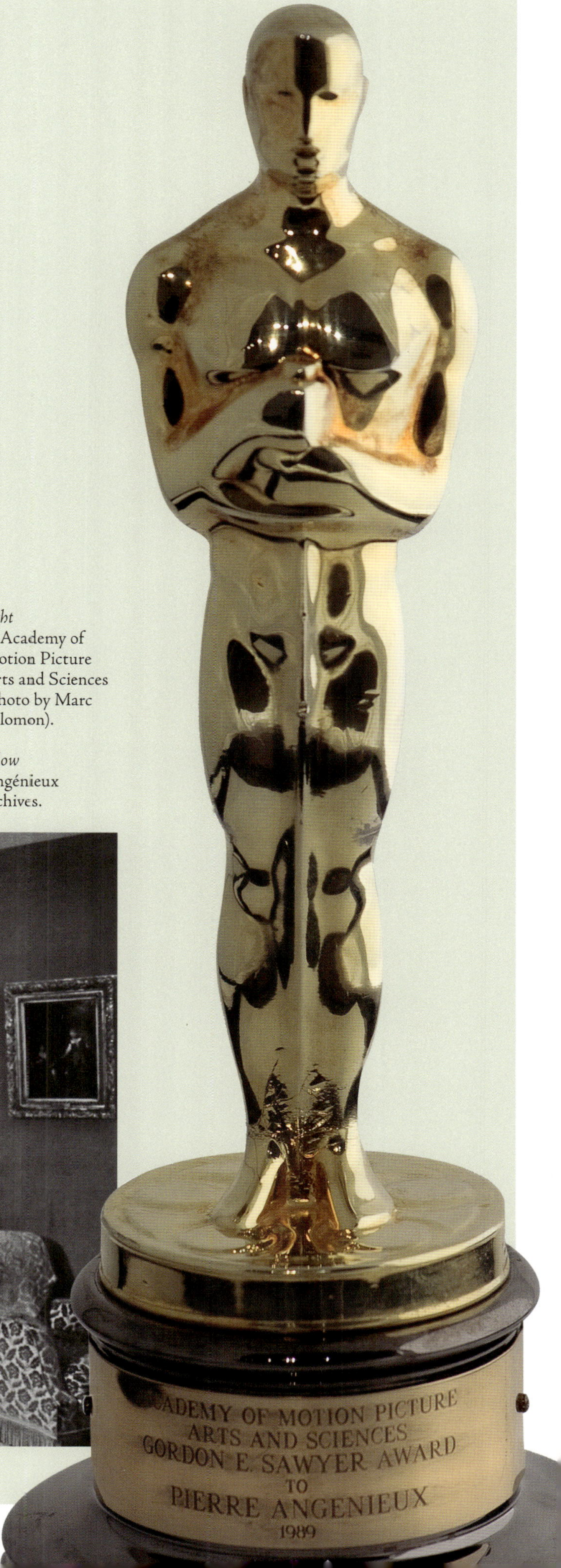

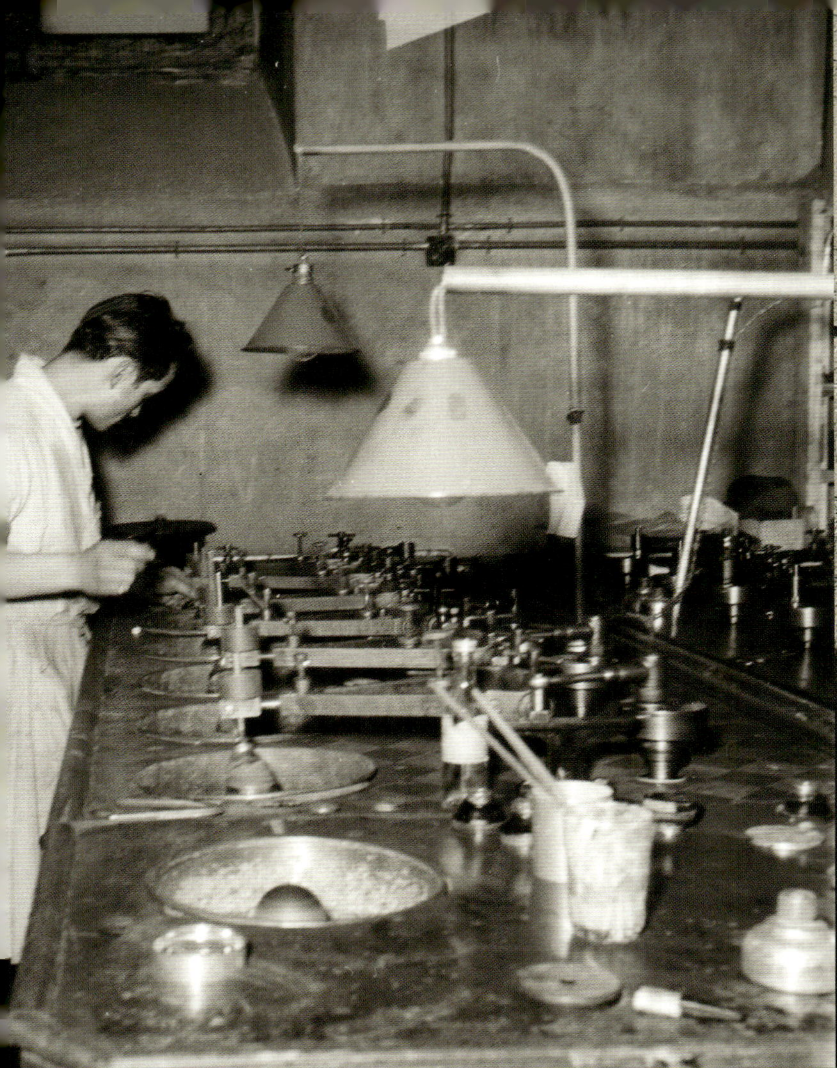

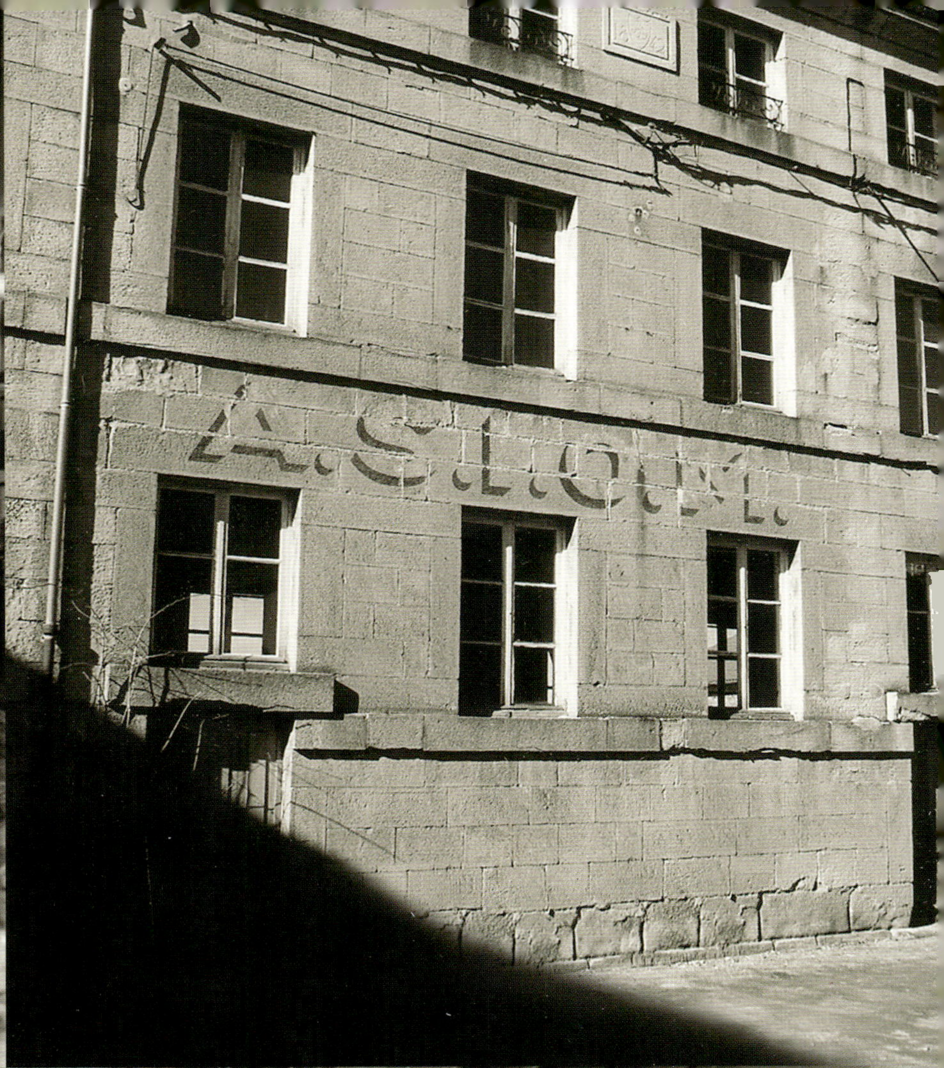

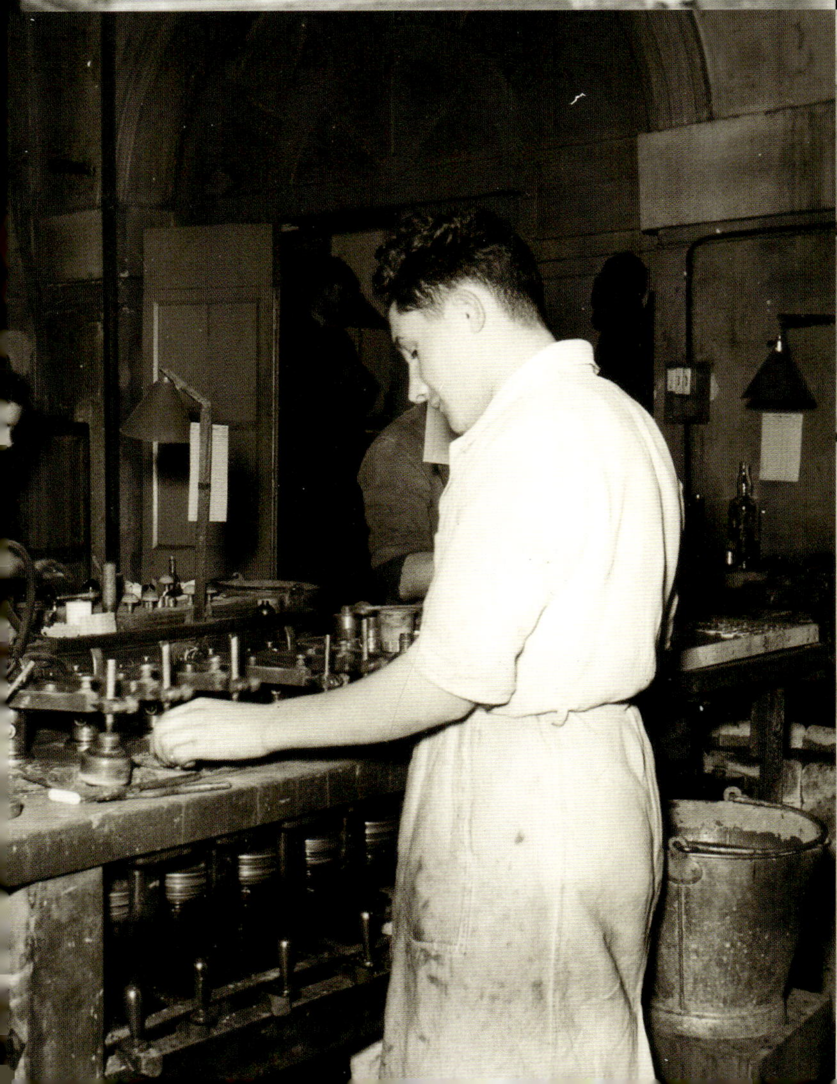

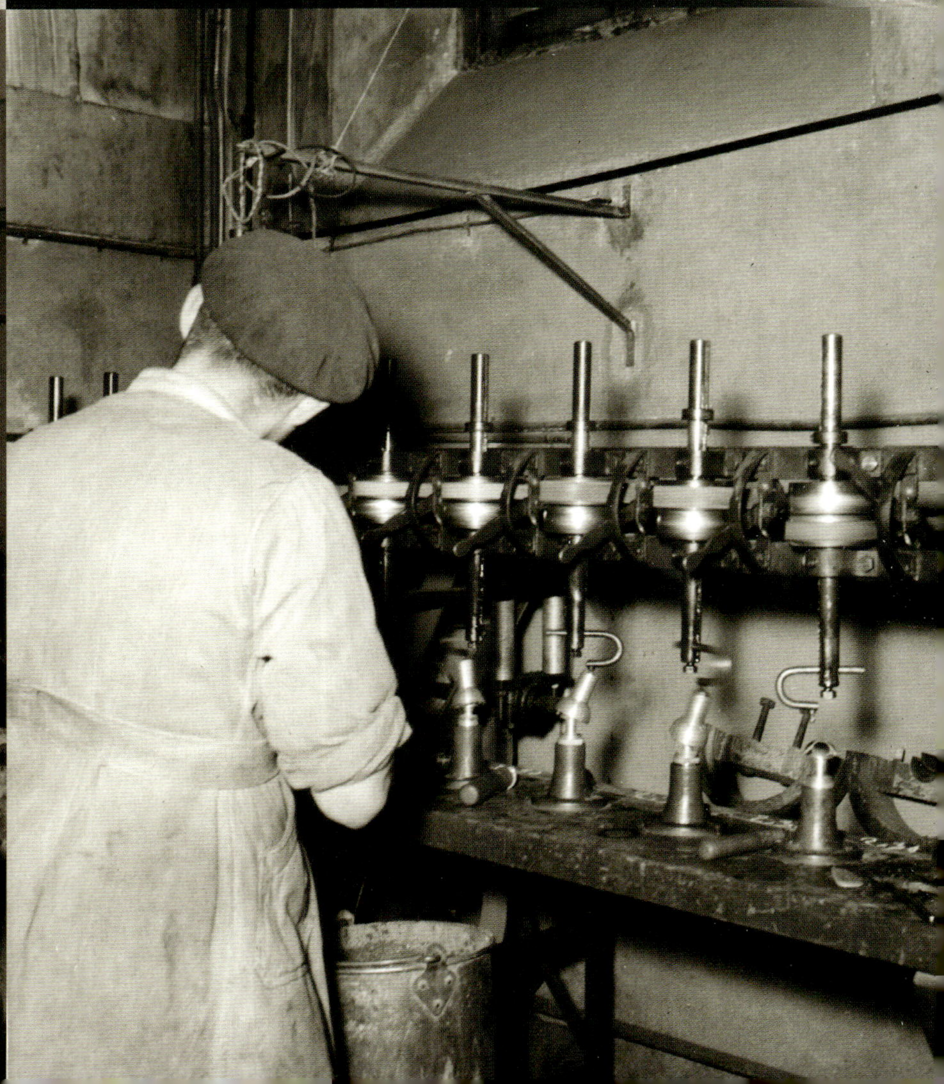

opposite
Facade of the ASIOM in the old school building in Saint-Héand.

The lens polishing workshop of the ASIOM, later Établissements Pierre Angénieux, in the old school (photo by Patrice-Hervé Pont).

The first polishing workshops in the Saint-Héand factory.

below
Headed notepaper of the Établissements Pierre Angénieux, June 1937, and an old testing laboratory.

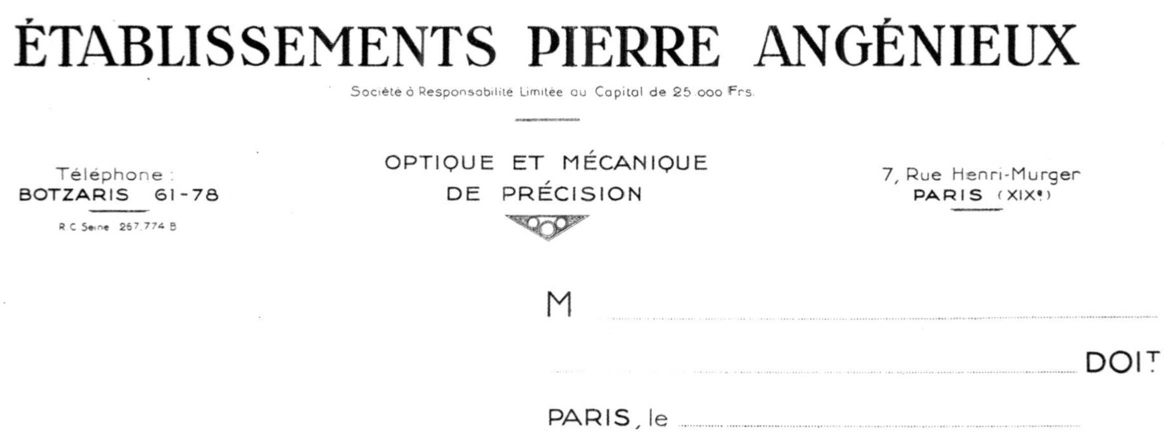
ÉTABLISSEMENTS PIERRE ANGÉNIEUX

Société à Responsabilité Limitée au Capital de 25.000 Frs.

Téléphone :
BOTZARIS 61-78
R C Seine 267.774 B

OPTIQUE ET MÉCANIQUE
DE PRÉCISION

7, Rue Henri-Murger
PARIS (XIXe)

M ....

.... DOIT

PARIS, le ....

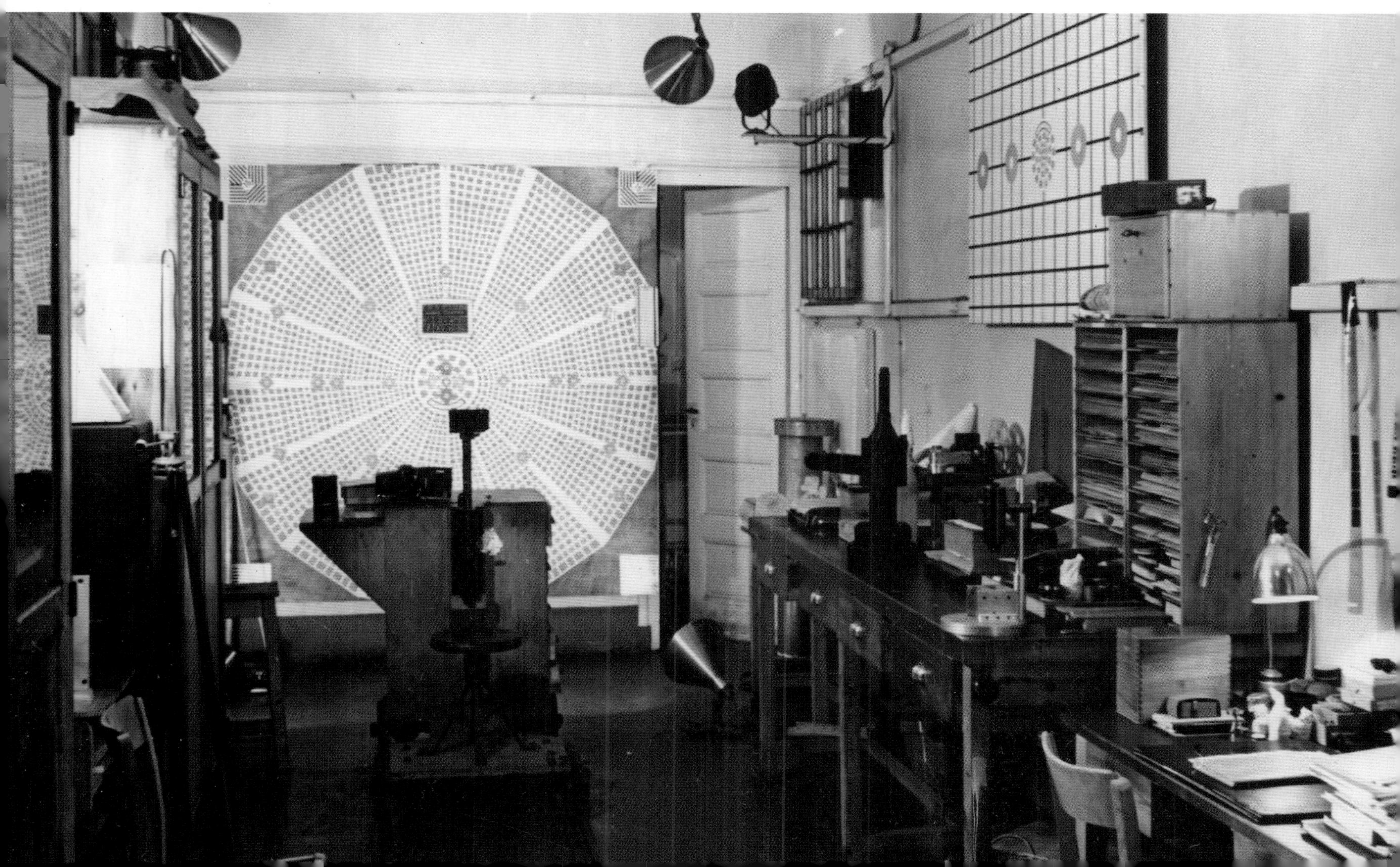

All those who worked with Pierre Angénieux are unanimous in their praise of his qualities as an exceptional man as well as his professional success.

« Pierre Angénieux's origins may help to explain his solid realism, refusal to show off, resistance to infatuation, and shrewd business sense.
All of his innovations were invented as a concrete response to the clearly identified needs of photographers or filmmakers.
All through his career, Pierre Angénieux succeeded in maintaining the close links established with the cinema world from the outset.
His intellectual capacity, mastery of the latest technology, and ability to listen enabled him to foresee trends and present the market with relevant products. »

André Masson

André Masson
Scientific director then chief executive of Angénieux from 1954 to 1988.

Jacques Debize
Debize graduated from the École Supérieure d'Optique in 1967 and joined Angénieux in 1969. One of his first calculations was for the NASA zoom. He left the company in 1972 only to return in 1973 and become Pierre Angénieux's right-hand man. Together they calculated the 42x16, which set the world record at the time, as well as the fixed focal lenses and zooms for the 24x36. Jacques Debize had a hand in all Angénieux's strategic developments. He retired in 2007 but continued to train young engineers.
Jacques Debize died in September, 2018.

Bernard Philibert
Philibert joined Angénieux in July 1961, initially as a draughtsman in the research department and then as head of the after-sales service until 1998, the year of his retirement.
Parallel to his professional career, Bernard Philibert was mayor of Saint-Héand from 1995 to 2014 and became an elected member of the Departmental Council.

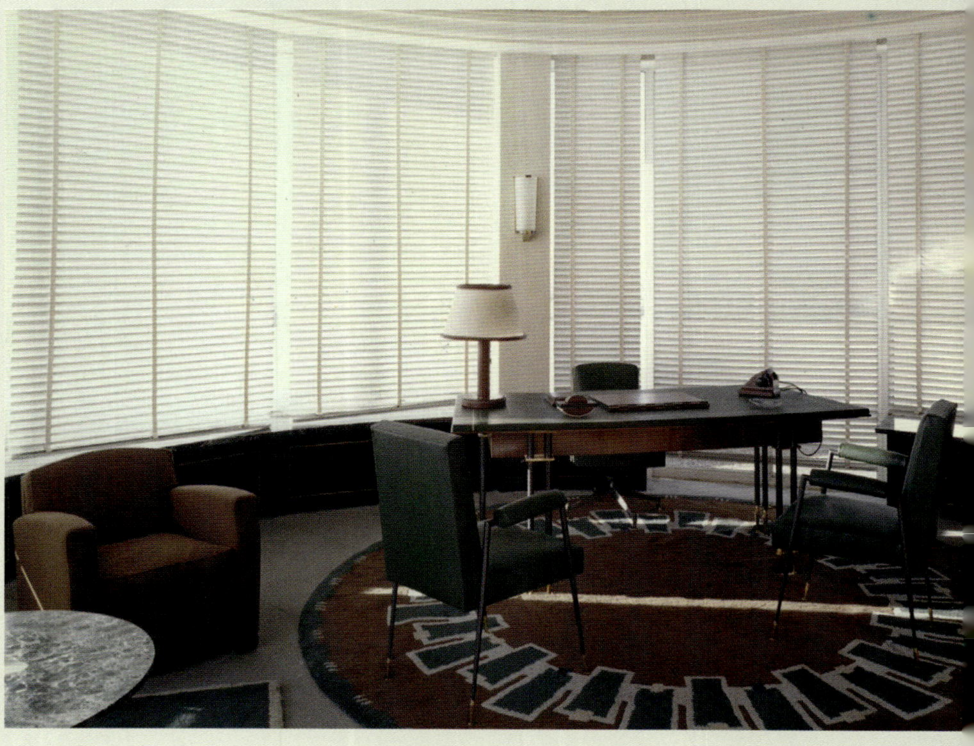

« Though very rigorous and demanding, Pierre Angénieux was also ready to show gratitude.
I truly appreciated his way of working. We did not always agree about the reason for a difference in calculation. In such cases, he would say, "I bet you a case of wine I'm right. Let's say that if you're right, I'll give you a case, but if I'm right you give me one bottle. That's fairer".
Pierre Angénieux loved fishing for pike, which he did regularly in his ponds in Forez. He was a simple man and hated being honored. He left it to his daughter Martine to go and receive the Technical Oscar in Hollywood in 1964 and would not even go to pick up his personal award in 1989.
On a more anecdotal level, I can also tell you that I once calculated the trajectories (curvature of bends, angle of tracks, optimal speeds) of the magnificent electric train set he had in his home in Geneva. »

Jacques Debize

A sophisticated art lover, Pierre Angénieux commissioned Jules Leleu, the most fashionable interior decorator of the time, to design his office (photo by Ito Josué, 1960s).

« [...] Pierre Angénieux was an exceptional man, one of those who stimulate the desire to work, to do a good job, to achieve the set aims.
Throughout his career, he succeeded in seeking out the brightest talents and persuading them to take part in the Angénieux adventure. He had the ability to recognize everyone's work and efforts.
Pierre Angénieux also believed in the virtues of practicing sport. Through the Saint-Héand athletics club, he founded the famous Saint-Héand thousand-meter race for youngsters aged 16–18. Each year 2,000 young people from the entire department took part, some of whom went on to become great champions. Pierre Angénieux made the name of Saint-Héand known all over the world and generated an entire economy in the town.
It is my great pride and honor to inaugurate a Pierre Angénieux street today. »

Bernard Philibert

Extract from the speech delivered in May 2003 at the inauguration of the street named after Pierre Angénieux in Saint-Héand

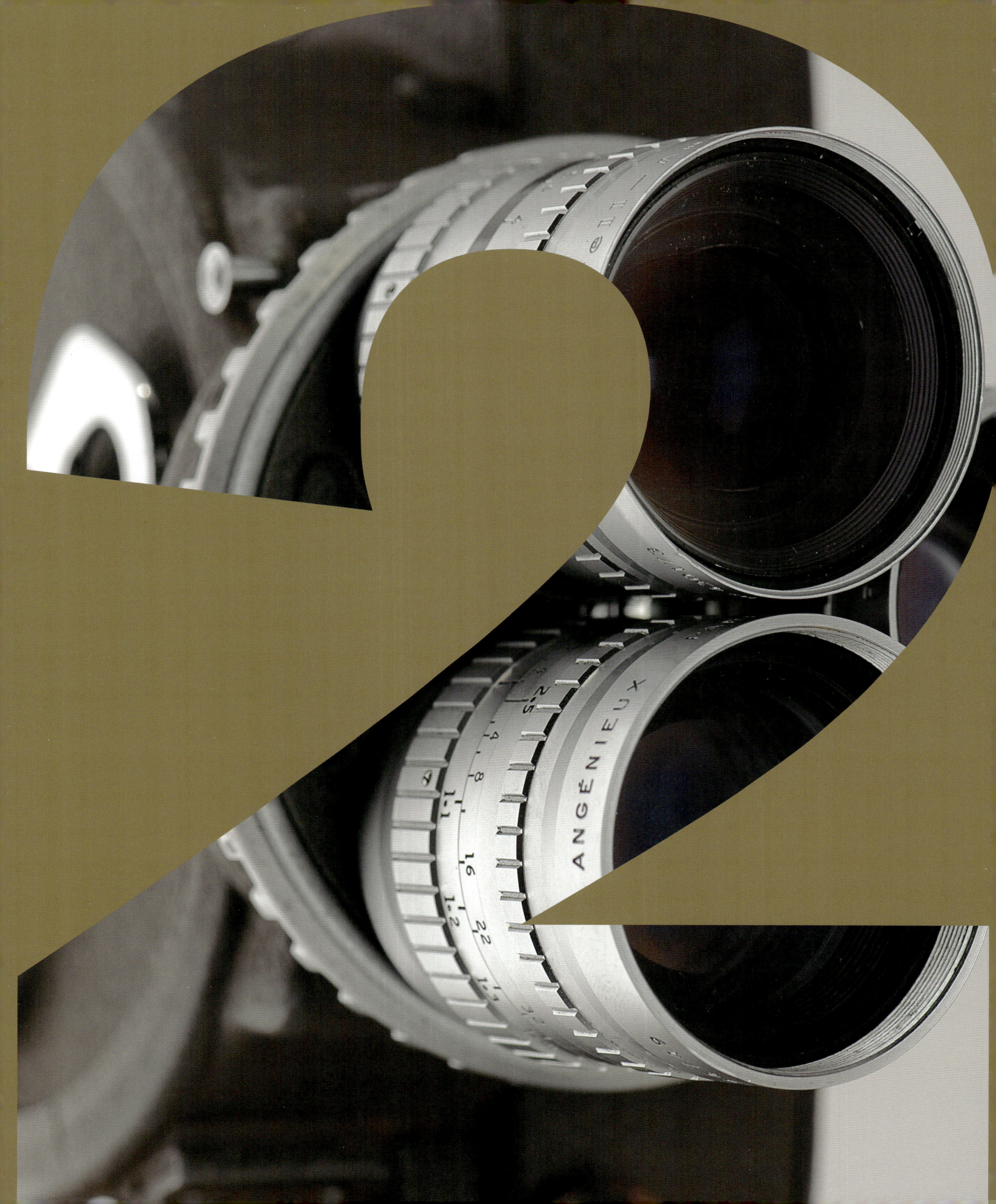
2
ANGÉNIEUX

THE PIONEERING YEARS
1935-1955

In August 1935, at the age of 28, Pierre Angénieux decided to set up in business for himself.
Photography was flourishing in France and the efforts of Pathé marked a heyday for home movies.
Pierre Angénieux opened his first **factory in Paris** near Buttes-Chaumont, initially manufacturing parts as a subcontractor.
Mechanical and optical items were produced respectively in Paris and **Saint-Héand**, where a factory was opened in 1937.
The **first lenses engraved** *P. Angénieux Ing. Paris*, an inscription devised with a view to the export trade, appeared in 1938.
The company had workforce of thirty employees by the eve of World War II.
The early 1950s saw a series of major innovations that were to establish the international reputation of the brand Pierre Angénieux.

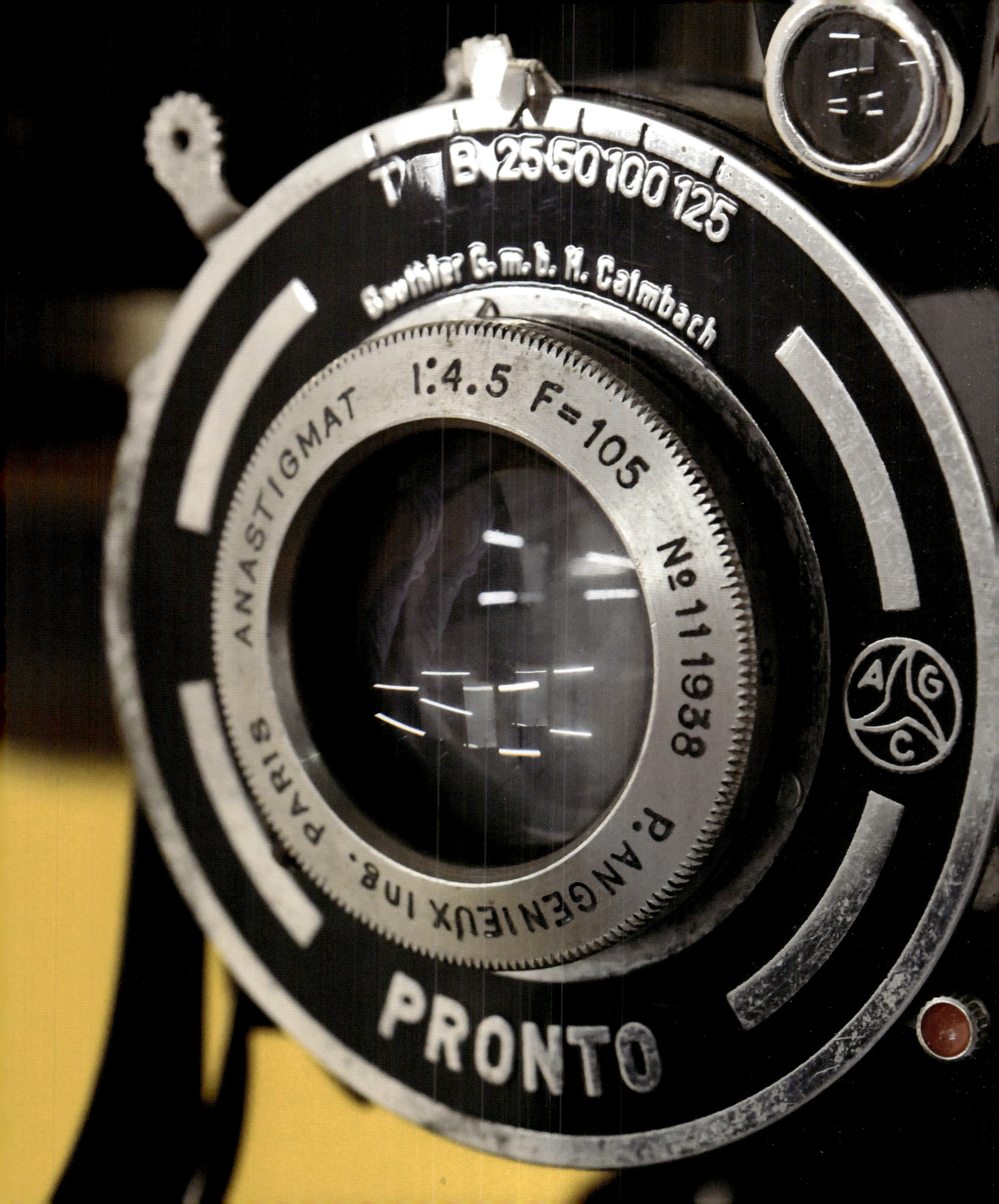

T B 25 50 100 125
Gauthier G.m.b.H. Calmbach
ANASTIGMAT 1:4.5 F=105
№ 111938
P. ANGENIEUX ing. PARIS
AGC
PRONTO

1939-1945
Pierre Angénieux Revolutionizes Optical Calculation

French cinema remained active throughout the war at the Studios de la Victorine in Nice, where Carné made *Les Visiteurs du Soirs* (1942) and *Les Enfants du Paradis* (1945), Cocteau *L'Éternel retour* (1943), Bresson *Les Dames du Bois de Boulogne* (1945), Clouzot *L'Assassin habite au 21* (1942) and *Le Corbeau* (1943), and Renoir *Falbalas* (1945).

Pierre Angénieux used these years of comparative inactivity to prepare for the future, concentrating on the development of new methods for the calculation of optical combinations.
A keen supporter of the German pionners' works in Scientific optics (Carl Zeiss, Ernst Abbe), he opted for the trigonometrical calculation of the trajectory of light rays, which he revolutionized by identifying a number of "pertinent" rays. The hours needed to calculate a lens were considerably reduced, thereby enabling him to test a larger number of solutions. Pierre Angénieux thus joined the very limited ranks of optical engineers creating new lenses. At the end of the war, with the German industry destroyed and demand booming, new markets opened up for French firms.

Against a Backdrop of War
Reclassified as fit for military service, Angénieux was specially assigned for an indefinite period as an engineer to his own firm, the Établissements Pierre Angénieux, on March 14, 1940.
The firm worked for the Ministry of Defense from September 1939 to June 1940. During the German occupation, Angénieux decided to close the factory in Paris and withdraw to Saint-Héand in the unoccupied zone, where he reorganized his activities around the optical engineering plant.
The patents taken out by Angénieux for prism binoculars and projectors in 1941 soon aroused the interest of the occupying authorities. In 1944 he was on the point of being arrested for refusing to work for Germany.

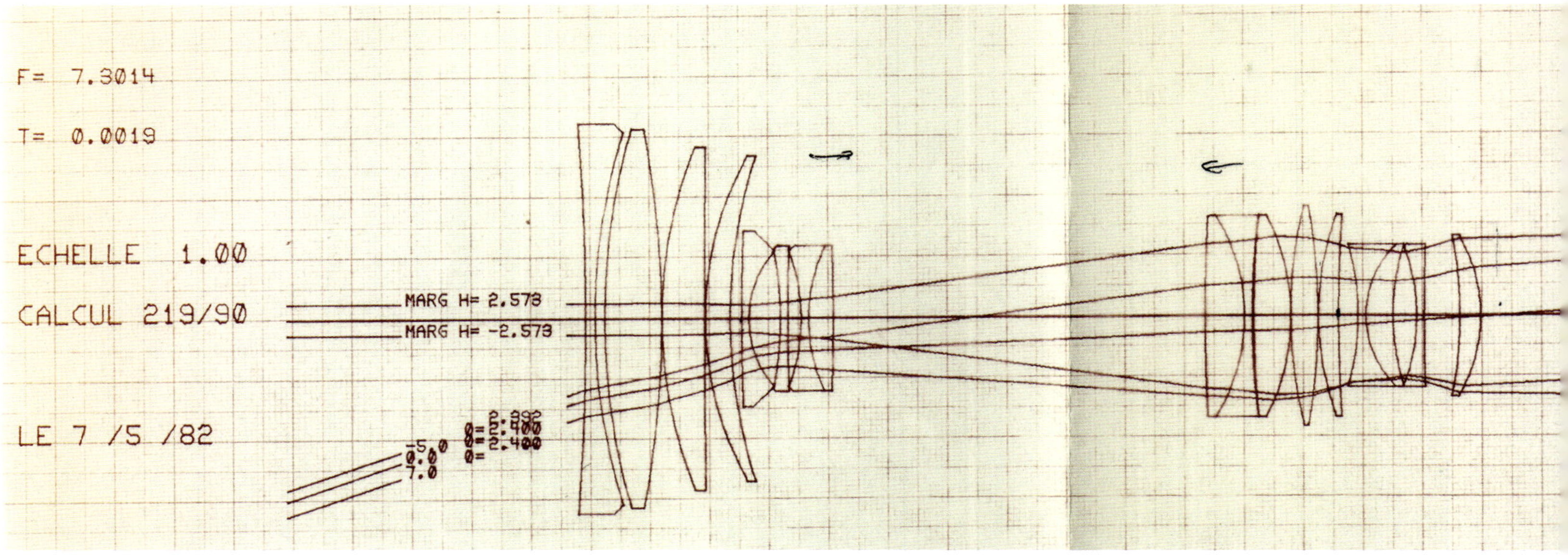

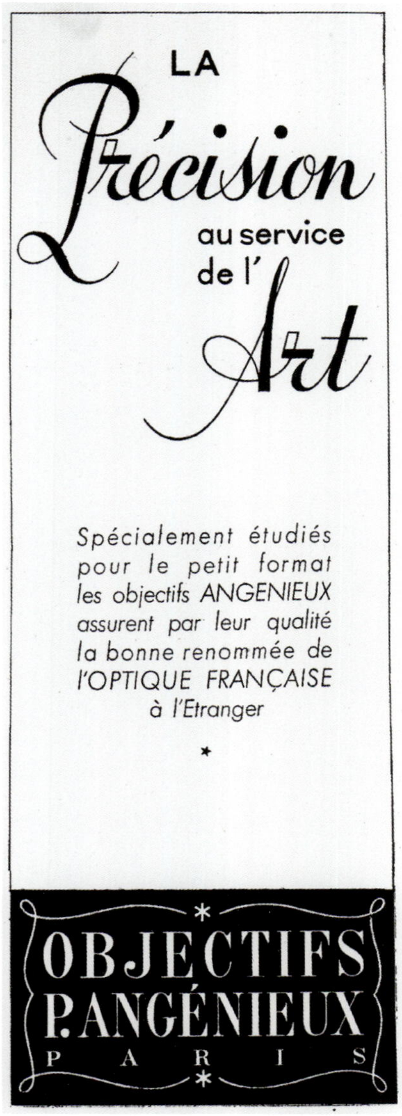

left
Advertisement: "Precision at the Service of Art," 1948.

« Pierre Angénieux spent the time at his disposal on establishing a more relevant method of calculation. While not actually new, this was applied with the same rigor as in Germany. There was and indeed is only one way to calculate an image: by following the rays of light and applying the known law of optics.
The only way is to follow all the rays emitted in an object-field and to do so through a certain number of lenses arranged so as to obtain the best possible image. It should be known that the calculation of the path of a ray of light takes 8–10 minutes per surface. If you have three lenses, for example, that makes six surfaces and means one hour to track a single ray. As some 300 rays must be tracked for just one trial, the calculation of a simple lens would take almost 2 years for at least 2 engineers. »

André Masson

right
Établissements Pierre Angénieux headed notepaper, August 1937 and 1950 (mentioning the Saint-Héand plant).

left
Pierre Angénieux's optimization of optical calculation made it possible to develop the complex zoom with mechanical compensation, a process accelerated by the use of early computers. Left, the drawing from July 1982 shows the trajectory of pertinent light rays (collection of André Masson).

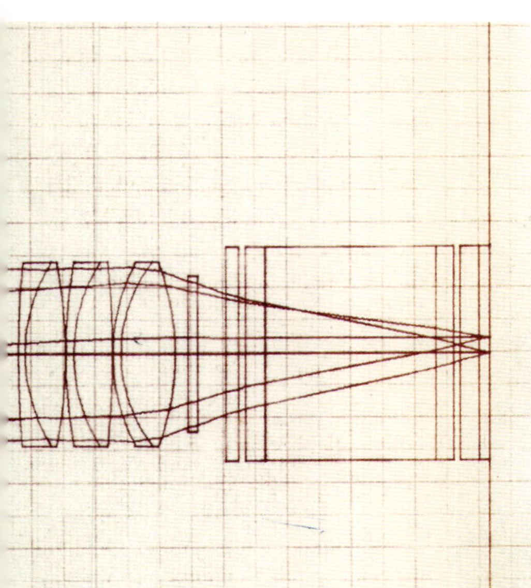

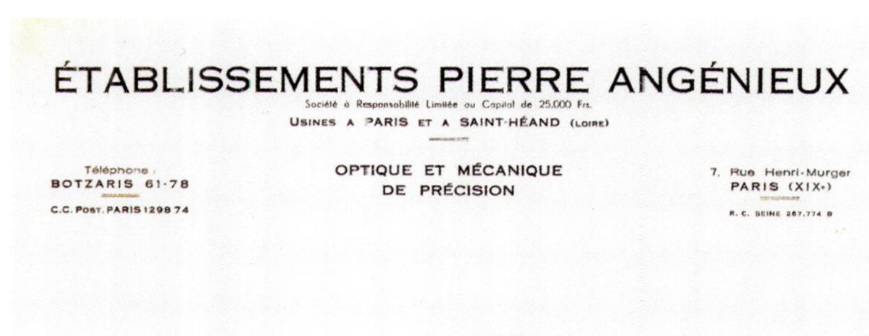
ÉTABLISSEMENTS PIERRE ANGÉNIEUX
Société à Responsabilité Limitée au Capital de 25.000 Frs.
USINES A PARIS ET A SAINT-HÉAND (LOIRE)

Téléphone :
BOTZARIS 61-78
C.C. POST. PARIS 1298 74

OPTIQUE ET MÉCANIQUE
DE PRÉCISION

7, Rue Henri-Murger
PARIS (XIXe)
R. C. SEINE 267.774 B

PARIS, LE

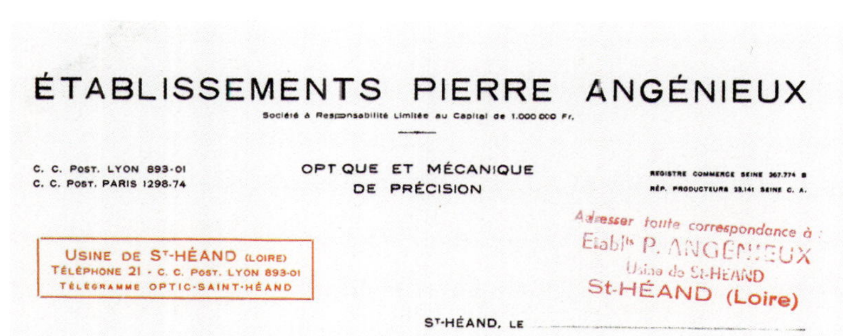
ÉTABLISSEMENTS PIERRE ANGÉNIEUX
Société à Responsabilité Limitée au Capital de 1.000 000 Fr.

C. C. POST. LYON 893-01
C. C. POST. PARIS 1298-74

OPTIQUE ET MÉCANIQUE
DE PRÉCISION

REGISTRE COMMERCE SEINE 267.774 B
RÉP. PRODUCTEURS 23.141 SEINE C. A.

USINE DE ST-HÉAND (LOIRE)
TÉLÉPHONE 21 - C. C. POST. LYON 893-01
TÉLÉGRAMME OPTIC-SAINT-HÉAND

toute correspondance à :
St-HÉAND (Loire)

ST-HÉAND, LE

1946
Construction Begins on a New Factory

Everything was now in place for the industrial launch. Pierre Angénieux gave himself the means to achieve his ambitions, increasing his workforce to 80 and commissioning the architectural firm of Ferraz and Seignol in Saint-Étienne to design a modern factory. He also recruited Lucien Granet, an engineer and fellow student at the Arts et Métiers in Cluny, to handle the construction of the new factory, and his brother-in-law Louis Moulard, a lawyer.

right
First overall color photo of the new factory in 1959, after a series of extensions. The building was based on the Kodak multistory factories in the USA. (Angénieux archives).

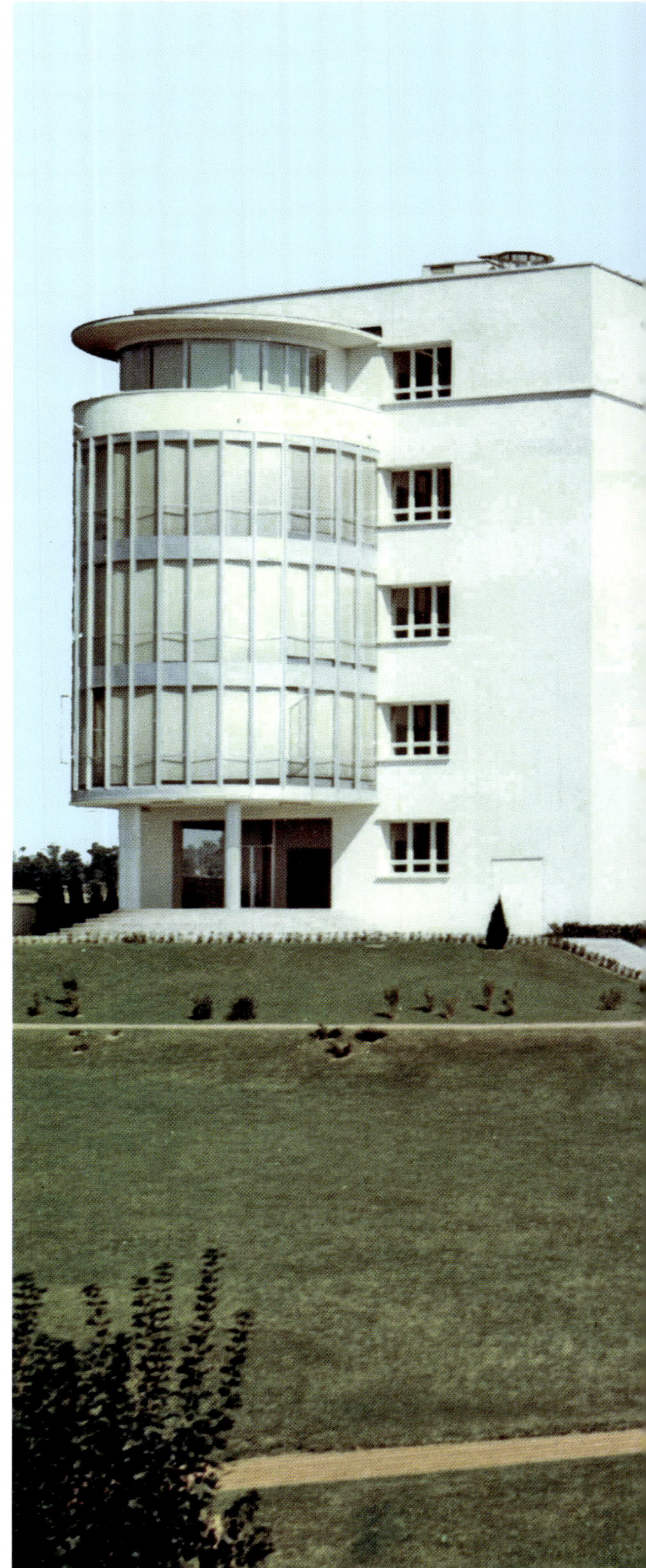

Lucien Granet
At the end of 1945, Pierre Angénieux asked Lucien Granet, a fellow student in Cluny, to join him and take charge of building the factory in Saint-Héand.
While Pierre Angénieux continued his research activities, Lucien Granet became the industrial director. After many years of highly fruitful collaboration, they retired at the same time in the mid-1970s.

Louis Moulard
Louis Moulard studied law with a view to becoming a notary. He was called up in 1939, taken prisoner, and held in Germany until 1945.
On his return, he married Marie-Louise Angénieux, Pierre's sister, in January 1946 and became head of the company's commercial department before taking over the administrative department.

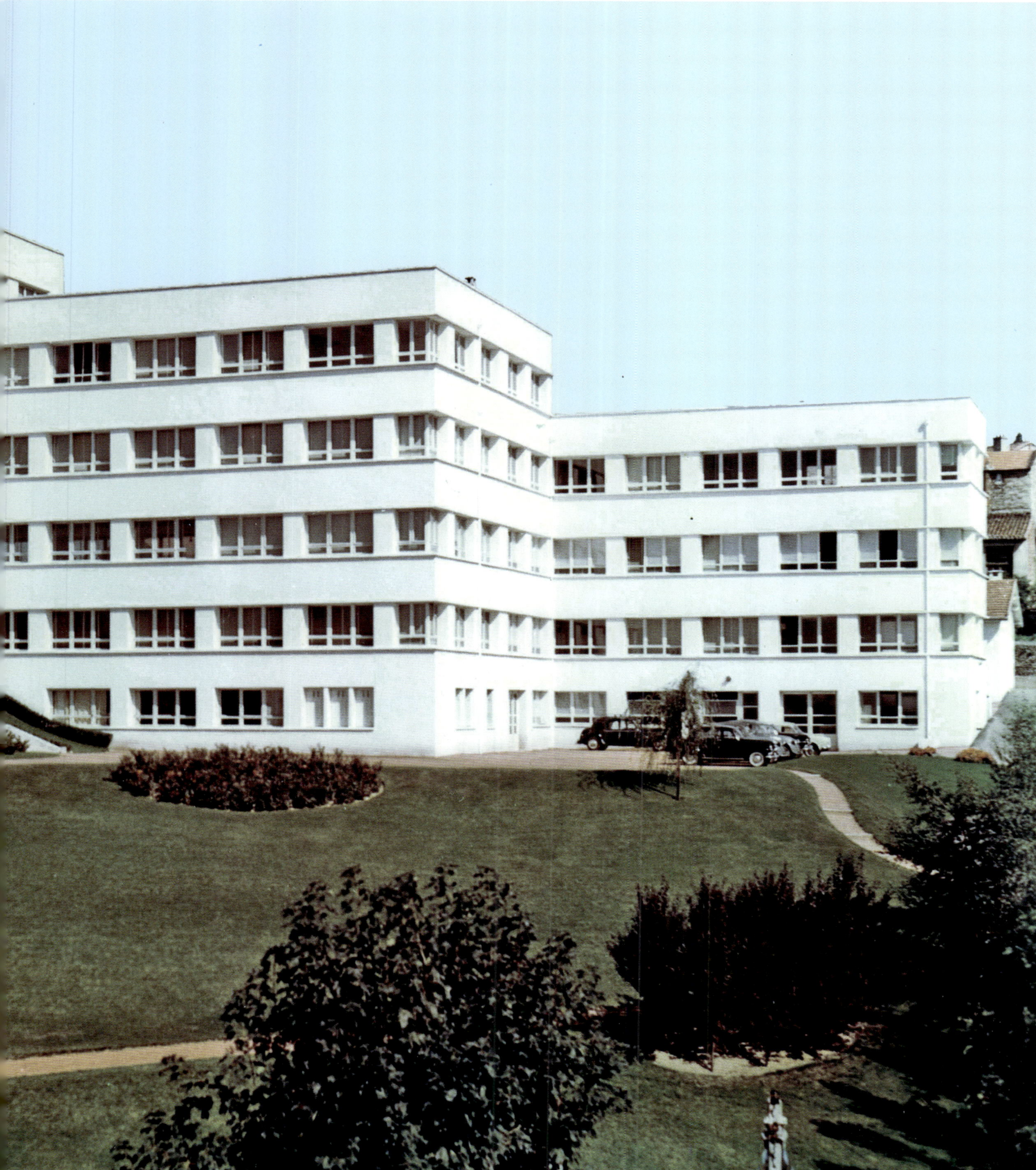

1935-1958
The Adventure of the Pictograph and S.E.B.A.G.A.

Fig.2.

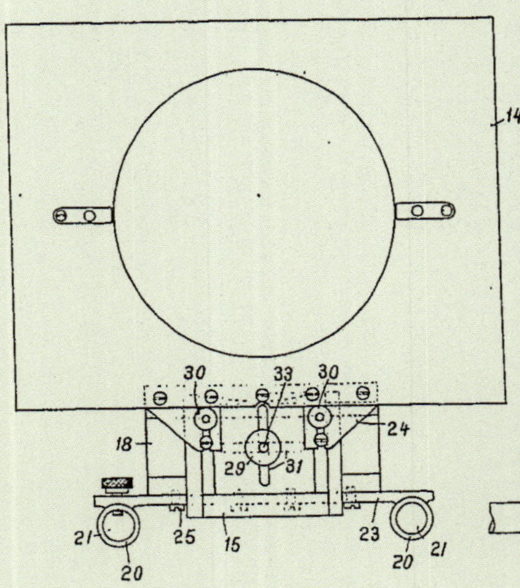

Fig.3.

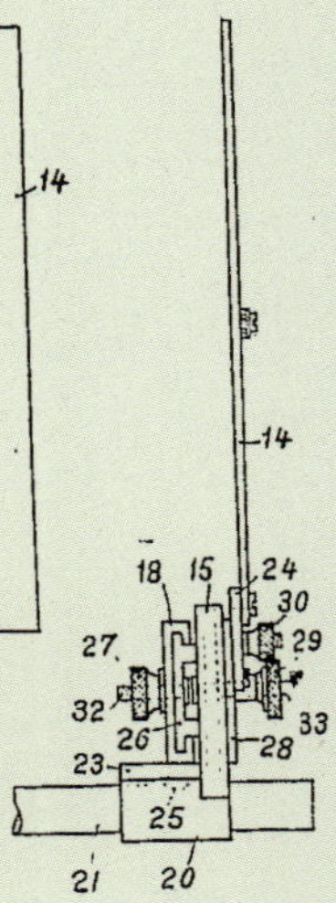

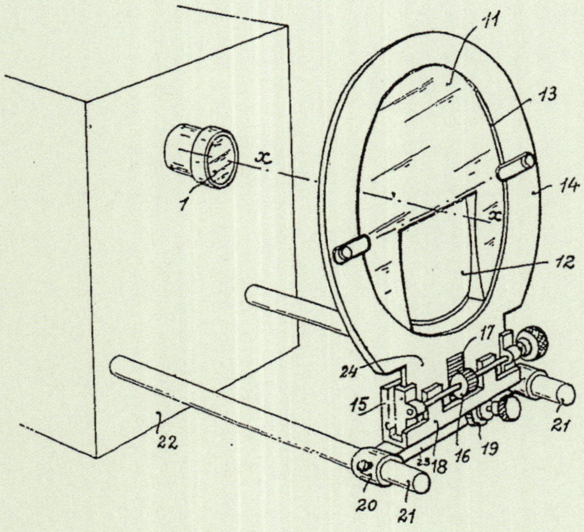

Fig. 1

Fig.4.

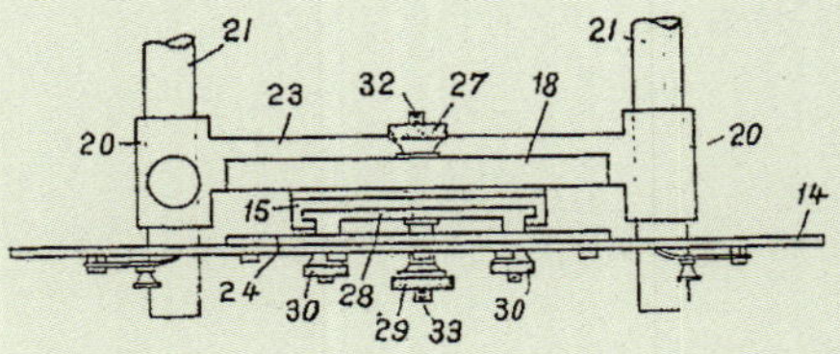

Pierre Angénieux met the greatest filmmakers of the time, including Abel Gance (1889–1981), who urged him to engage in optical research to further the progress of cinema.

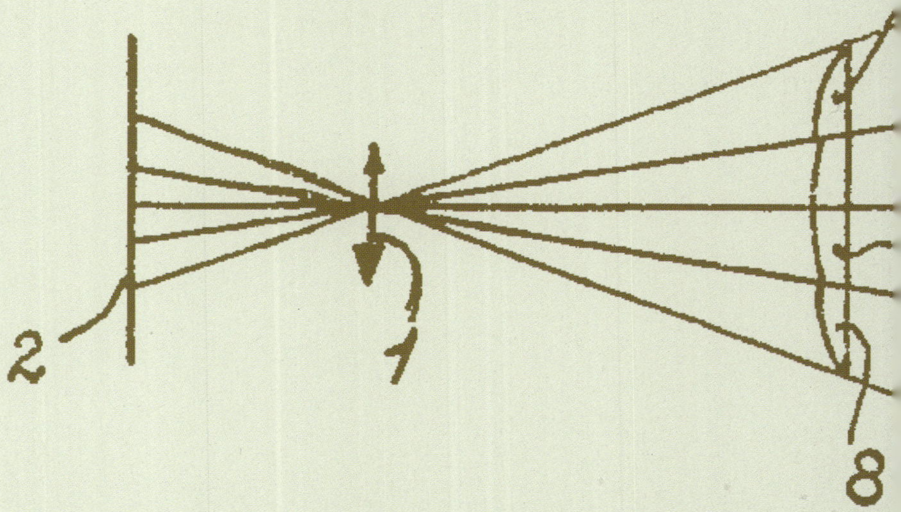

Abel Gance is one of the very few directors of the history of cinema to have invented and created the tools he required to capture his visions on film. In 1925, he and André Debrie invented a process of filming with three cameras next to one another, thus obtaining three times the standard width and making it possible to develop a narrative in three different perspectives. Abel Gance was to use this new process, called Polyvision, in his *Napoleon* (1927). Freeing himself from the traditional frame, he sought to give his film a more epic and indeed mythic dimension.
In 1929–1932, again with André Debrie, he applied for a patent for *perspective sonore* or "sound perspective", a forerunner of stereophonic sound, and brought out a sound version of *Napoléon* retitled *Napoléon Bonaparte*. In January 1937, prompted by a number of Primitive paintings seen in a museum in Amsterdam, he came up with the idea of a process that would make clear focusing possible simultaneously at different focal depths of the field of vision. He contacted Pierre Angénieux, who had founded his company just two years earlier, and they applied for an initial patent on July 5, 1937. This marked the birth of the Pictograph.

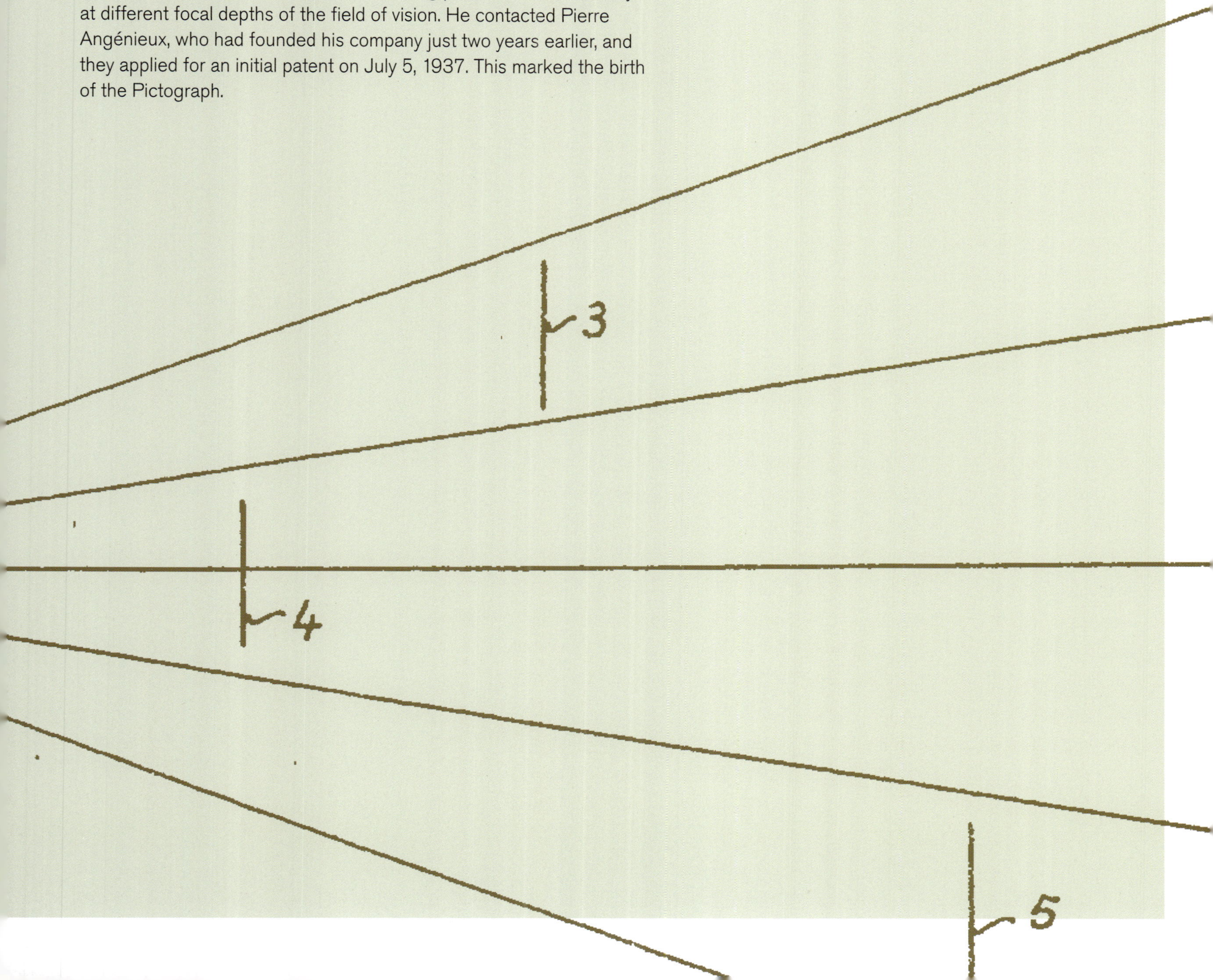

Patent of July 5, 1937 for the "improvement of optical systems for photography" consisting of "lens elements assembled and placed before a lens to divide its field into different zones with different focusing distances".

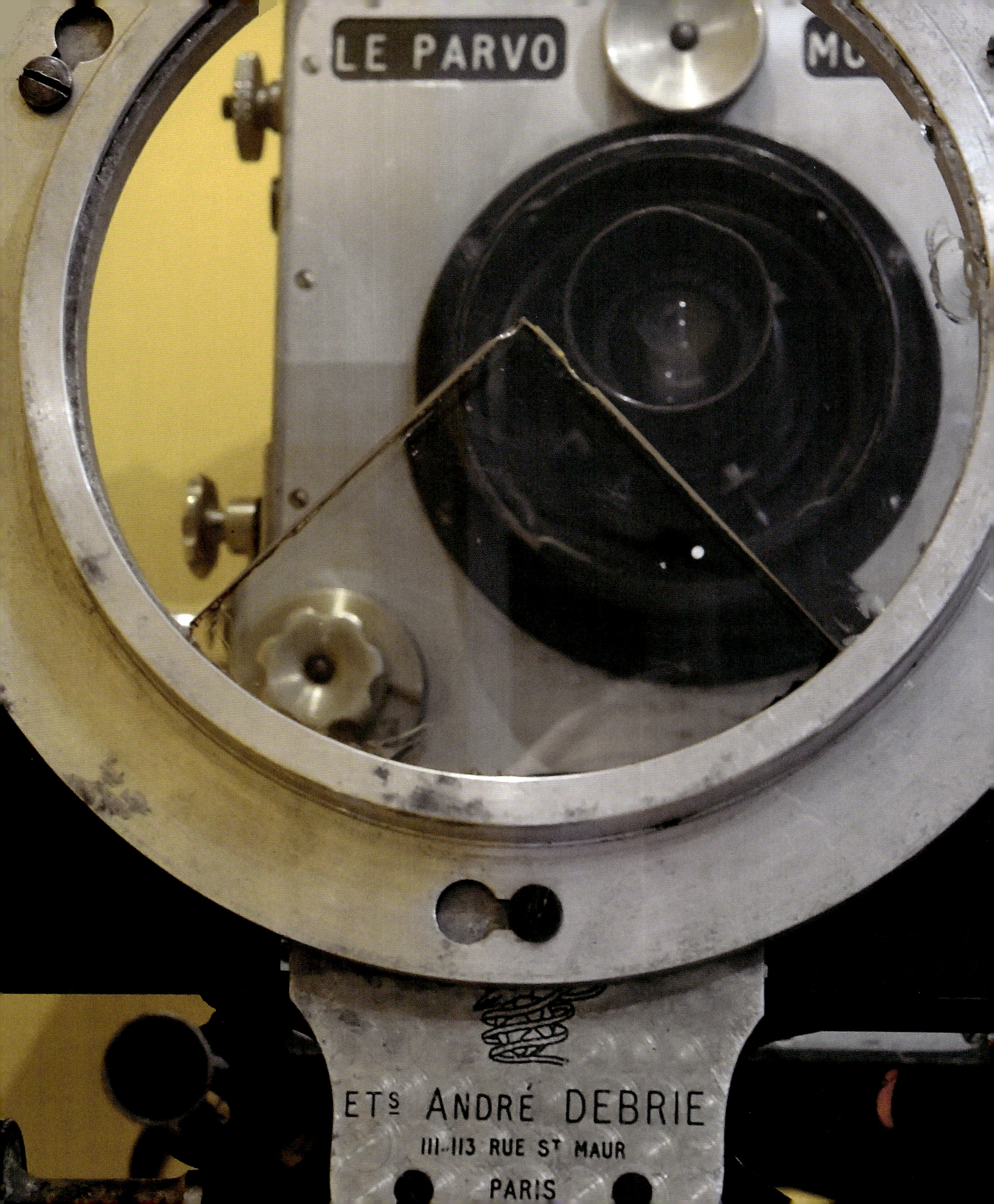
LE PARVO
Ets ANDRÉ DEBRIE
111-113 RUE St MAUR
PARIS

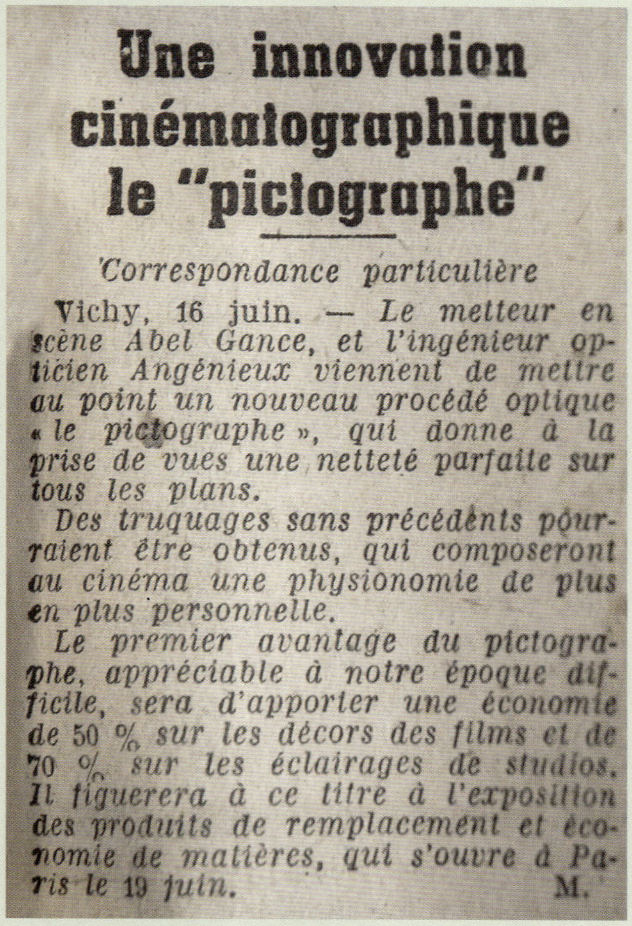

Une innovation cinématographique le "pictographe"

Correspondance particulière

Vichy, 16 juin. — *Le metteur en scène Abel Gance, et l'ingénieur opticien Angénieux viennent de mettre au point un nouveau procédé optique « le pictographe », qui donne à la prise de vues une netteté parfaite sur tous les plans.*

Des truquages sans précédents pourraient être obtenus, qui composeront au cinéma une physionomie de plus en plus personnelle.

Le premier avantage du pictographe, appréciable à notre époque difficile, sera d'apporter une économie de 50 % sur les décors des films et de 70 % sur les éclairages de studios. Il figurera à ce titre à l'exposition des produits de remplacement et économie de matières, qui s'ouvre à Paris le 19 juin. M.

The Pictograph is a system of lenses, each of which can be made to measure according to the desired division of the image. It thus became possible to replace scenery with simple models or photographs, looking forward to the later processes of visual superimposition.

The **S.E.B.A.G.A.** (Société d'Exploitation des Brevets Abel Gance et Pierre Angénieux) was founded on August 5, 1937 by Abel Gance, Pierre Angénieux and the cinematographer Roger Hubert in partnership with Forrester-Parant Productions. By contract, the purpose of the company was the commercialization, sale and licensing of the Pictograph and derivatives in France and other countries. The Pictograph was first used for Abel Gance's *J'accuse* in May 1937 and then for *Capitaine Fracasse* in 1942.

Patent applications were submitted in Germany and the United States in 1937, the United Kingdom in 1938, and Spain, Holland, Italy, Portugal, Sweden, and Switzerland in 1943. The US patent was not definitively granted until 1952, having been delayed until 1941 and then confiscated for military purposes for the duration of the war.

Meanwhile, unbeknown to the S.E.B.A.G.A., the Pictograph was used in the United States from June to October 1940 in the making of Orson Welles's film *Citizen Kane*. The director of photography Gregg Toland had indeed been present at the testing of the system in 1939. No legal action could be taken, however, as the patent had not yet been granted in the United States.

In spite of the German occupation, the scarcity of materials and electricity, and the departure of Abel Gance for Spain, Pierre Angénieux was unsparing in his efforts. While the new system did arouse the interest of the major US production companies, the S.E.B.A.G.A. failed to obtain the expected revenues and was wound up in 1958.

The event was even mentioned in the local press of Saint-Étienne. Above, copy of the *Mémorial de Saint-Étienne* of June 17, 1943 (Municipal Archives of Saint-Étienne 7C7 275).

An innovation for motion pictures The "Pictograph" Special report Vichy, June 16. Director Abel Gance and optics engineer Angénieux have just developed a new optical procedure, the "Pictograph", which offers a perfectly sharp image at every distance. Never-before-seen illusions can be achieved, which will give motion pictures a much more personal look. The first advantage of the Pictograph, which we can certainly appreciate in these difficult times, will be to reduce the costs of film sets by 50%, and of studio lighting by 70%. As such, it will appear in the exhibition of products that replace or save materials, which opens in Paris on June 19.

Angénieux restored the set of Pictograph lenses held by the Cinémathèque Française in 2009.

right and opposite Parvo Debrie camera, model K, series Y, equipped with a Gance-Angénieux Pictograph lens (Cinémathèque Française Collection, photographs by Éric Perrin and Marc Salomon).

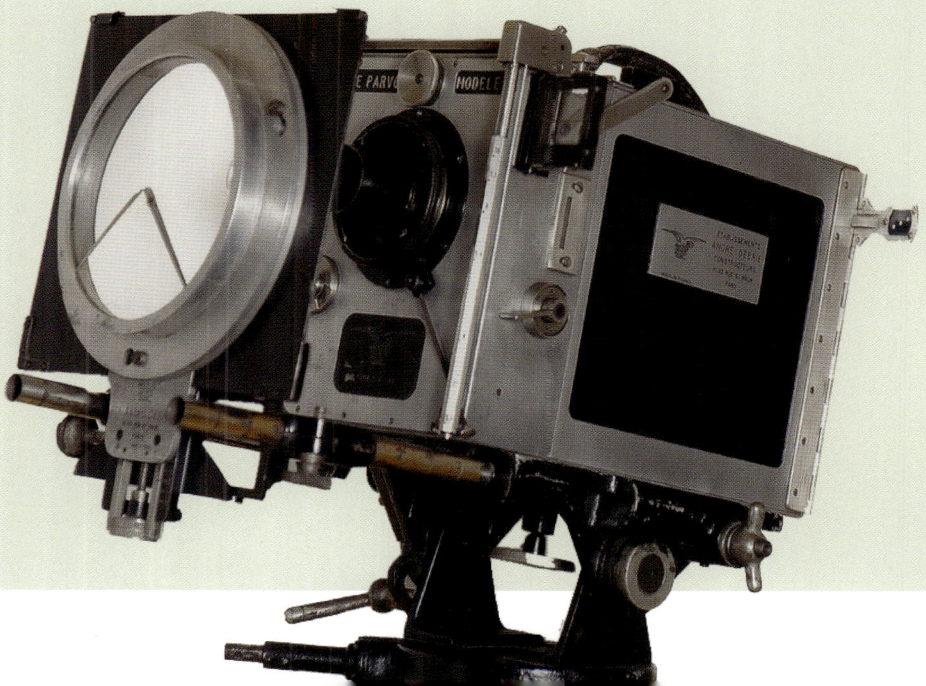

Advertising brochure for the Pictograph printed for an exhibition on economizing the use of materials and replacement products held at the Grand Palais in Paris (June 19–July 4, 1943). The event was accompanied by the making of a promotional demo film entitled *Pathé présente le Pictographe d'Abel Gance et Angénieux* (Cinémathèque Française Collection, Abel Gance archive).

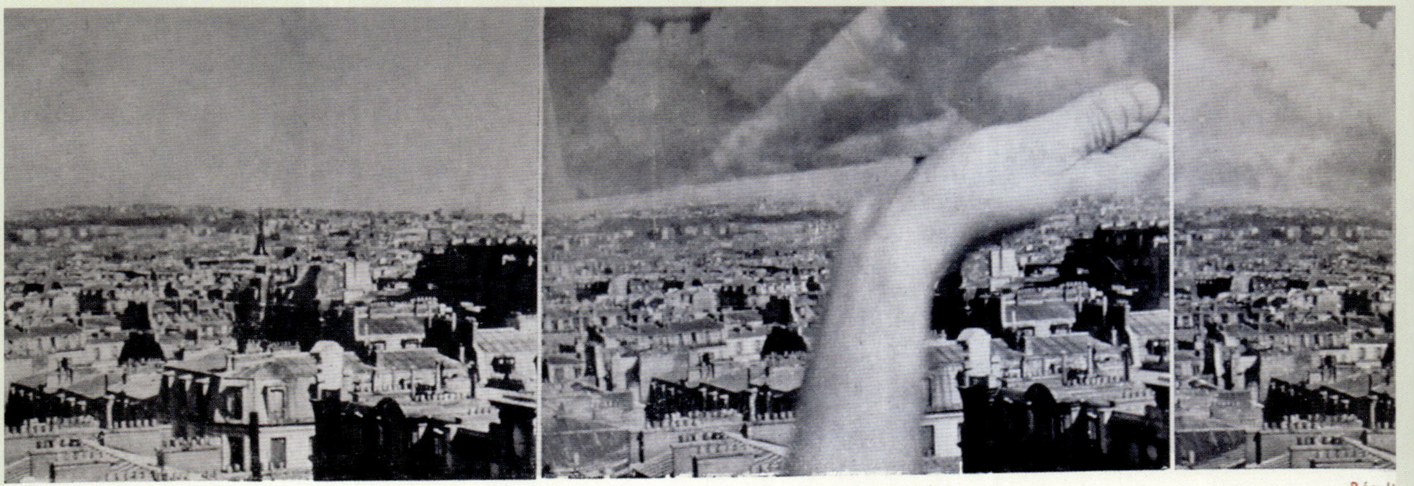

Panorama de Paris.

L'operateur ajoute une photo représentant des nuages.

Résult

Quelques mots sur :

LE PICTOGRAPHE

Le Cinéma nous a habitués aux miracles.

Du muet au parlant, de la couleur au relief, des truquages primitifs de Méliès à ceux de « L'Homme invisible », chaque fois on a fait un bond en avant dans ce métier, sans cesse, d'après son étymologie même, en « mouvement ».

Aujourd'hui, un metteur en scène célèbre : Abel GANCE, et un opticien réputé, Pierre ANGÉNIEUX, lancent une invention sensationnelle qu'ils ont appelée le **Pictographe.**

De quoi s'agit-il?

Tout simplement de supprimer une partie importante des décors, dont le devis sera baissé de 50 %, avec une économie de temps et de lumière de près de 75 %, tout en conférant aux images une valeur picturale et artistique infiniment plus grande.

∴

N'oublions pas, tout d'abord, que c'est à Abel GANCE que le Cinéma Français doit, entre autres grandes œuvres marquées de sa profonde personnalité (*J'Accuse, La Roue, Napoléon, Un Grand Amour de Beethoven, etc...*), le **Tryptique,** qui fut à l'origine de l'**Ecran Magnascopique,** et la **Perspective Sonore,** qui permit de donner un plus juste équilibre aux valeurs auditives et visuelles, laquelle perspective fut d'ailleurs fort habilement « empruntée » par Walt Disney, pour son *Fantasia.*

Voilà pour le passé. Restons-y un instant encore pour parler de la naissance de l'idée qui devait devenir, avec la collaboration de Pierre ANGÉNIEUX : le **Pictographe.**

∴

GANCE, de passage à Amsterdam, alla visiter le célèbre Musée Hollandais et tomba en arrêt devant les toiles étonnantes des Primitifs.

Emotion artistique, bien entendu, mais aussi méditation qui allait le conduire sur des voies nouvelles.

Derrière les personnages d'Holbein ou de Memling, l'auteur de *Mater Dolorosa* réfléchit à l'importance que les peintres d'alors donnaient aux fonds de leurs toiles, synthétisant dans un paysage ou dans un décor très fouillé et très net tout ce que ces portraits comportaient de résonnances picturales, les faisant ainsi participer, par cet équilibre instinctif des valeurs, à la réussite totale de leurs tableaux.

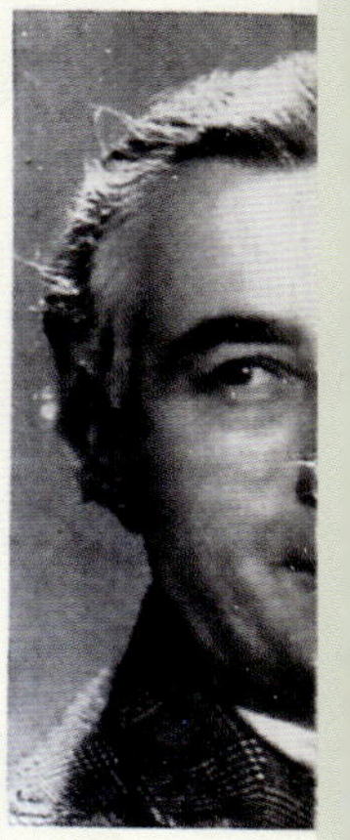

Abel G

Un parallèle se fit dans son esprit : Pourquoi ne pas essayer d'appliquer le même procédé au cinéma commandé, jusqu'à présent, par des objectifs indociles et toujours optiquement arbitraires?

Ce jour-là, il nota sur ses feuillets ces quelques lignes brèves : « Pour gros plans, adopter idée de l'Ecole Primitive. Fond synthétique en réduction derrière un personnage. Echelle de 1/3 réduite pour paysages ou constructions. »

Il ajoutait : « Cela doit former un grand pas en avant dans la technique de notre métier. »

Le **Pictographe** était né.

∴

Mais il fallait l'élever, le faire grandir, lui donner, si j'ose dire, une éducation.

Pour ce faire Abel GANCE rendit visite à son

Photo prise avec l'objectif habituel. Mise au point sur le premier plan.

La même scène avec le "Pictographe" qui permet la mise au point totale sur tous les plans.

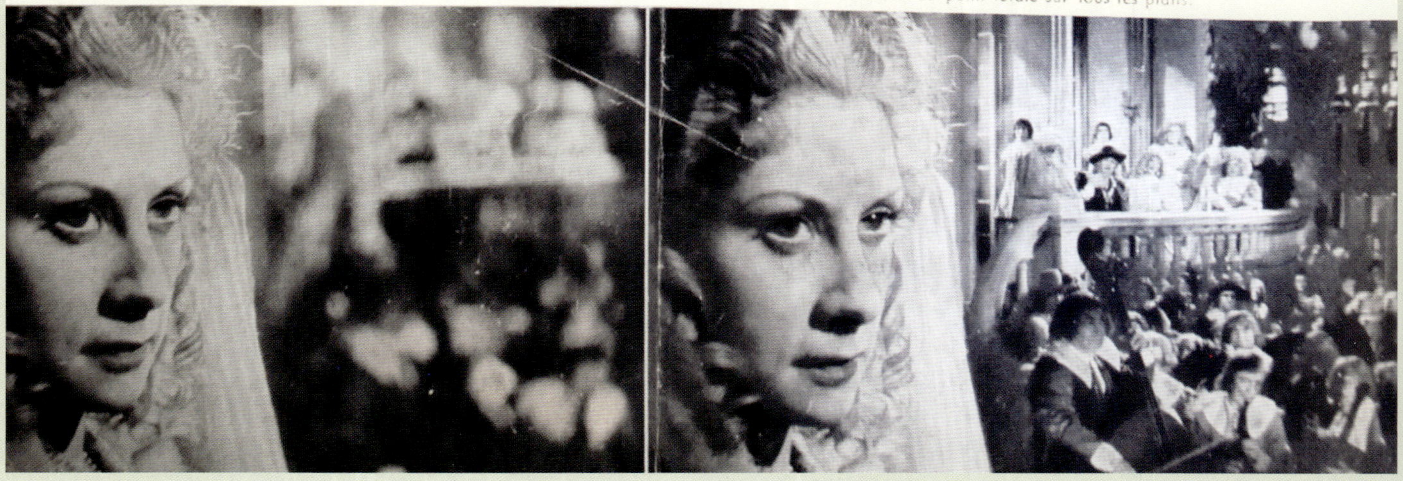

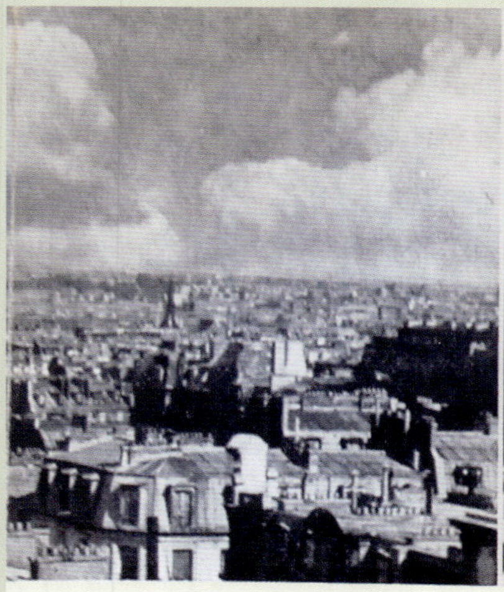

...nt avec le "Pictographe".
...ans commentaire...

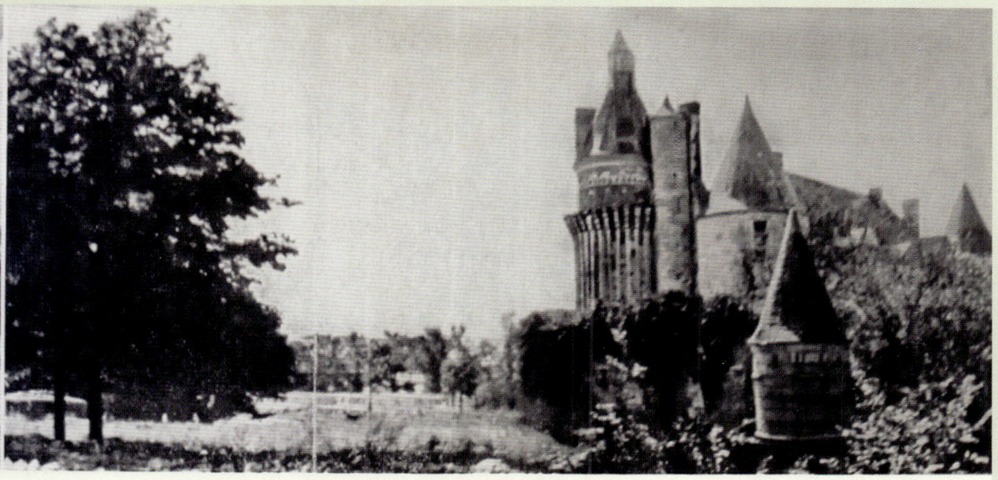

Un magnifique Château de la Loire... qui n'est qu'une simple photo 6×9. L'extérieur de gauche seul est réel. Les 2 vues "pictographiées" donnent cette étonnante photo d'ensemble.

...ami l'opticien ANGÉNIEUX et, tous deux, penchés sur le même problème pendant des mois, trouvèrent enfin la clef du problème.

Grâce à un système de lentilles différentes, placées à une distance X de l'appareil, on allait pouvoir filmer *en très Gros Plan* un sujet, ayant derrière lui, soit d'autres interprètes, soit un décor dans le lointain, en ayant la possibilité de rendre à l'écran une scène d'ensemble parfaitement nette.

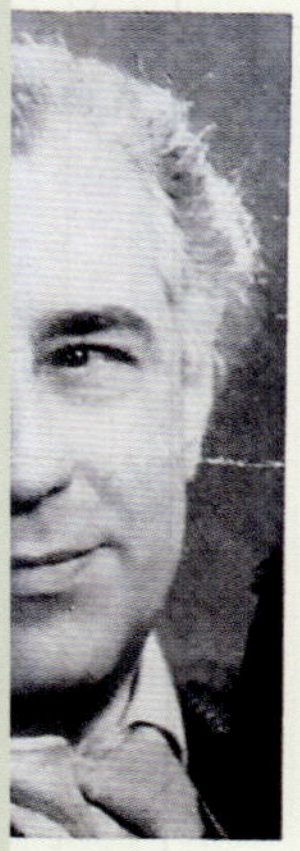

...ANCE

C'était **la mise au point totale sur tous les plans.**

Jusqu'alors, les réalisateurs et les opérateurs étaient esclaves des objectifs qui ne « consentaient » à prendre qu'un champ avarement délimité en laissant dans l'ombre, ou flous, les lointains qu'il fallait encadrer dans un second plan, puis un troisième, etc.., d'où une perte importante de temps, et pour les techniciens et pour les acteurs.

Heureuse et utile vengeance maintenant : les 28, 35, 40, 50, etc., allaient à leur tour *obéir*. L'Homme avait vaincu la Machine. Et si les objectifs n'acceptaient pas la chose « aveuglément », ils le faisaient, au contraire, d'une manière lumineuse, ce qui est bien le moins quand on est destiné à cet emploi.

Et c'est ainsi que l'on est arrivé à avoir dans un même plan un magnifique château, dans un Extérieur s'étendant sur des kilomètres, avec la facilité nouvelle pour les interprètes de jouer en venant *depuis le fond le plus éloigné jusqu'au très gros plan*, le tout dans une seule prise de vue directe.

∴

La belle affaire, direz-vous, et où voyez-vous l'économie : construire un château!

Que l'on se rassure. *Le Castel en question n'est autre qu'une simple pictographie.* Seul, le paysage est réel et, par le système optique du **Pictographe,** on obtient *l'illusion totale* que l'ensemble du Décor existe et vit, d'autant mieux, que la qualité photographique est supérieure à ce que l'on avait l'habitude de voir et permet d'obtenir un réalisme étonnant.

Ce Château peut être, d'ailleurs, aussi bien une Ville, un Pays, une Mansarde, un Palais, etc..., et c'est là que nous touchons au miracle, tous ces éléments pouvant devenir des images *mobiles.* C'est ainsi que nous pourrons avoir la mer soulevée par les flots d'une violente tempête, le frissonnement du vent dans les arbres, un port avec des bateaux sous la houle, — tout est maintenant possible et les exemples iraient à l'infini...

C'est là, on l'admettra, un progrès inappréciable, un pas énorme fait dans le domaine multiple de la technique, dans l'économie d'une production, et, par conséquent, dans son financement.

L'invention bouleverse toute la prise de vue et, par contre coup, le style même des scenarii, ce qui a pu faire dire à Marcel ACHARD qu'avec le **Pictographe**, *il faudrait réécrire les découpages.*

∴

Ce nouveau procédé sensationnel a été breveté en France *le 1er août* 1938 et il l'est actuellement en Allemagne, en Amérique, en Angleterre et dans la plupart des Pays.

Le Gouvernement Français, en la haute personnalité de M. GALEY, Directeur Général de la Cinématographie Nationale, a bien voulu comprendre, le premier, l'énorme retentissement du **Pictographe** et a tenu à faire subventionner un *Film de Démonstration* à l'usage des techniciens.

Ce film, le public ne le verra pas, car il ne servira d'abord que d'études et de bancs d'essais, mais les spectateurs n'en seront pas lésés pour autant et, dans quelque temps, ils pourront, à leur tour, apprécier les résultats probants de l'invention dans *tous* les films à venir.

Pour une fois, un proverbe a menti, mais d'une façon constructive, car si Abel GANCE et Pierre ANGÉNIEUX ont été Prophètes dans leur pays, ce dernier aura, grâce à eux, un rayonnement nouveau à travers le Monde, et le **Pictographe** ira porter au-delà des frontières l'esprit inventif de la France, dira le labeur patient de ses enfants et montrera que nos Trois Couleurs sont encore flamboyantes et claquent toujours aussi fort dans le vent...

Ce personnage semble visiblement attendre quelque chose.

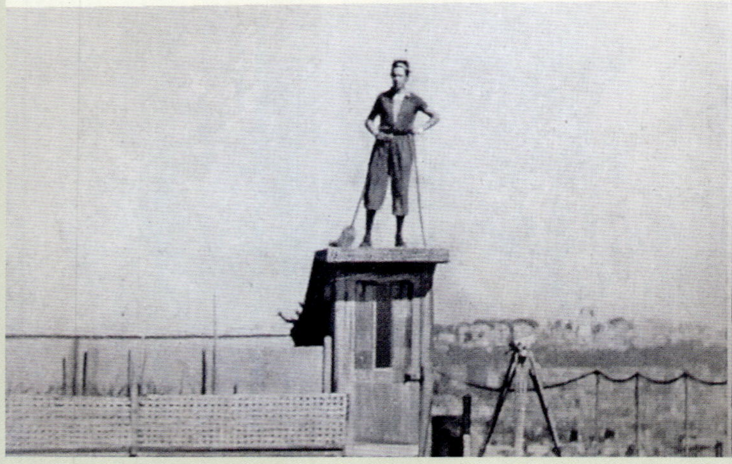

La voici : en une seconde, grâce au "Pictographe" il a été transporté du toit des Studios Pathé au milieu d'un merveilleux paysage de neige.

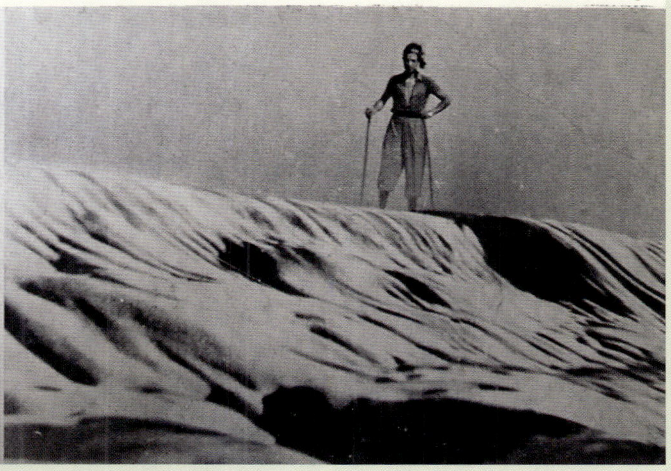

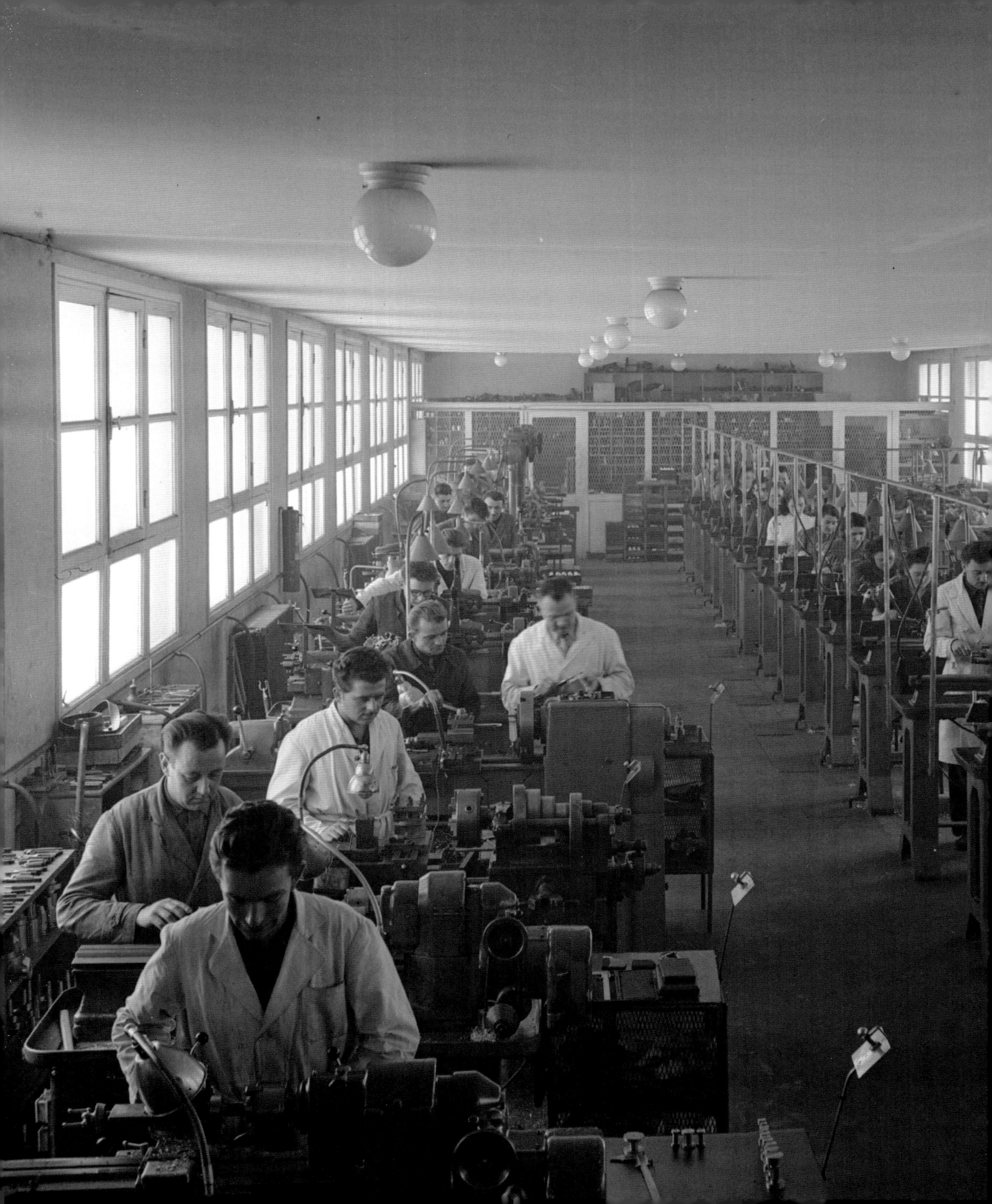

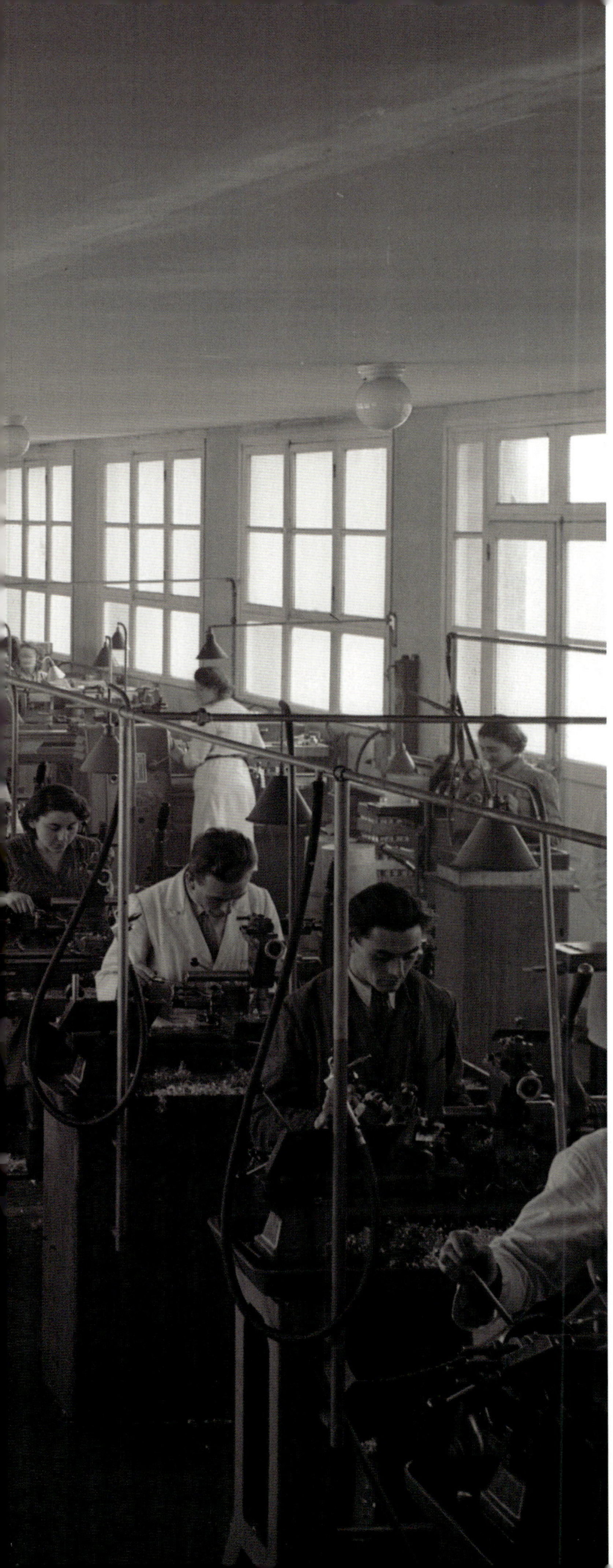

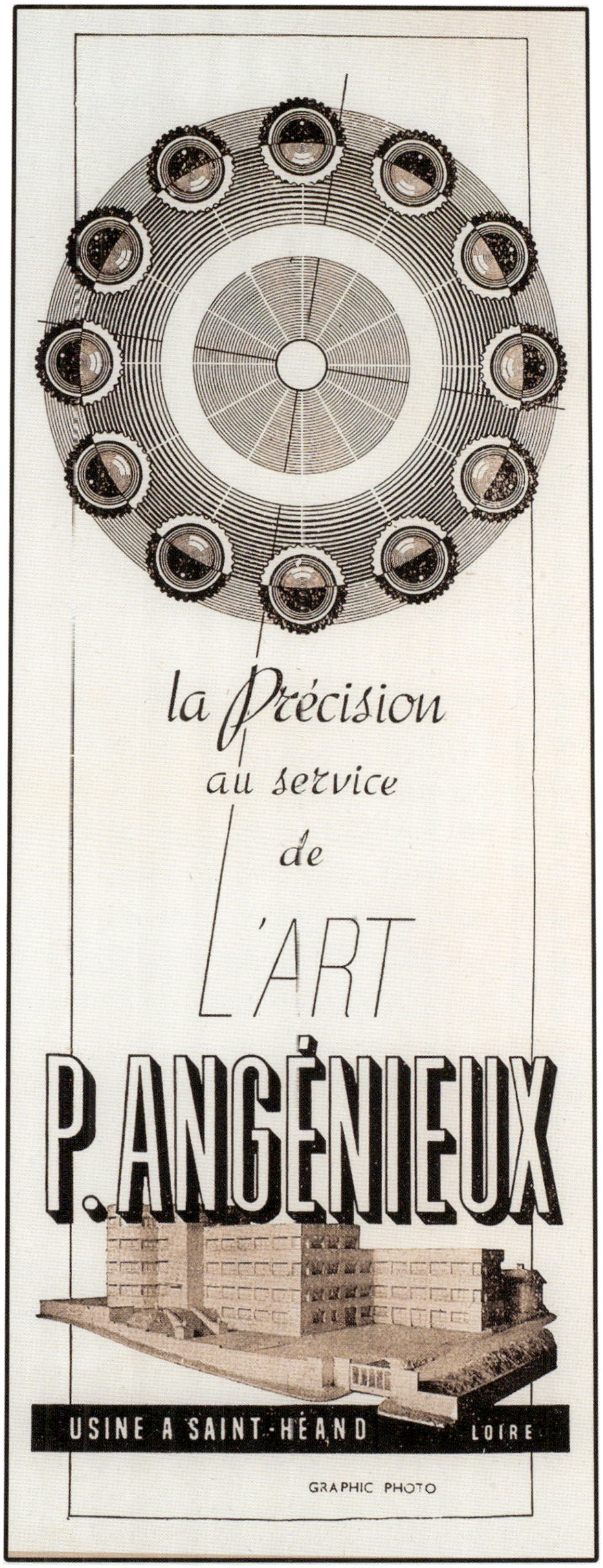

View of the machine workshop, ca. 1950.

Advertisement "Precision at the Service of Art" (1955), including a model of the new factory.

During the post-war period, everything that had been prepared or produced in the shadows emerged into broad daylight.

Just after the war, having no time as yet to apply his new method of calculation to the design of radically new lenses, Angénieux sought only to optimize the classic formulas and focused on producing more directly marketable photographic optics solutions.
Angénieux 24x36 lenses were produced for the many French owners of pre-war Leica and Contax cameras, for which it was impossible to obtain original lenses, and for those buying the brand-new Swiss Alpa Reflex, to complete their set of equipment.
Angénieux presented a broad range of lenses: 50mm (in three versions, 2.9, 1.8 and 1.5), 75mm, 90mm and 135mm as well as a 35mm/3.5, the widest focal length that could be built at that time for a reflex camera.
On this market, which was still dominated by the 6x9 and 6x6 formats, the three focal lengths most required by camera manufacturers were 100/105mm for the 6x9 folding camera, 75mm for the dual lens 6x6 reflex, and 45/50mm for the 24x36.

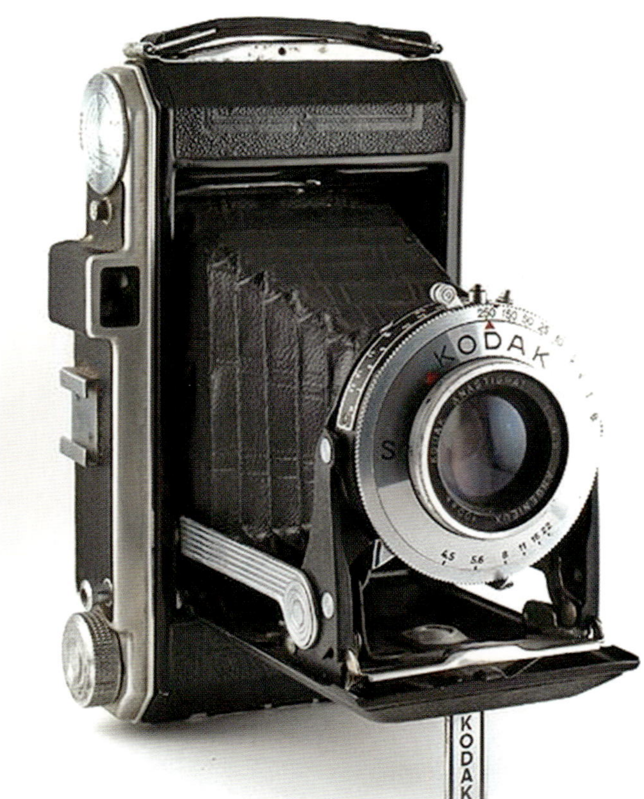

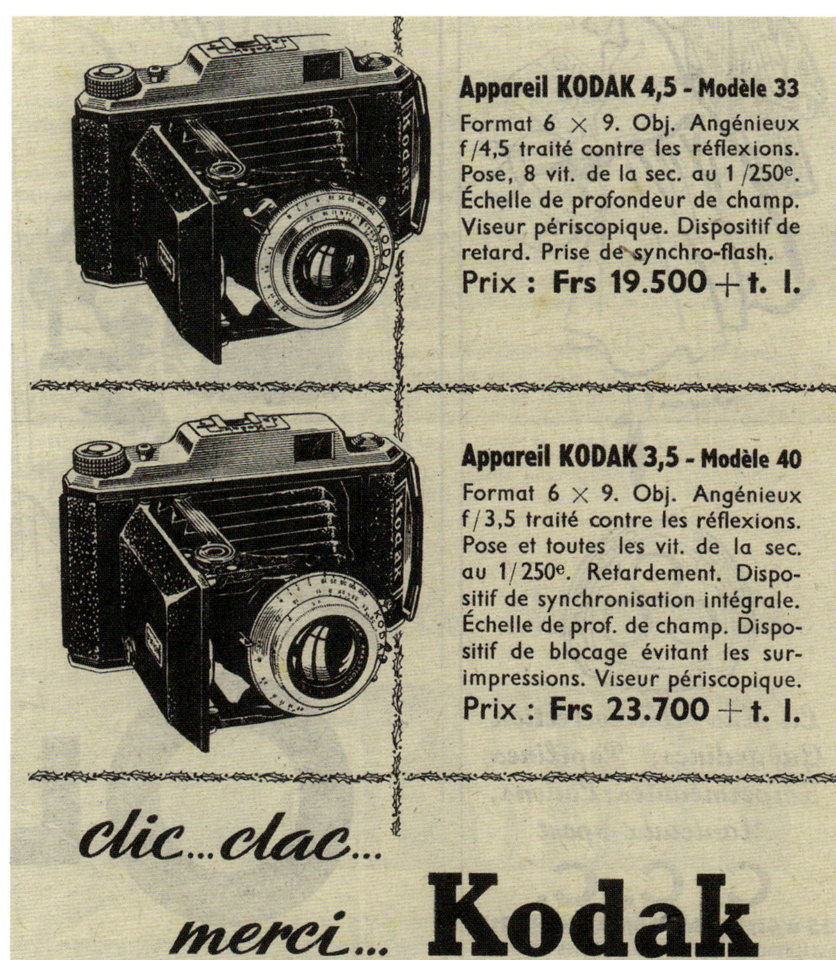

opposite
Kodak Pathé 6x9 folding camera with an Angénieux 100mm, 1:4.5 lens (collection of Jean-Paul Bouchet, author of the photo).

Advertisements for the Kodak film cameras assembled by Angénieux and equipped with its lenses, *Paris-Match,* December 27, 1952.

below
Alpa reflex camera. Two Angénieux lenses: 35mm, no. 85576, and 50mm, no. 57450 (photo by J.L. Princelle).

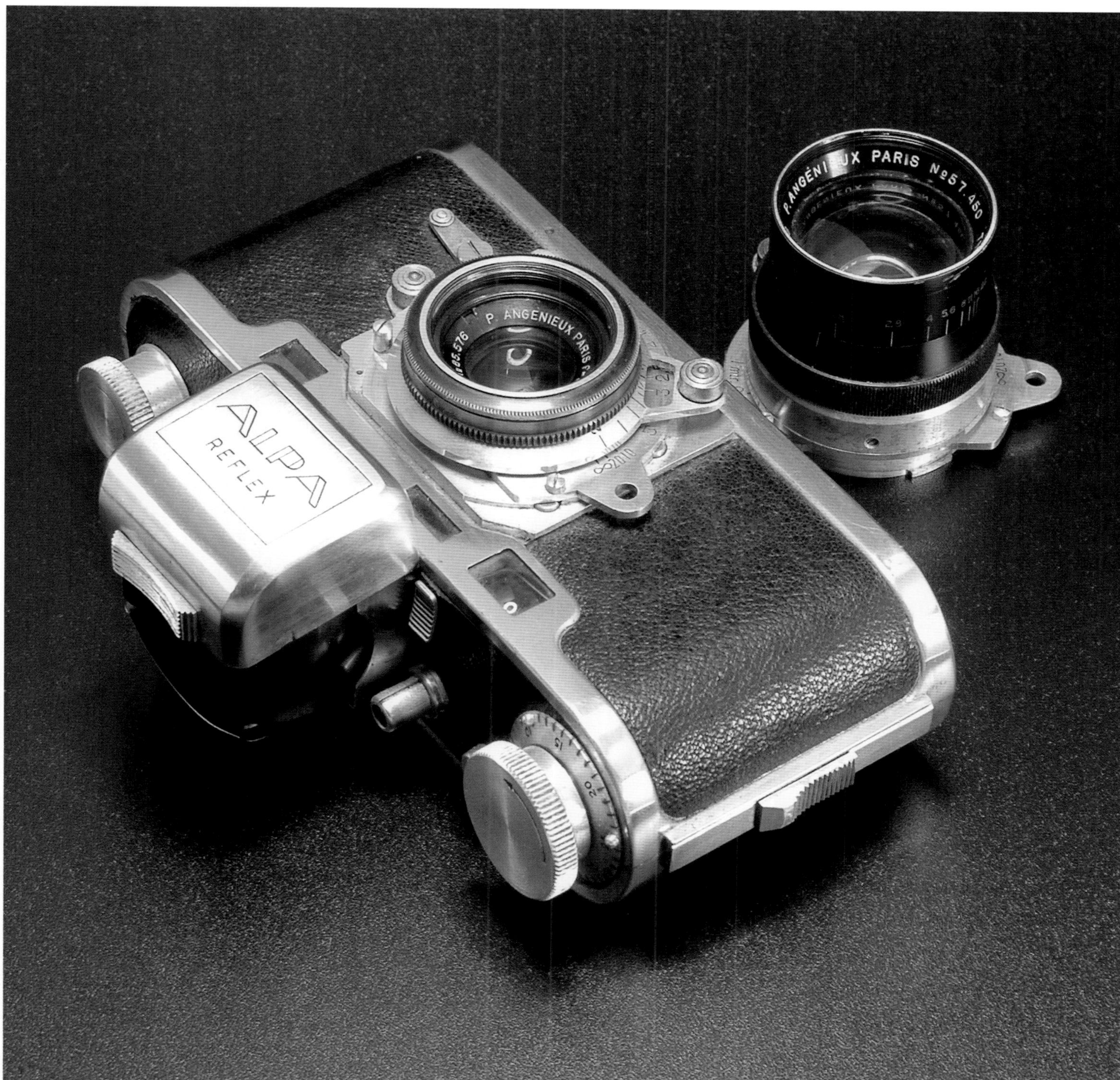

1950
Invention of the Retrofocus

FIG.1

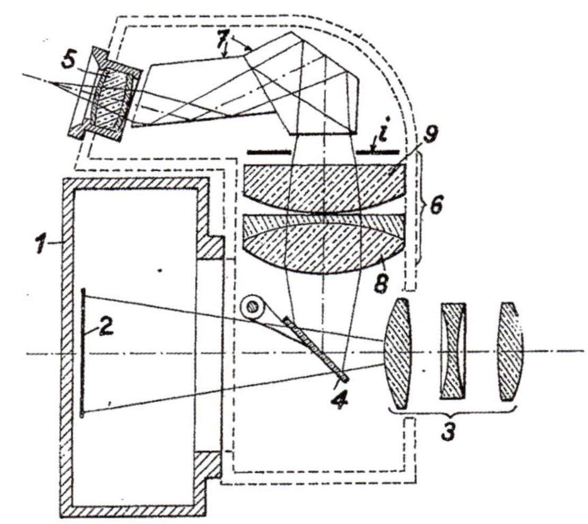

Pierre Angénieux offered photographers and filmmakers new optical possibilities in 1950 with the Retrofocus.
The invention arose out of the simple fact that the moving mirror system of the new reflex cameras made it impossible for photographers to use lenses with a focal length of less than 40mm.
Pierre Angénieux worked initially on improvements to the viewfinders of reflex cameras, for which he applied for several patents between 1943 and 1950.
His invention of the Retrofocus, a telephoto configuration leaving sufficient room behind the last lens, then made it possible to use a wide-angle lens on a reflex camera.

This major innovation lay at the origin of the universal development of 24x36 SLR (Single Lens Reflex) camera with interchangeable lenses.

above
First addition made by Pierre Angénieux to the patent of June 30, 1947, for an "improvement to camera viewfinders".

below
February 17, 1950 Pierre Angénieux applied for a patent for a wide-angle lens: the Retrofocus.

opposite
Rectaflex camera fitted with an Angénieux R1, 35mm Retrofocus lens, no. 166387, 1949 (collection of the Granet family, photo by Éric Perrin).

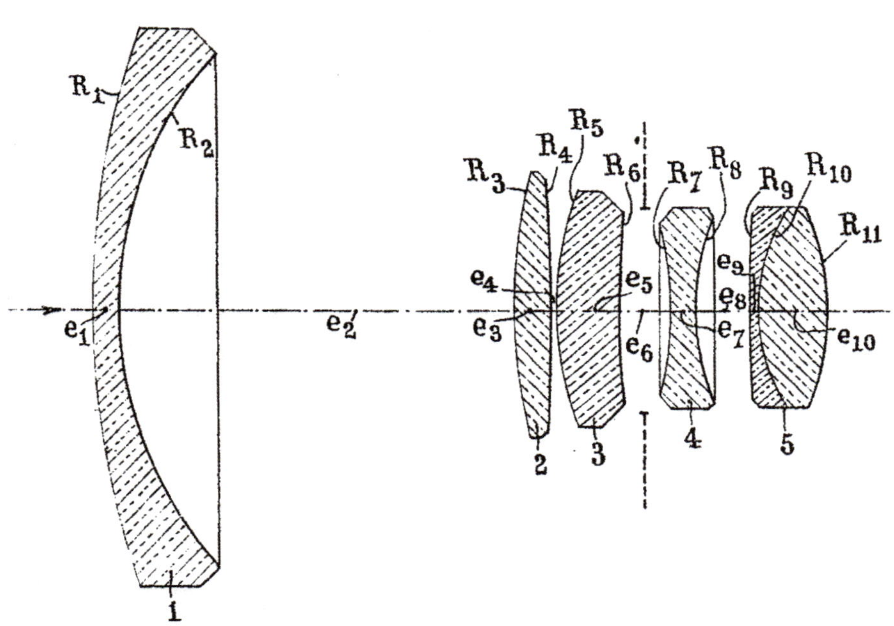

« The first stroke of genius came in the field of photographic lenses with the invention of the Retrofocus.
It was impossible to make a lens with a focal length of below 40mm, as the minimum flange focal distance was much smaller than the focal length. Angénieux's invention was the production of a lens system with a flange focal distance far greater than the focal length. »

André Masson

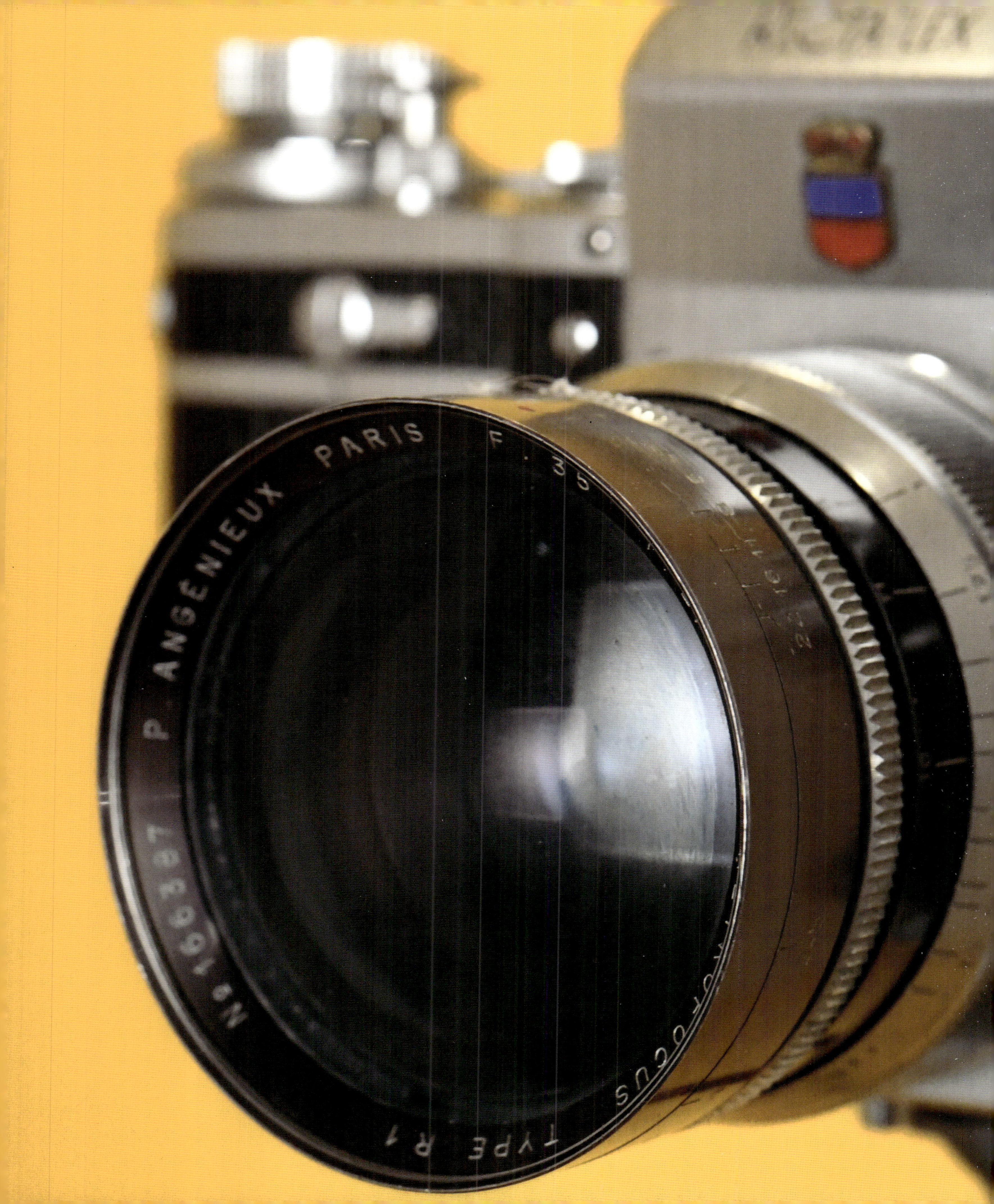
P. ANGÉNIEUX PARIS F. 35
Nº 166387
TYPE R1

The R1, the very first Angénieux **Retrofocus** lens, was a 35mm with an aperture of f/2.5, a world record at the time. First presented to the public at the 21st Salon de la Photographie et du Cinéma in Paris, it was ideal for reflex cameras but also aroused the interest of owners of 24x36 rangefinders by virtue of its remarkable brightness. More than 50,000 units were sold. It was followed in 1953 by the 28mm f/3.5 Retrofocus, a wide-angle lens opening the field of view up to 75°, and in 1957 by a 24mm f/3.5 reaching 83°.
While SOM-Berthiot, the other French manufacturer in the market of 24x36 prime lenses, took great pride in offering an exhaustive range, Angénieux concentrated on the bestselling lenses, which explains its lesser presence in the sector of longer focal length.

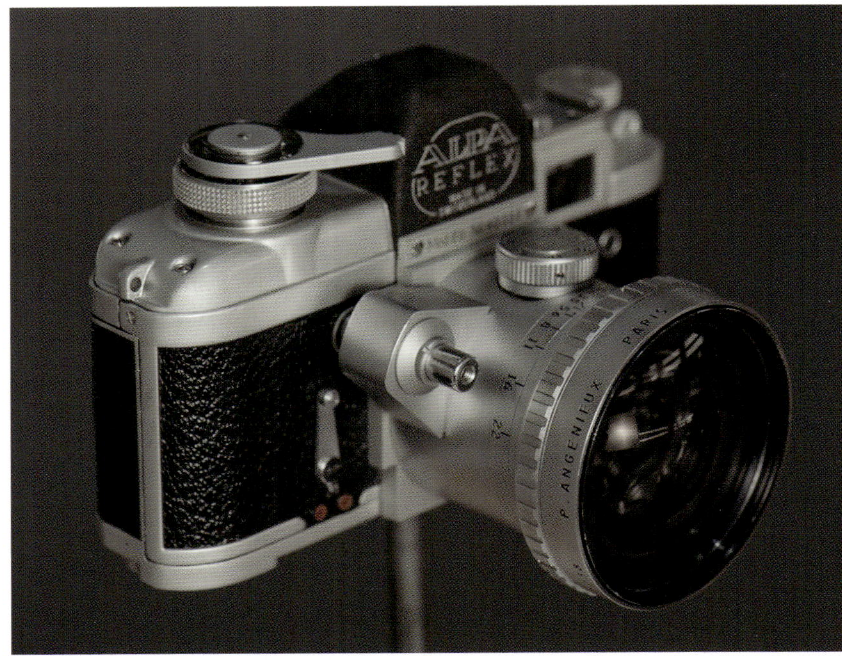

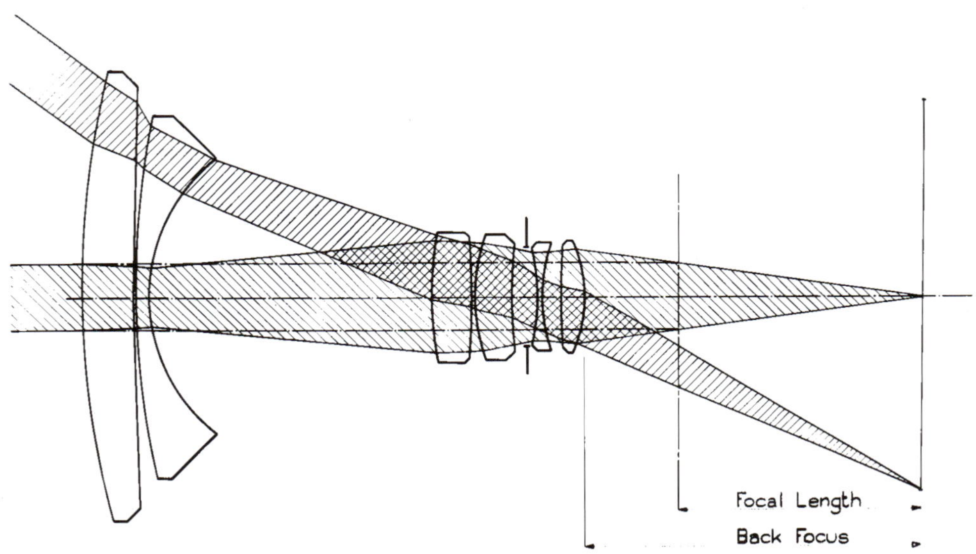

below left
Angénieux R11, 28mm, 1:3.5 Retrofocus lens no. 1235135 (collection of the Granet family, photo by Éric Perrin).

right
Optical diagrams and sectional view of the Angénieux R11, 28mm Retrofocus lens.

above
Swiss Alpa reflex camera, 6B, no. 40358. Angénieux R61 24mm, 1:3.5, Retrofocus lens no. 791238 (collection of Jean-Paul Bouchet, photo by Marc Salomon).

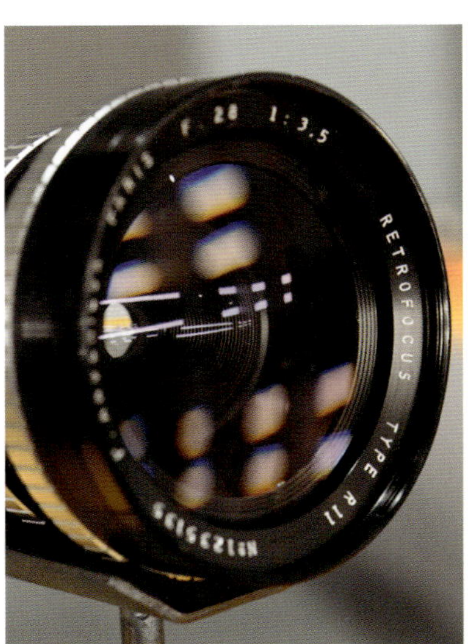

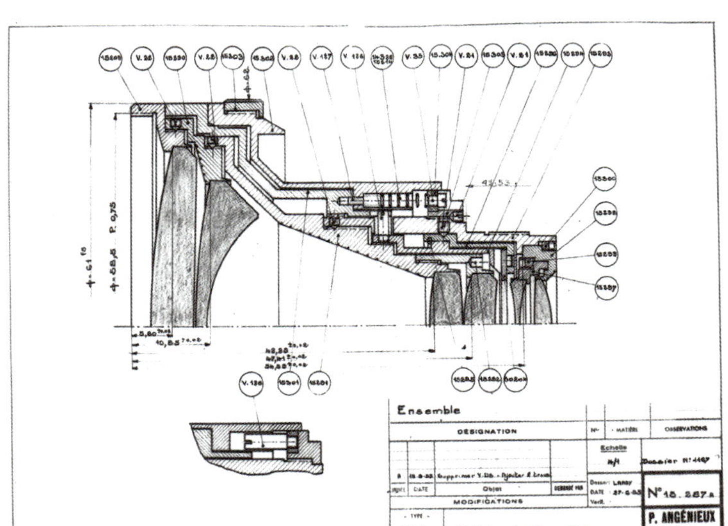

1951
First Fixed Lenses for Cinema

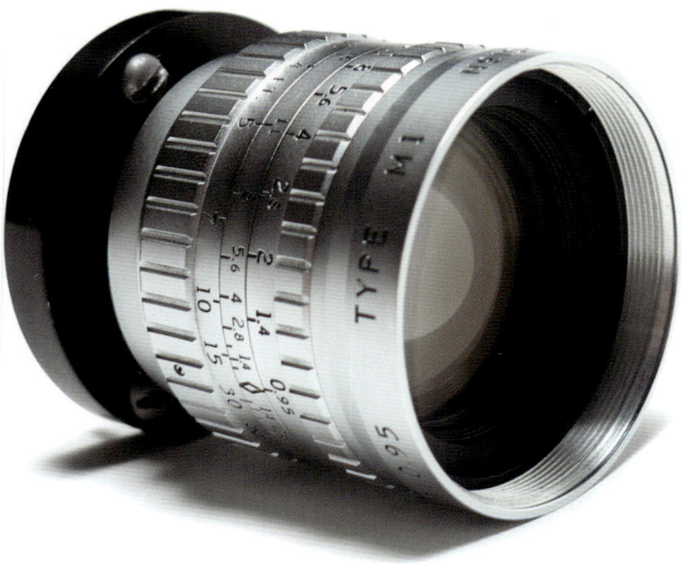

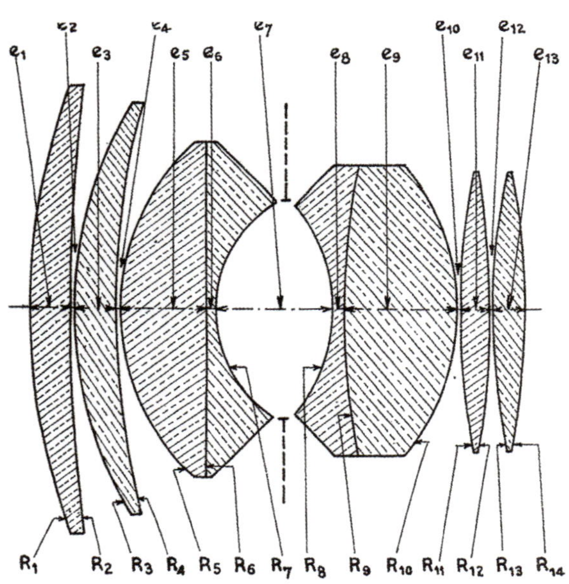

Angénieux 25mm (wide- angle) f/0.95 lens for 16mm cameras.

On March 26, 1953, Pierre Angénieux applied for a patent for a "lens with a very large aperture." The new architecture developed enabled him to reach 0.95.

Someone as passionately interested in the seventh art as Pierre Angénieux could hardly fail to develop optical systems for cine cameras, where interest in the reflex view principle was all the greater. As the first cameras had the view finder on the top or the side, the point of view differed from that of the lens. This difference in framing, or parallax error, was all the more problematic in cinema, where the image projected was the one actually taken with no possibility of realignment.

Moreover, interest soon arose in changing the lens so as to vary the angle of field of view. On a turret-front camera, these lenses must also meet further requirements, the diameter of the forward lens being limited by the need to fit in with the others.

The first cine camera lenses were officially presented at the 22nd Salon de la Photographie et du Cinéma in March 1951. While the amateur market (8mm) was in the hands of SOM-Berthiot, Angénieux aimed rather at the semi-professional (16mm) and professional (35mm) motion picture camera market.

1953, achievement of an f/0.95 aperture

Angénieux mastered another fundamental characteristic of lenses in 1953 in response to the requests of directors wishing to film in available light in dark places like the corridors of the Paris subway. The previous limit of 1.00 for an aperture was superseded to reach f/0.95, thus increasing consequently the amount of light reaching the sensitive surface of the lens.

A total of ten fixed lenses were to be developed **for the 35mm motion picture format**: 14.5mm/3.5 (Retrofocus), 18.5mm/2.2 (Retrofocus), 24mm/2.2 (Retrofocus), 28mm/1.8, 32mm/1.8, 40mm/1.8, 50mm/0.95 (ultra-bright), 50mm/1.8, 75mm/1.8, and 100mm/2;

twelve **for the 16 mm motion picture format**: 5.9mm/1.8 (Retrofocus), 9.5mm/2.2 (Retrofocus), 10mm/1.8 (Retrofocus), 12.5mm/2.2 (Retrofocus), 15mm/1.3 (Retrofocus), 25mm/0.95 (ultra-bright), 25mm/1.4, 28mm/1.1 (ultra-bright), 50mm/1.5, 75mm/2.5, 100mm/2.5, and 150mm/2.7;

and only a few **for the 8mm format**: 6.5mm/1.8, 12.5mm/1.8, 12.5mm/1.9, and 35mm/1.8. SOM-Berthiot remained the top supplier of French manufacturers on the cine camera lens market.

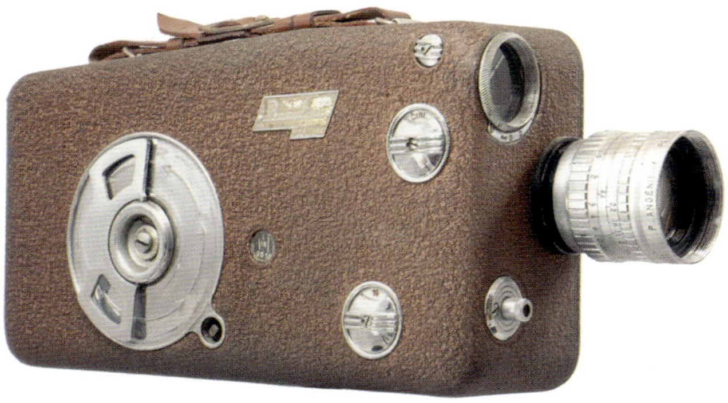

above
Beaulieu M16 16mm camera (1951), Angénieux lens (photograph © J.L. Princelle).

right
Pathé Webo M Reflex16 16mm camera, turret with three Angénieux lenses: 25, 50, and 75mm (photograph © J.L. Princelle).

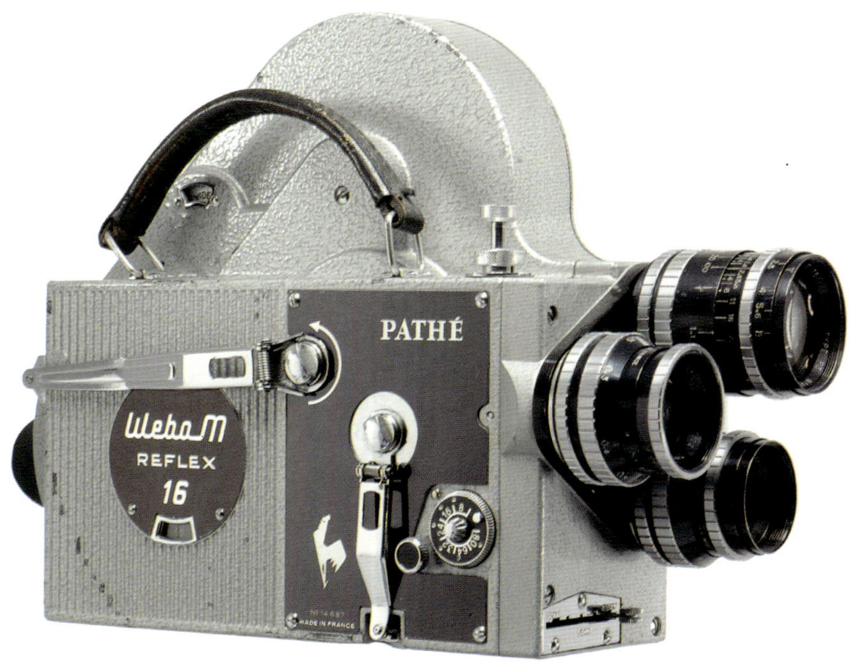

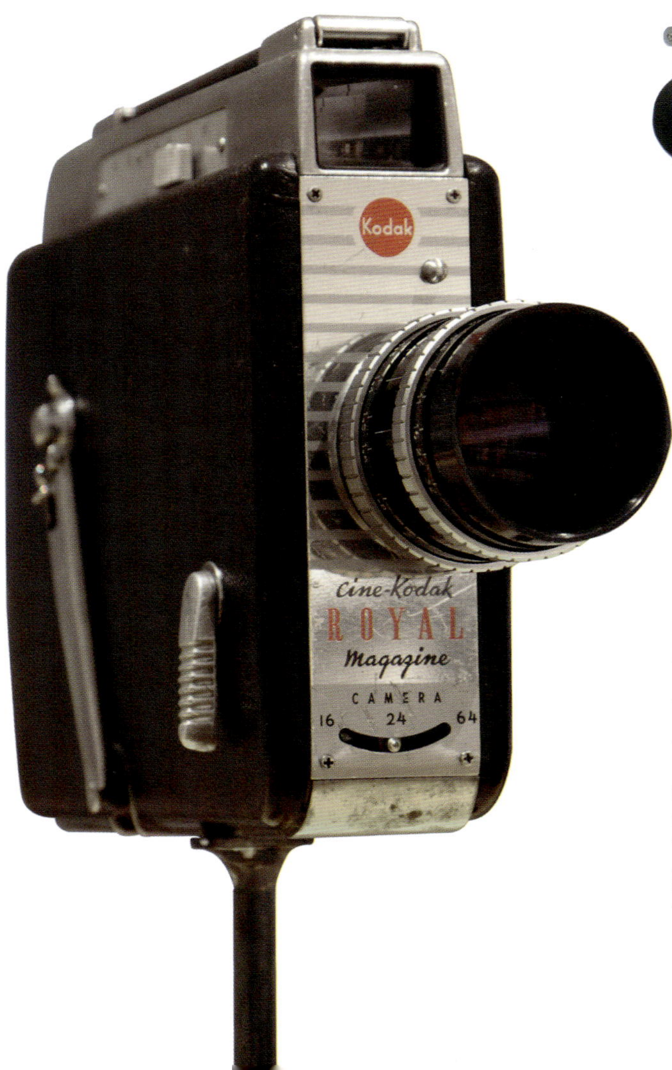

Kodak Royal Magazine camera, 16mm, no. 55365, United States, Angénieux 75mm, 1:2.5 lens, no. 390626, 1955 (collection of the Granet family, photo by Éric Perrin).

opposite
F40mm f/1.8 fixed lens for the 35mm format (Cinémathèque Française collection, Photo Stéphane Dabrowski).

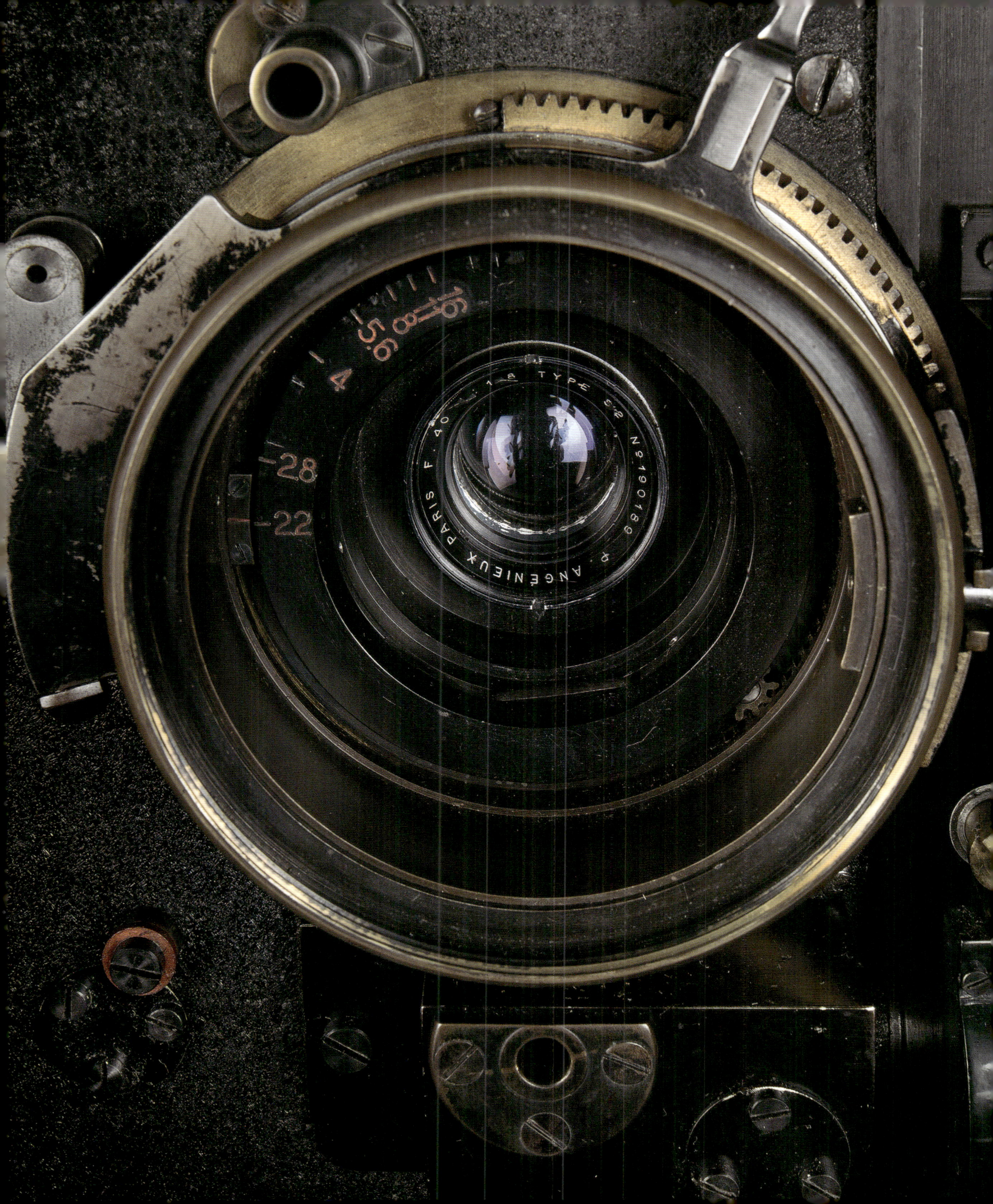
16
11
8
5.6
4
2.8
2.2
ANGÉNIEUX PARIS F. 40
TYPE S2
Nº190189

The Angénieux Brand Acquires an International Reputation

None of the leading manufacturers of reflex cameras, namely Exakta, Praktika, Rectaflex and Alpa, were lens makers. Angénieux filled this gap with a complete range of classic lenses for each in addition to its own Retrofocus. These lenses are characterized by a splendid esthetics, rigorously identical from a focal distance to another.
Angénieux produced lenses (primarily 45mm/f/3.5) for all of the **Kodak** cameras sold in Europe by agreement with the company's management in France from 1950 to 1956. Pierre Angénieux and his teams also worked on the design and manufacturing of a package consisting of a 24x36 camera and a slide projector. Kodak even contracted Angénieux to handle the complete manufacturing process of its Pony cameras, which accounted for up to 60% of Angénieux sales in the 1950s

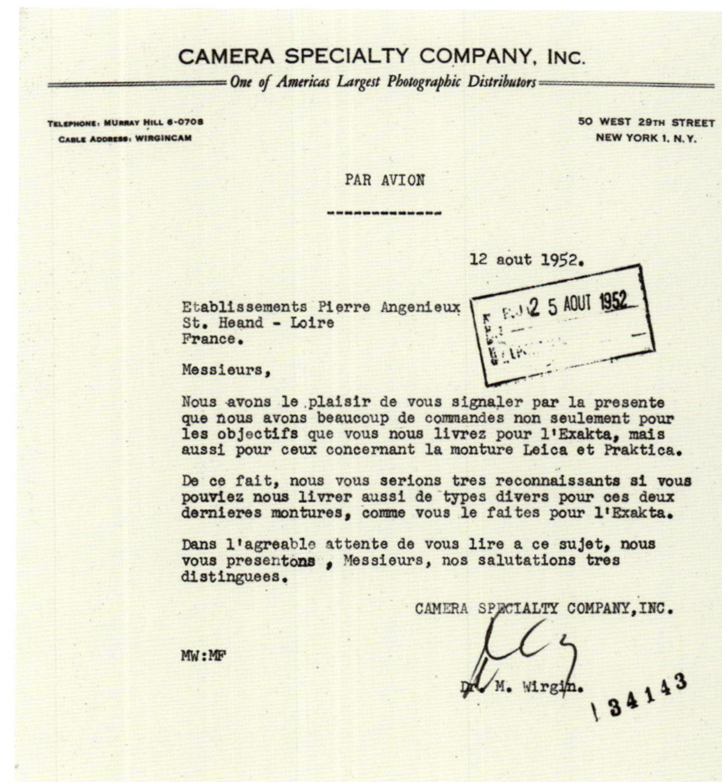

CAMERA SPECIALTY COMPANY, INC.
One of Americas Largest Photographic Distributors

TELEPHONE: MURRAY HILL 6-0708
CABLE ADDRESS: WIRGINCAM

50 WEST 29TH STREET
NEW YORK 1, N.Y.

PAR AVION

12 aout 1952.

25 AOUT 1952

Etablissements Pierre Angenieux
St. Heand - Loire
France.

Messieurs,

Nous avons le plaisir de vous signaler par la presente que nous avons beaucoup de commandes non seulement pour les objectifs que vous nous livrez pour l'Exakta, mais aussi pour ceux concernant la monture Leica et Praktica.

De ce fait, nous vous serions tres reconnaissants si vous pouviez nous livrer aussi de types divers pour ces deux dernieres montures, comme vous le faites pour l'Exakta.

Dans l'agreable attente de vous lire a ce sujet, nous vous presentons , Messieurs, nos salutations tres distinguees.

CAMERA SPECIALTY COMPANY, INC.

MW:MF

Dr. M. Wirgin.

134143

below
Kodak Pony Flash camera, Saint-Héand, Angénieux 45mm, 1:3.5 lens (Angénieux archives, photo by Éric Perrin).

Kodak Senior no. 1 projector, 1956–1961, Saint-Héand. Angénieux 100mm lens (collection of the Granet family, photo by Éric Perrin).

right
Letter ordering Angénieux lenses for the Camera Specialty Company, New York, 1952.

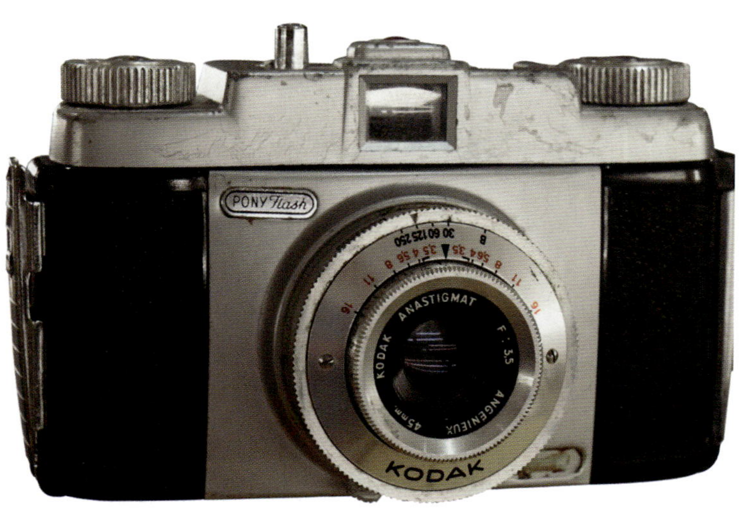

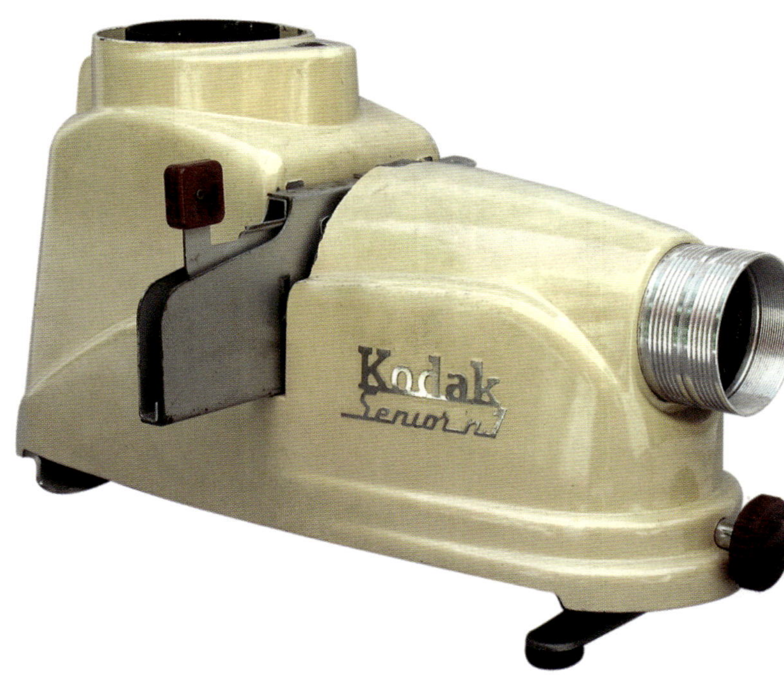

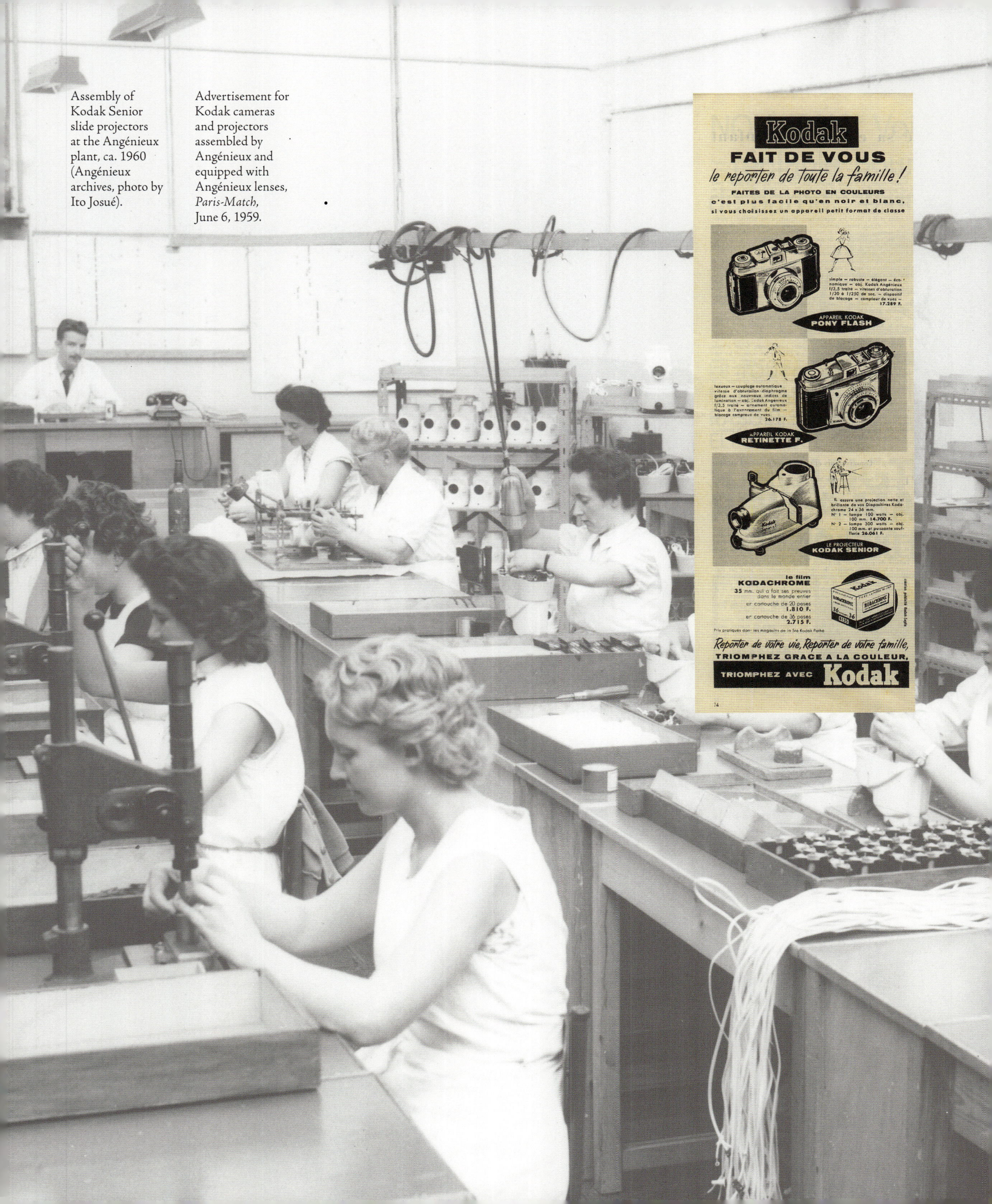

Assembly of Kodak Senior slide projectors at the Angénieux plant, ca. 1960 (Angénieux archives, photo by Ito Josué).

Advertisement for Kodak cameras and projectors assembled by Angénieux and equipped with Angénieux lenses, *Paris-Match*, June 6, 1959.

Kodak
FAIT DE VOUS
le reporter de toute la famille!
FAITES DE LA PHOTO EN COULEURS
c'est plus facile qu'en noir et blanc,
si vous choisissez un appareil petit format de classe

simple — robuste — élégant — économique — obj. Kodak Angénieux f/3,5 traité — vitesses d'obturation 1/30 à 1/250 de sec. — dispositif de blocage — compteur de vues — 17.289 F.

APPAREIL KODAK
PONY FLASH

luxueux — couplage automatique vitesse d'obturation diaphragme grâce aux nouveaux indices de lumination — obj. Kodak Angénieux f/3,5 traité — armement automatique à l'avancement du film — blocage compteur de vues. 26.178 F.

APPAREIL KODAK
RETINETTE F.

Il assure une projection nette et brillante de vos Diapositives Kodachrome 24 x 36 mm.
N° 1 — lampe 100 watts — obj. 100 mm. 14.700 F.
N° 2 — lampe 300 watts — obj. 100 mm. et puissante soufflerie 26.061 F.

LE PROJECTEUR
KODAK SENIOR

le film
KODACHROME
35 mm. qui a fait ses preuves dans le monde entier
en cartouche de 20 poses 1.810 F.
en cartouche de 36 poses 2.715 F.

Prix pratiqués dans les magasins de la Sté Kodak Pathé

création publicité Kodak-Sepia

Reporter de votre vie, Reporter de votre famille,
TRIOMPHEZ GRACE A LA COULEUR,
TRIOMPHEZ AVEC Kodak

74

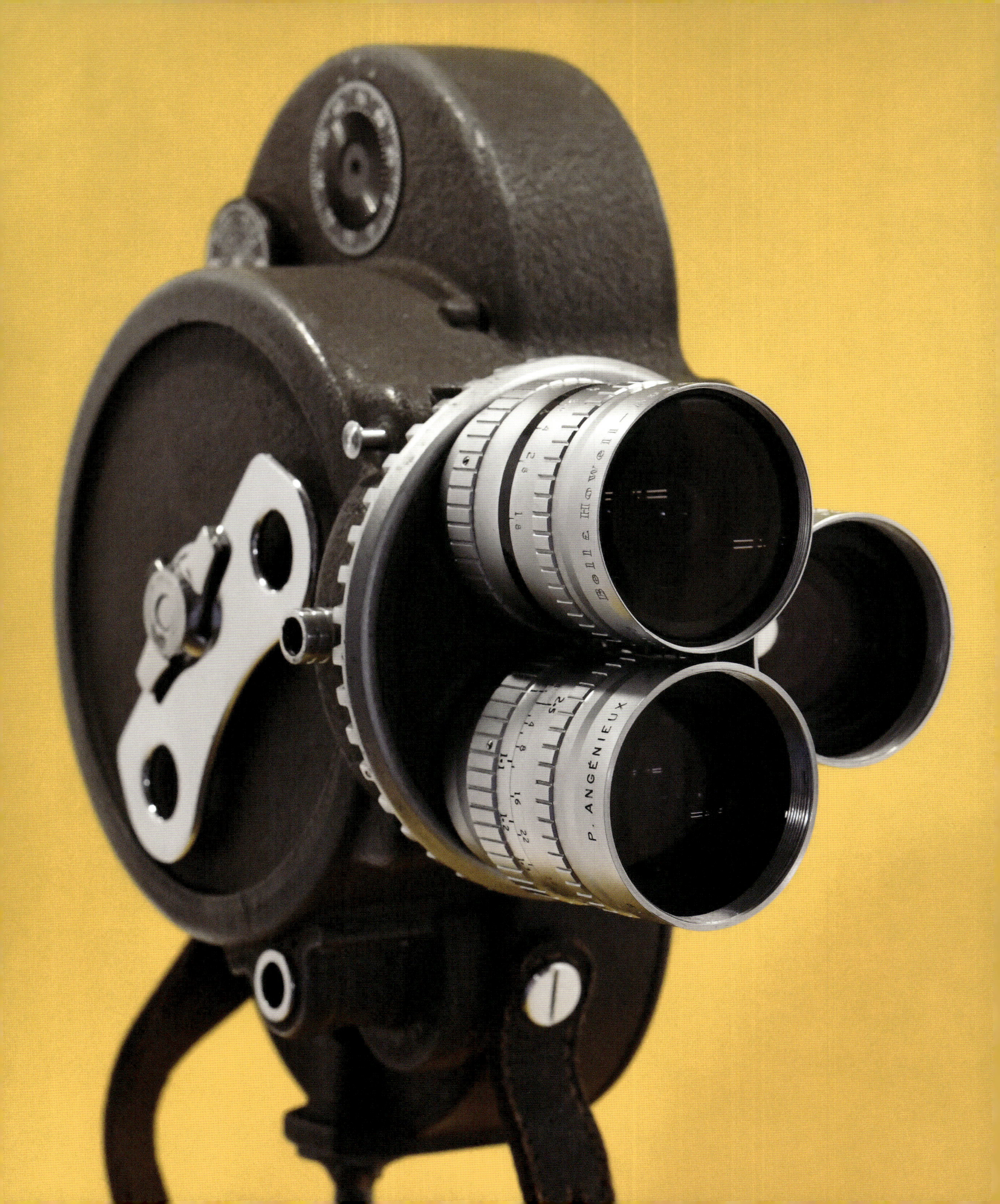
P. ANGÉNIEUX

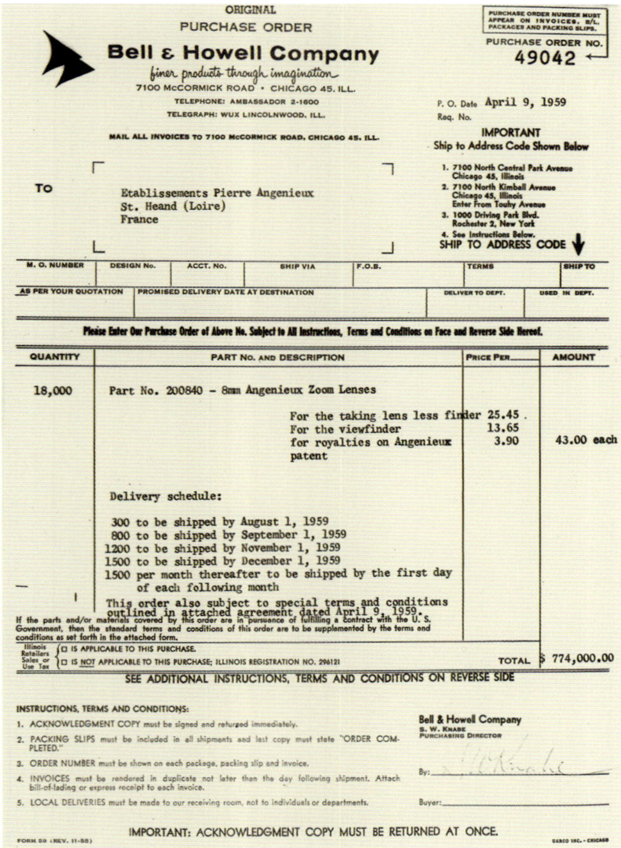

ORIGINAL
PURCHASE ORDER

Bell & Howell Company
finer products through imagination
7100 McCORMICK ROAD · CHICAGO 45. ILL.
TELEPHONE: AMBASSADOR 2-1600
TELEGRAPH: WUX LINCOLNWOOD. ILL.

MAIL ALL INVOICES TO 7100 McCORMICK ROAD. CHICAGO 45. ILL.

PURCHASE ORDER NUMBER MUST APPEAR ON INVOICES, B/L, PACKAGES AND PACKING SLIPS.
PURCHASE ORDER NO. 49042

P. O. Date April 9, 1959
Req. No.

IMPORTANT
Ship to Address Code Shown Below
1. 7100 North Central Park Avenue Chicago 45, Illinois
2. 7100 North Kimball Avenue Chicago 45, Illinois Enter From Touhy Avenue
3. 1000 Driving Park Blvd. Rochester 2, New York
4. See Instructions Below.
SHIP TO ADDRESS CODE

TO
Etablissements Pierre Angenieux
St. Heand (Loire)
France

M. O. NUMBER	DESIGN No.	ACCT. No.	SHIP VIA	F.O.B.	TERMS	SHIP TO

AS PER YOUR QUOTATION	PROMISED DELIVERY DATE AT DESTINATION	DELIVER TO DEPT.	USED IN DEPT.

Please Enter Our Purchase Order of Above No. Subject to All Instructions, Terms and Conditions on Face and Reverse Side Hereof.

QUANTITY	PART No. AND DESCRIPTION	PRICE PER	AMOUNT
18,000	Part No. 200840 - 8mm Angenieux Zoom Lenses		
	For the taking lens less finder	25.45	
	For the viewfinder	13.65	
	for royalties on Angenieux patent	3.90	43.00 each
	Delivery schedule: 300 to be shipped by August 1, 1959 800 to be shipped by September 1, 1959 1200 to be shipped by November 1, 1959 1500 to be shipped by December 1, 1959 1500 per month thereafter to be shipped by the first day of each following month This order also subject to special terms and conditions outlined in attached agreement dated April 9, 1959.		

If the parts and/or materials covered by this order are in pursuance of fulfilling a contract with the U. S. Government, then the standard terms and conditions of this order are to be supplemented by the terms and conditions as set forth in the attached form.

Illinois Retailers Sales or Use Tax
☐ IS APPLICABLE TO THIS PURCHASE.
☐ IS NOT APPLICABLE TO THIS PURCHASE; ILLINOIS REGISTRATION NO. 296121

TOTAL $ 774,000.00

SEE ADDITIONAL INSTRUCTIONS, TERMS AND CONDITIONS ON REVERSE SIDE

INSTRUCTIONS, TERMS AND CONDITIONS:
1. ACKNOWLEDGMENT COPY must be signed and returned immediately.
2. PACKING SLIPS must be included in all shipments and last copy must state "ORDER COMPLETED."
3. ORDER NUMBER must be shown on each package, packing slip and invoice.
4. INVOICES must be rendered in duplicate not later than the day following shipment. Attach bill-of-lading or express receipt to each invoice.
5. LOCAL DELIVERIES must be made to our receiving room, not to individuals or departments.

Bell & Howell Company
S. W. Knabe
Purchasing Director
By:
Buyer:

IMPORTANT: ACKNOWLEDGMENT COPY MUST BE RETURNED AT ONCE.

FORM 59 (REV. 11-58)

above
Order from Bell & Howell, 1959.

The early 1950s saw the heyday of home movies in 16mm, 8mm, and then Super 8. The performance of Angénieux fixed lenses aroused the interest of the major American camera manufacturer Bell & Howell, who used them for their legendary B&H 70 16mm turret cameras, widely used all over the world. Over 35 years, Angénieux supplied nearly 30,000 sets consisting of three lenses – 10mm wide-angle, standard 25mm/0.95, and 75mm telephoto – a combination superior in performance to the Taylor-Hobson lenses with which the camera had previously been equipped (12mm, 25mm/1.4, 50mm).

In 1953 Angénieux exported 40% of its output to the United States and had a workforce of 300.

The reputation established by Angénieux in the 1950s prompted NASA to use its ultra-bright lenses for the Gemini, Surveyor and Ranger reconnaissance missions that paved the way for the **Apollo** program, not least by mapping the moon landing areas and determining the solidity of the ground.
Three Angénieux 25mm f/0.95 ultra-bright lenses were used by Ranger 7 on July 31, 1964 to take the first close-range photographs of the moon's surface.

opposite
Bell & Howell BH 70 turret camera with three Angénieux lenses (Angénieux archives, photo by Éric Perrin).

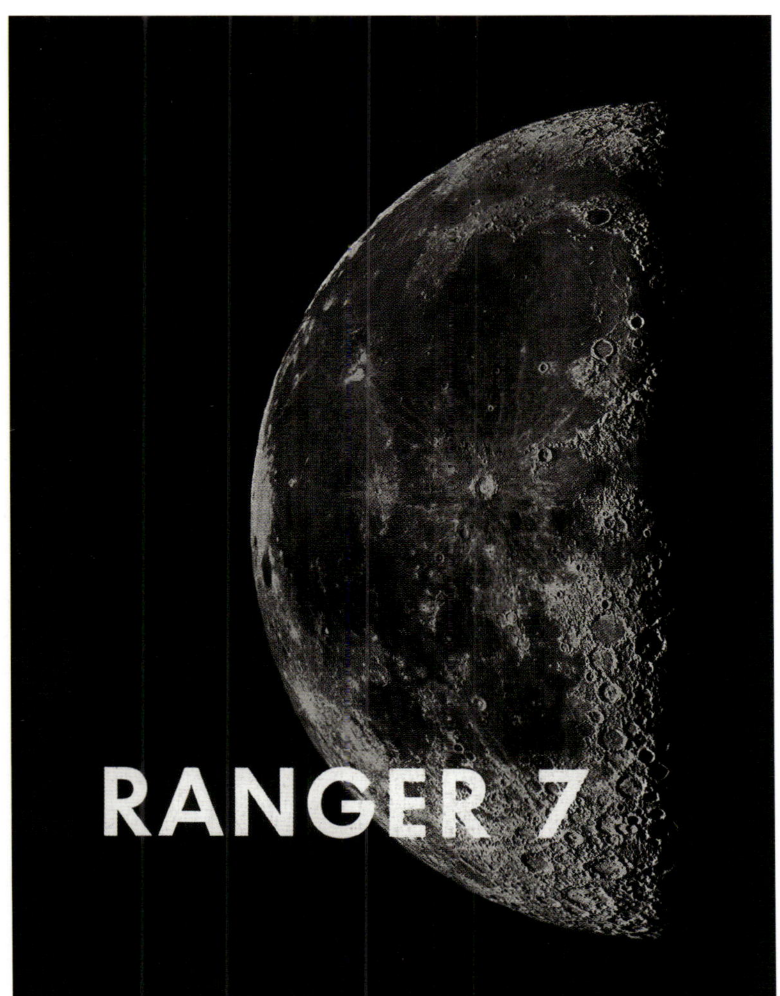

Cover of a NASA brochure presenting the Ranger 7 mission.

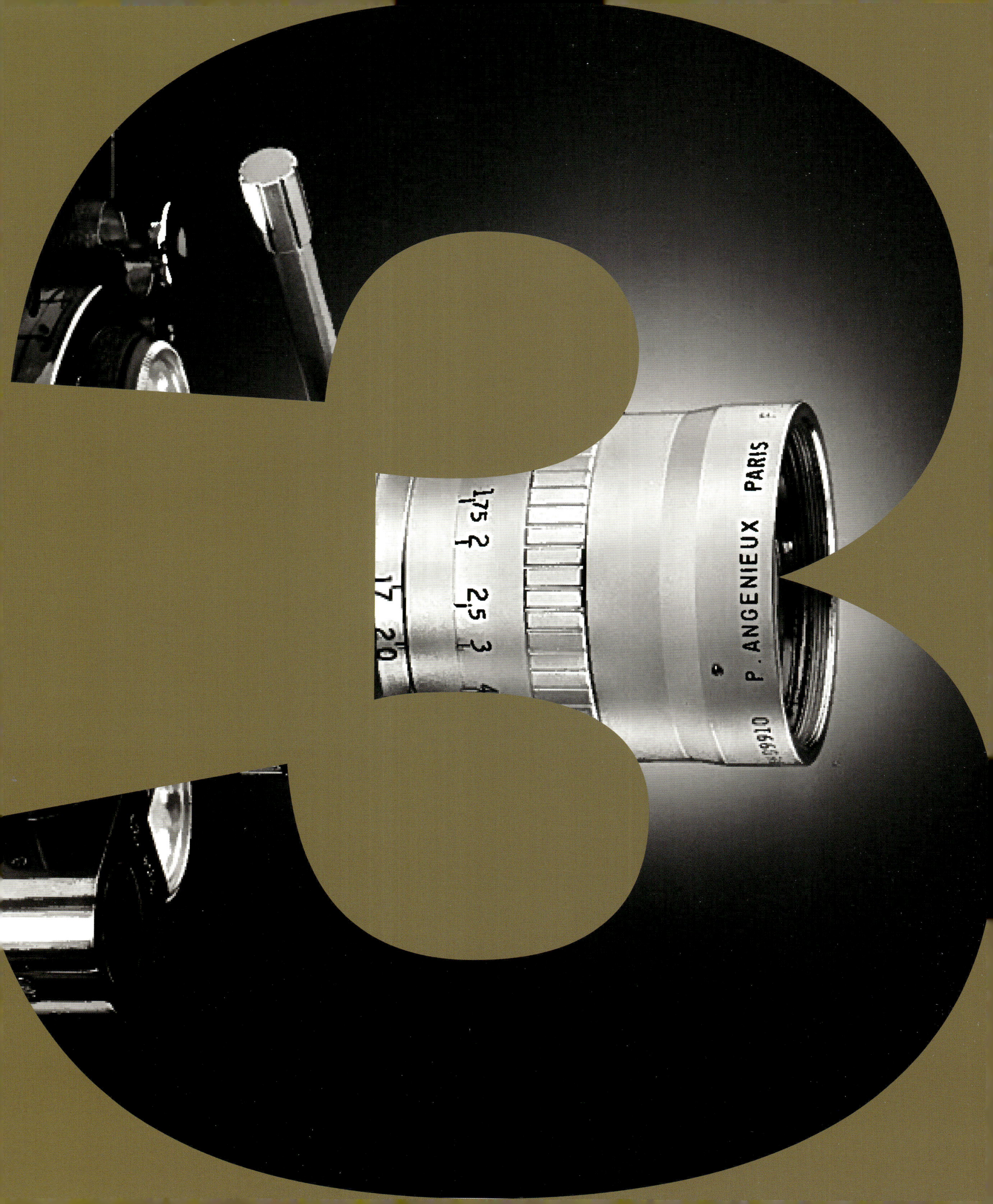
P. ANGENIEUX PARIS

THE ZOOM
TECHNOLOGY AT THE SERVICE OF ART

Throughout the history of cinema, directors of photography have placed the latest technical advances at the service of their **work** and **creativity**. The first Angénieux 16mm zooms were soon to triumph in the world of **television**, **news** and **information**, where their practical advantages were immediately obvious. Thanks to the choice of mechanical compensation, Angénieux zooms can attain variations in magnification at **still unequalled levels of optical performance**. With the first 10x zoom for 35mm cinema, Pierre Angénieux offered directors of photography a new tool. The zoom gradually found favor with cinematographers shooting fiction film in 35mm. A key invention in the history of the filmmaking, it has had a lasting influence on the **style** and **esthetics of the cinematographic image**.

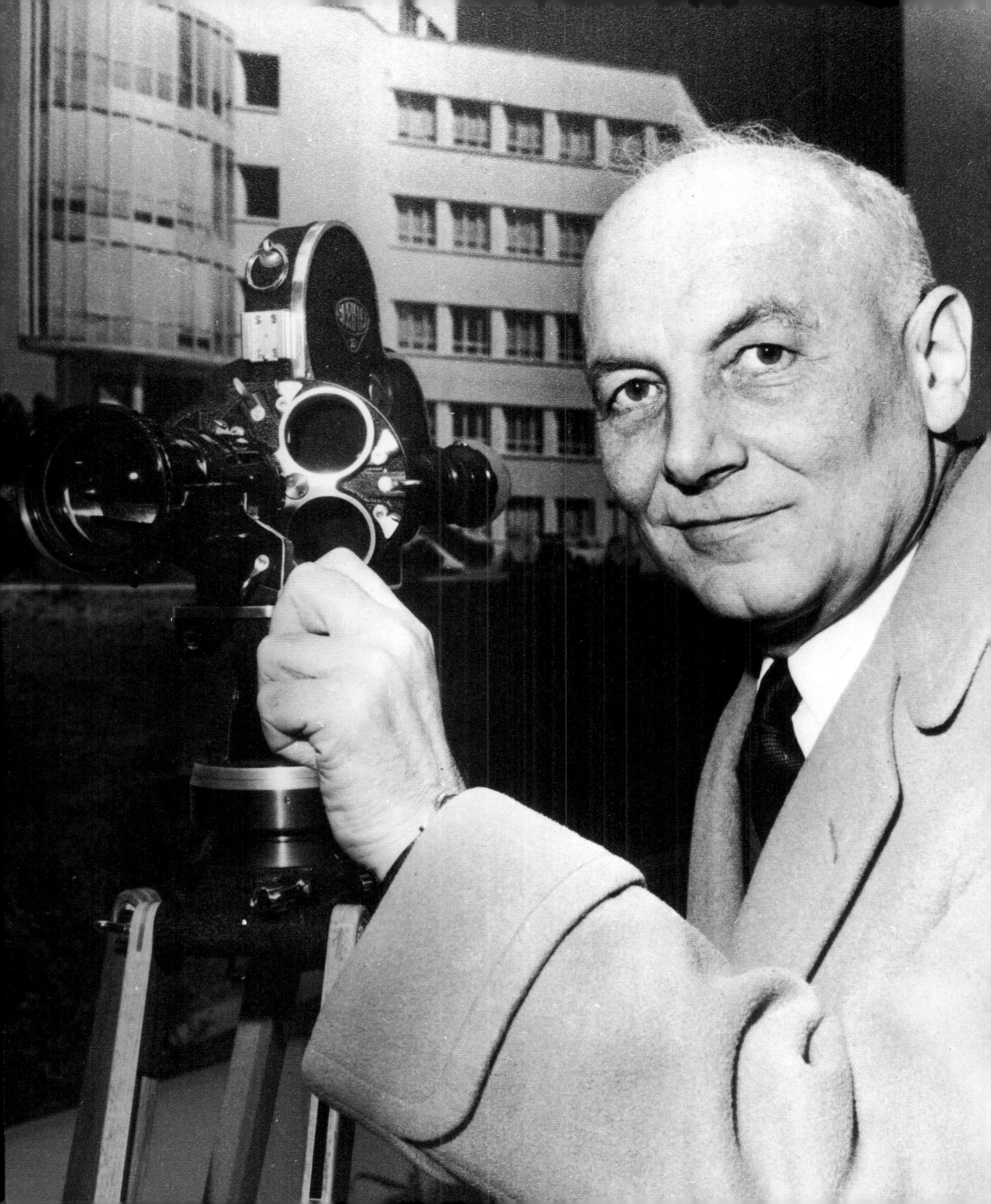

Looking back on the Invention of the Zoom

Extract from an article by Priska Morrissey in the French magazine *Positif*, February 2008

The zoom technology has existed since the end of the 1920s at least. Various Anglo-Saxon historians have helped to trace the technical and aesthetic history of these lenses[1] but the information remains scattered and sometimes contradictory. By comparing these works with the reviews and films of the time, I have endeavoured to develop a summary and an initial outline of a history of the zoom.

Some Definitions to Start With

Every camera lens, and hence every cine camera lens, is defined among other things by its focal length in millimeters (mm), i.e. the distance between the film and the optical center of the lens set to infinity. The focal length determines the angle of view. The shorter the focal length, the wider the angle and the larger the field covered. Conversely, the longer the focal length, the narrower the angle of view.

The lenses used in cinematography are of fixed focal length ranging from short or very short focus (roughly between 12 and 35mm) to telephoto lenses (over 85mm). The zoom is instead a lens of variable focal length. The different models are designated in terms of their minimum and maximum focal lengths, e.g. 18-70mm or 80-200mm. A zoom is made up of several groups of lenses whose movement in relation to one another lengthens or shortens the focal length. A zoom therefore makes it possible to alter the angle of view and the size of the subject photographed. This change in focal length, called *zooming*, modifies the depth of field. Herein lies the difference between zooming and a tracking shot, where the camera moves in space and the depth of field remains the same. A zoom can, however, also be used only to change focal length more easily. Some operators prefer to work with a set of fixed focal lengths and others with zoom. Fixed focal lengths and zooms are very often both used in shooting a film.

Not all lenses with variable focal length can claim to be zooms, only those equipped with a system of mechanical or optical compensation making it possible to maintain focus on the object filmed throughout a change in focal length. While the addition of a compensation system has no great importance in photography, as the photographer can refocus for each image, it is crucial in cinema to preserve constant focus and clarity when zooming in or out. It is a matter both of simplifying and accelerating the change of fixed focal lengths and of making the magnification or contraction of the image visible on the screen. One of the challenges in manufacturing zooms was to design a lens offering perfect focus at all the intermediary positions of the lens. The early models of the late 1920s did not always achieve this, as demonstrated by the slight blurring to be found in films during the change in focal length as against the sharpness of the final image.

Priska Morrissey is a lecturer in cinema studies at the Université Rennes 2. Her work explores the history of film professions and techniques, as exemplified by the thesis, defended in 2008, on cameramen in France between 1895 and 1926. She has published various articles and edited two special issues of the magazine *1895* on professions in the film industry and color processes.
Together with Emmanuel Siety, she recently published *Filmer la peau* (Presses universitaires de Rennes, 2018), which combines scholarly studies and interviews with people involved in filmmaking.

The Archeology of the Photographic Zoom

The first real lens with variable focal length was produced in the field of precision optical instruments rather than photography. A telescope with a mobile concave achromatic lens making it possible to alter the magnification of the objects viewed while continuing to look through the instrument was presented to the Royal Society in 1834 by the British mathematician and physicist Peter Barlow. This was followed by other inventions of the same kind for astronomical and other telescopes. The first to apply the principles of variation in focal length to photography was the Italian optician Ignazio Porro, who invented a new lens with variable focal length for a telescope to study a solar eclipse in 1851, and then adapted the same for use in photography. While telephoto lenses and variable focal length naturally developed together, varifocal telephoto lenses did not appear on the market until the 1890s, whereupon they enjoyed great success. Photographers working in the fields of the natural sciences, architecture, medicine and of course astronomy immediately put them to practical use. Some even made deliberate use of the lack of sharp focus to produce supposedly artistic photographs. At the beginning of the 20th century, most industrialists opted for optical systems of fixed focal length and the lenses produced at the time cannot in fact truly be described as zooms, as a change in focal length entailed the loss of focus and necessitated manual refocusing. Photographers therefore had to wait until the early 1960s for a lens with variable focal length, which was finally an authentic zoom. The first zoom, even though this name had yet to be coined, dates from 1902, when an optician in Chicago named Clile C. Allen patented The Adjustable, a varifocal projection lens equipped with a system of mechanical compensation making it possible to preserve focus. The invention was intended primarily for the stereopticon, a kind of improved magic lantern with two projectors making it possible to fade out and in. The birth of the photographic zoom dates from 1914, when the American Lodewick J.R. Holst developed and produced his "telephotographic apparatus" (patent no. 1138217, May 1915). He did not, however, manufacture it on the industrial scale and the patent remained an isolated case.

The Interwar Period: a Specific and often Spectacular Effect

The lens with variable focal length and compensation system was the subject of experiments for filming in the late 1920s. The camera had come of age in the meantime and was capable of increasingly complex movements. The early 1910s saw the first models equipped with a turret offering a choice of fixed lenses, a device designed to facilitate and accelerate the changing of focal lengths in a shot or between one shot and another was based on the one already found on microscopes. The zoom was naturally involved in this attempt to combine different lenses, as it makes it possible to change focal length without having to rotate the turret and hence with no break in filming.
The birth of the cinema zoom lenses took place within the broader context of the major

[1] See Rudolf Kingslake, 'The Development of the Zoom Lens,' *Journal of the SMPTE*, vol. 69, August 1960, and *A History of Photographic Lens*, Academic Press, London, 1989; Paul Joannides, 'The Aesthetics of the Zoom Lens,' *Sight and Sound*, vol. 40, no. 1, winter 1970–1971; A.D. Clark, 'Zoom lenses,' *Applied Optics*, vol. 7, 1973; John Belton, 'The Bionic Eye Zoom Esthetics,' *Cineaste*, vol. 11, no. 1, winter 1980–1981; Carlo Montarano, 'Nascita (probabile) dello zoom', *Immagine*, no. 16, 1990–1991; Barry Salt, *Film Style and Technology: History and Analysis*, Starword, London, 2nd edition, 1993.

technical innovations of the late 1920s; namely color, widescreen images, and of course, sound. Amongst the three major innovations, two of them can undoubtedly be related to the invention of cinema zoom lenses. On one hand, the transition to sound movies requires the camera to remain more static. Perhaps the zoom lens could play a role where people were searching for tools to simulate camera movements? On the other hand, the moment we start imposing larger screen sizes, suddenly the projected image size becomes larger. Did these new processes, based on optical research, foster the invention of the zoom? Probably.

What is certain is that various engineers and cinematographers were at work in the late 1920s on a zoom for filming. Historians refer almost exclusively to American patents but it is highly likely that similar patents can also be found in Europe, at least in Germany, where various zooms for 16mm film came onto the market in the mid-1930s.[2] The Americans who took out patents include Leon F. Douglas (1928), I.G. Ries and D.G. Shearer (1928), Otto B. Durholz (1931), and the cinematographer Joseph Walker, renowned for his long and fruitful collaboration with Frank Capra in the 1930s, who patented his Traveling Telephoto in 1929. This variable-length telephoto lens served as a basis for work on the Walker ElectraZoom, which he began to produce after the end of World War II for TV cameras.[3]
Studios like Paramount equipped themselves with zooms at this point. According to the cinematographer Milton Krasner, who worked with Lang, Hawks, Wilder, and above all Minnelli in the early 1960s and used a zoom for the first time on Ralph Murphy's *Golden Harvest* (1933), Paramount used Walker's original model.[4] According to other sources,[5] however, the Paramount zoom was produced by Lorenzo Del Riccio as an adaptation of his work on Magnascope, one of the processes making it possible to magnify the image during projection.[6]
The first industrially produced zooms for professional 35mm cinema included the Cooke Varo (40-120mm), developed by the British inventor Arthur Warmisham and manufactured by Bell & Howell at its plant in Chicago as from 1932. The Varo was on the market up to World War II but its success was limited.
The name "zoom" was borrowed by American cinema from the field of aviation, where it meant rapid ascent or descent, in the early 1930s. The term then covered two different effects, namely a very fast tracking movement on a dolly and any movement carried out by this lens with variable focal length, and soon extended to the instrument that made the latter possible. This shift encapsulates the early use of the zoom lens, confined precisely to optical tracking shots. The zoom made it possible to home in suddenly on a potentially dramatic or revealing detail of the image.

Clarence Badger's romantic comedy *It*, starring Clara Bow and produced by Paramount in 1927, appears to have been the first fiction movie filmed with a zoom.[7] It contains a zoom

[2] The Busch Vario-Glaukar (25-80mm) lens of the Astro-Transfokator company, invented in 1932, adapted for cameras produced by Siemens.
[3] **For the testimony of the operator on his contribution to the history of the zoom,** see Joseph Walker and Juanita Walker, *The Light on Her Face*, ASC press, Los Angeles, 1984, pp. 266–272.
[4] See George Turner, 'Milton Krasner, ASC,' *American Cinematographer*, vol. 67, no. 9, September 1986.
[5] Rick Mitchell, *Films in Review*, vol. 38, no. 6-7, June-July 1987, p. 383 (Letters to the Editor).
[6] The rivals of Magnascope included one invented by George Palmer, chief projectionist of the Marble Arch Cinema in London. His Giant Expanding Pictures system was inaugurated in July 1930 and bought by several London cinemas the same year despite its poor quality. Rudolf Kingslake also speaks in the above-mentioned article of 1960 about the existence of a German zoom for film projection called the Busch Vario-Neokino.
[7] Another candidate for this honor is Paramount's war film *Wings* (1927), but the director William Wellman denied any use of a zoom. See Scott Eyman, '"Wild Bill": William A. Wellman,' *Focus on Film*, no. 29, March 1978.

out, shot from the roof of a neighboring building, a tilt down to establish the setting in the centre of an American metropolis, and a zoom in to focus attention on a group of people clustered around the entrance to a department store, where the action takes place. The zoom replaces a dolly shot, which would have been impossible or too complicated to stage, and links the sign of the store with its entrance, the emblem of success for its young salesgirls. This effect reappears in Erle C. Kenton's *Island of Lost Souls* (1932), where the camera zooms in on the astonishing discovery by the male lead of the animal-like ears of one of Moreau's creatures, lying unconscious on the deck of the freighter.
A certain number of the films known for their use of the zoom in the late 1920s and early 1930s were produced by Paramount, which was then famous for its technical innovations. The list includes *The Four Feathers* (1929) by Merian C. Cooper, Lothar Mendes and Ernest B. Schoedsack, *Rango* (1931) by Schoedsack, *Love Me Tonight* (1932) by Rouben Mamoulian, and *Thunder Below* (1932) by Richard Wallace, renowned for a vertical zoom imitating the fall of a body. The use of a zoom is, however, perhaps related not so much in general to the signature of a studio as to the involvement of certain figures. The presence of zooming in Capra's *Dirigible* (1931) and *American Madness* (1932) is unquestionably due to the fact that the cinematographer listed in the credits is the above-mentioned Joseph Walker. *Private Worlds* (1935), directed by Gregory La Cava and photographed by Leon Shamroy, where the zoom is briefly used in a scene of madness, was produced by Walter Wanger, who displayed a keen interest in the latest technical inventions during his seven years with Paramount (1923–1931) and subsequent career as an independent producer. Merian C. Cooper also produced a number of films for RKO including one or more zoom effects, such as *She* (1935) by Irving Pichel and Lansing C. Holden.

A falling body, the discovery of a surprising detail or an object of desire, a fit of madness or a trajectory guided by the rhythm of a melody (*Love Me Tonight*), the zoom was used precisely for the "non-normal" effect produced by its characteristic movement. The poor quality of its optics undoubtedly explains its limited success in the 1930s and above all the fact that it was not yet used as a fixed lens. The early zooms had in fact numerous defects, including poor resolution and limited aperture of the diaphragm. The use of this lens therefore required a great deal of light and its aberrations were insufficiently corrected, especially chromatic aberration due to the dispersal of rays of light.

The Zoom in the Post-war Period: from Television to 35mm Movies

The zoom was established as an essential tool after World War II, first in television and then in amateur and professional filmmaking.
The late 1940s marked a crucial turning point in the history of the zoom with the appearance of two new and improved lenses for 16mm cameras in 1946 and 1950. Their optical compensation system involved the combination and separation of lens elements, thus

avoiding the use of a cam shaft to maintain focus and the possible flaws connected with its manipulation by the cameraman.
The first was the American Zoomar lens (17.5-70mm) developed by Frank Back, the version of which for 35mm cameras, covering the focal lengths from 36 to 82mm, did not appear until 1954. The second is the renowned French Pan Cinor lens, whose name refers to its ability to replace the entire set of Cinor lenses developed by SOM-Berthiot for 16mm cameras. This proved superior to the Zoomar in its correction of aberrations. Patented in 1949 and first presented in shows and conferences in Paris and Milan in 1950, it enjoyed immediate success. The Pan Cinor was adapted for 8mm cameras in 1952 and over 30,000 had been sold by 1956, 80% of production being marketed outside France through the Swiss campany Paillard.
The use of electronic calculators to perform the calculations required for lens production made the transition from 16 to 35mm possible and the French firm SOM-Berthiot brought out a Pan-Cinor for 35mm cameras (38.5-154mm f/3.8) in 1956. The lens had already been tried out in 1955 at a Christmas soccer match between France and Belgium in Brussels.
It was again in 1956 and in France that Angénieux presented its first Zoom (the name under which it was registered), a 17-68mm lens for 16mm cameras. Its success was ensured by the fact that Pierre Angénieux opted for a mechanical compensation system making it possible to increase sharpness of focus and magnification. It is thanks to Angénieux that the term *zoom* came into general use in non-English-speaking countries, replacing the more or less evocative names of earlier models. [...] Due to the requirements of live broadcasts, television soon became the major sphere of use. The zoom became the lens *par excellence* of TV cameras for coverage of live events, especially sports, and for shooting fiction films live in the studio. Practical applications were obviously found in amateur filmmaking as well. The zoom made its grand entrance into cinema in the late 1950s. Another key milestone came in the following decade with the appearance of the Angénieux 10x, 25-250, which received a Scientific and Technical Oscar in 1964.

Bernard Bastide, Jean-Pierre Berthomé, Richard Cornu, John Hayes, and Jean-Jacques Meusy are thanked for their assistance.

Pierre Angénieux, Inventor of the Modern Zoom

« After the Retrofocus, Pierre Angénieux embarked on the calculations for a lens of variable focal length, the dream of opticians in that period (...) This is what Pierre Angénieux accomplished: lenses of variable focal length with mechanical compensation. The first Angénieux 4x zoom was presented in 1956 and work began immediately on a magnification of 10x. »

André Masson

1956, the First Angénieux Zoom, the 17-68mm f/2.2 for 16mm Cameras

Pierre Angénieux's greatest innovation remains the creation and industrial production of the first zoom with mechanical compensation, which was to revolutionize practice in television, cinema, and photography. The first prototype was a 4x zoom for 16mm cameras presented in 1956 and brought into production in 1957, initially for Bell & Howell 16mm cameras for amateurs. This zoom replaced the turret with its three lenses of fixed focal length. This surpassed the zooms with optical compensation in power and performance, especially the Pan Cinor of SOM-Berthiot developed by Roger Cuvillier. Pierre Angénieux gave his lens a new dimension, namely the ability to vary the magnification of the image with a device of limited size and no impairment of focus during change in focal length, a great challenge regarded as impossible by German optical engineers. The list of requirements was fully met by the F=17-68 f/2.2, the first Angénieux zoom, deliveries of which began in 1958. Over 70,000 of these lenses were sold on the market between 1957 and the end of the 1970s.

This new zoom technology of the late 1950s was a major event for the movie industry heralding the end of turret cameras and the widespread use of reflex viewfinding. Pierre Angénieux registered Zoom as a brand name in France, after which the word entered the French language.

While the Angénieux range featured just one zoom among a multitude of fixed focal lengths in 1958, the only fixed focal length to survive in 2001 was the super 5.9mm wide-angle for 16mm film cameras.

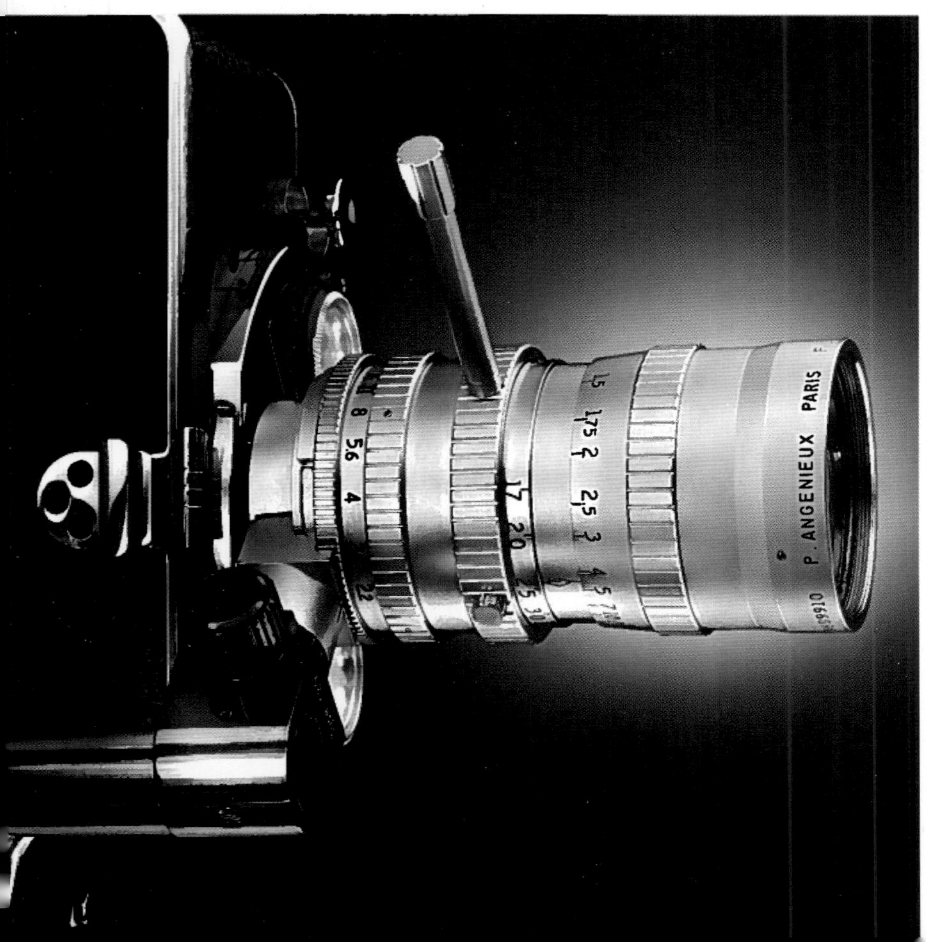

The first Angénieux 17-68mm zoom (Angénieux archives).

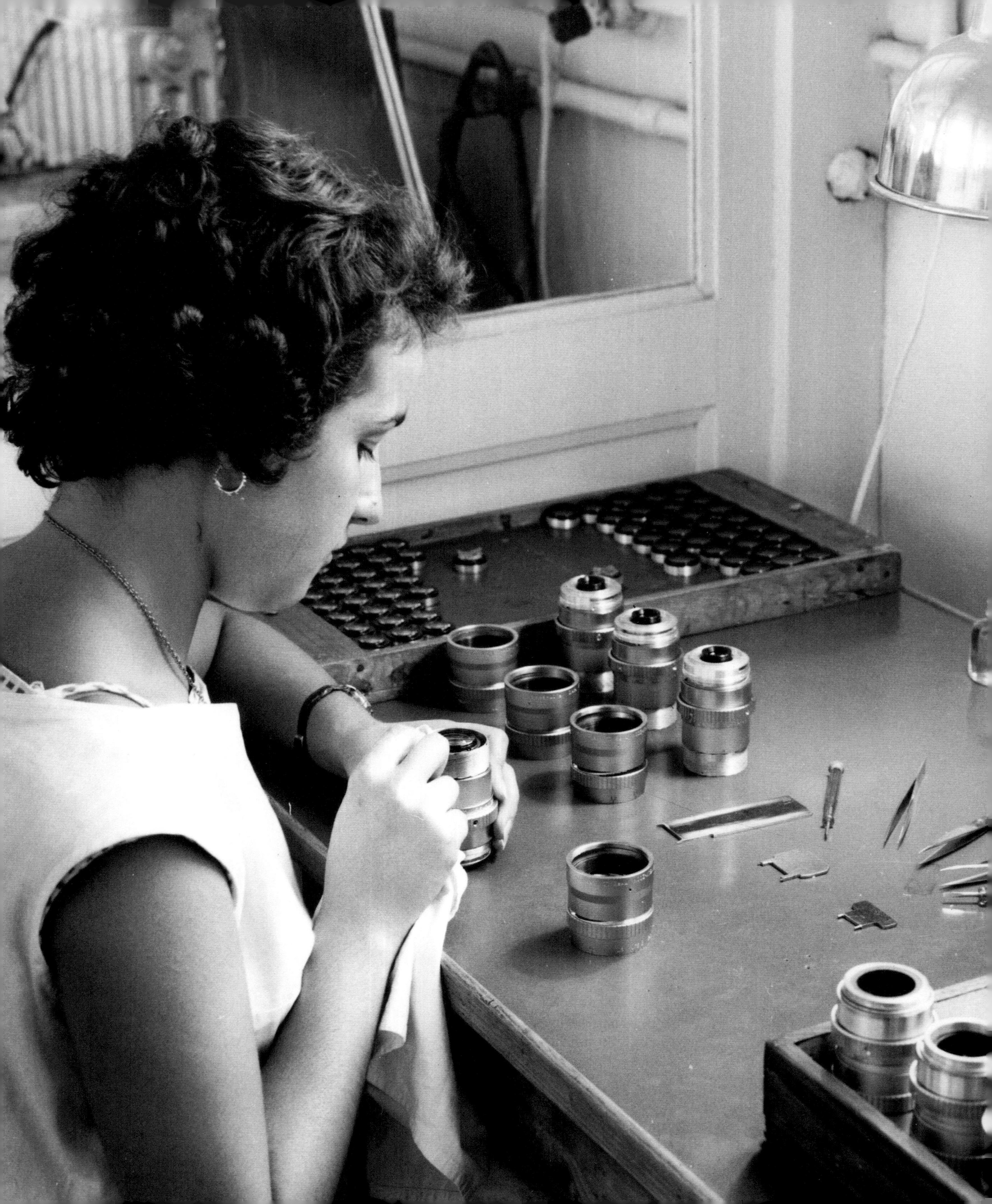

The Challenge of the Zoom

Jacques Debize

A zoom must be calculated so that all the rays of light emitted from a point of the object to be filmed will converge, regardless of the number of lens elements to be passed through, on a single point so as to obtain a sharply defined image with the minimum of optical defects. As is known, however, rays of light differ in terms of their point of impact and incidence on a lens, their color, the shape of the lens, and quality of the glass. It is only the rays very close to the optical axis that preserve their direction. Away from this axis, the light is scattered. The calculation must therefore follow the trajectory of each ray of light and simulate its passage through a lens so as to obtain the right optical combination (succession of lens elements) in compliance with the given constraints as regards technical specifications, cost, size, and weight.

All of Pierre Angénieux's genius lay in developing a system of calculations involving only the rays essential to the formation of an image and thus reducing the time required to calculate a lens by ninety percent. He considered only the rays in a cross-sectional plane of the lens, 10–20 rays per lens, no more, on the optical axis in the center of the image for three colors and three different apertures and then two or three points of the image in the field and always in this meridian plane. Intuition and experience play a key part in the selection of rays and appraisal of the final image.

If the image was no good in the end, it was necessary to revise the choice of optical combination by adding a lens element or changing its shape and so on. Each new study considered acceptable involved the creation of a prototype to assess optical and mechanical performance and decide on modifications wherever necessary.

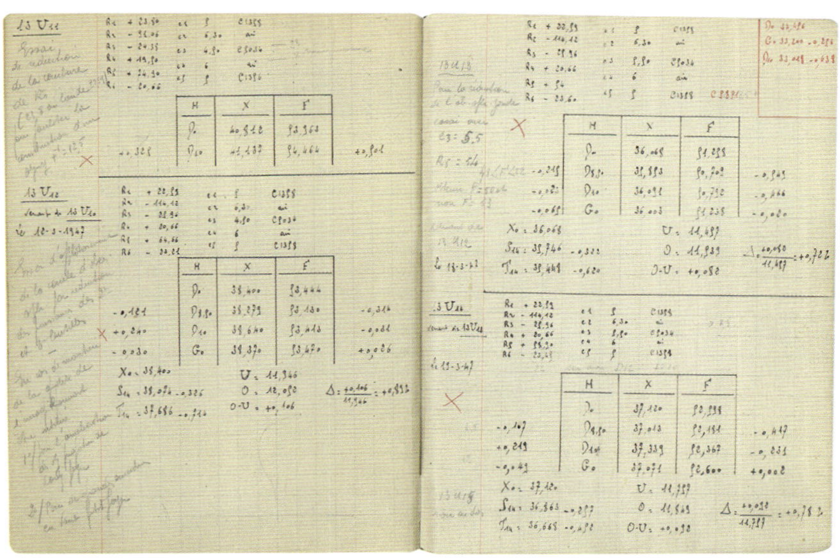

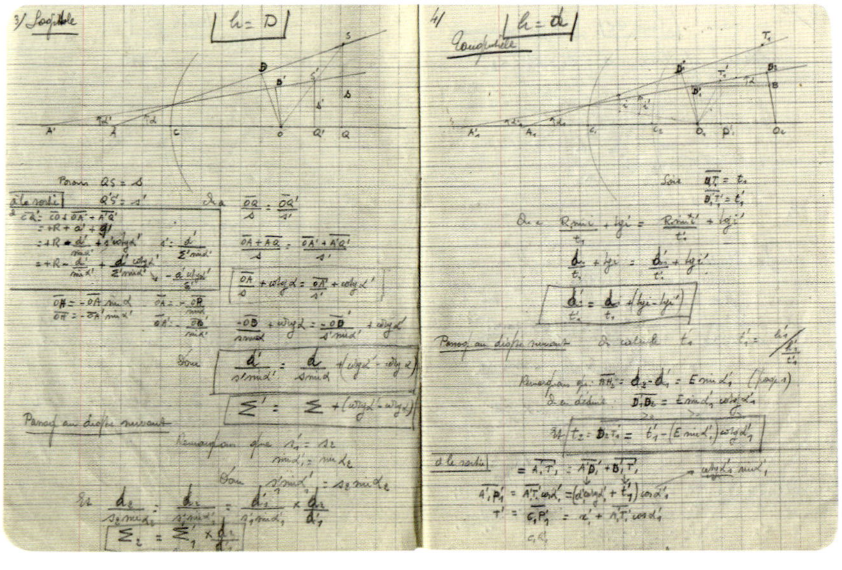

opposite
Assembly of the 17-68, the first Angénieux 16mm zoom (Angénieux archives. photo by Ito Josué).

left
Above, notebook of optical calculations by Pierre Angénieux. (André Masson collection). Below, tables of intermediate results of optical calculations by Marcel Moulin annotated by Pierre Angénieux, 1947–1963. (Angénieux archives).

Before the first computers, the calculation of a lens of fixed focal length with four or five lens elements took two engineers two years of work. And it was all work done by hand. For these daunting calculations, Pierre Angénieux selected the members of his team that made the fewest mistakes. It was an accountant named Moulin that had the privilege of spending his time compiling tables of figures, thus relieving the engineers of this burdensome task.
Pierre Angénieux bought his first electronic calculator in 1956. He was one of the first in the world of optical engineering to do so. It was a Bull Gamma 3. I started work at Angénieux in 1969, just six months after the company bought its first computer, an IBM 1130 as big as a sideboard. And this was only a sort of super calculator that we fed with one punch card for each light ray trajectory to be calculated. This immediately reduced the time taken to about one thirty-fifth. Where it had once taken seven minutes to calculate the passage through one lens element, it now took twelve seconds. While it thus became possible to calculate a lens of fixed focal length in a matter of months, it still took years for a zoom. Log tables and slide rules were still indispensable, as indeed was Monsieur Moulin. This early computerization of calculations made it possible to take giant strides in the development of increasingly complex and precise zooms, while the engineer still played a crucial part in the choice of solutions.

In the case of a lens of fixed focal length, in order to maintain quality in the center of the image and in different points of the field (as the position of the object filmed is liable to change), it is necessary for the optical elements inside the lens to move so as to correct the focus. With the zoom, there is also the movement of optical elements making it possible to expand or contract the visual field and hence to make the image of the objects smaller or larger. Crucial importance therefore attaches to the mechanical part of a zoom. Everything hinges on ensuring perfect guidance of the optical elements so that the image does not move in zooming.
Pierre Angénieux's achievement once again was to design a mecnanical system so precise as to ensure perfect stability of image. He took out a patent for an isostatic system of

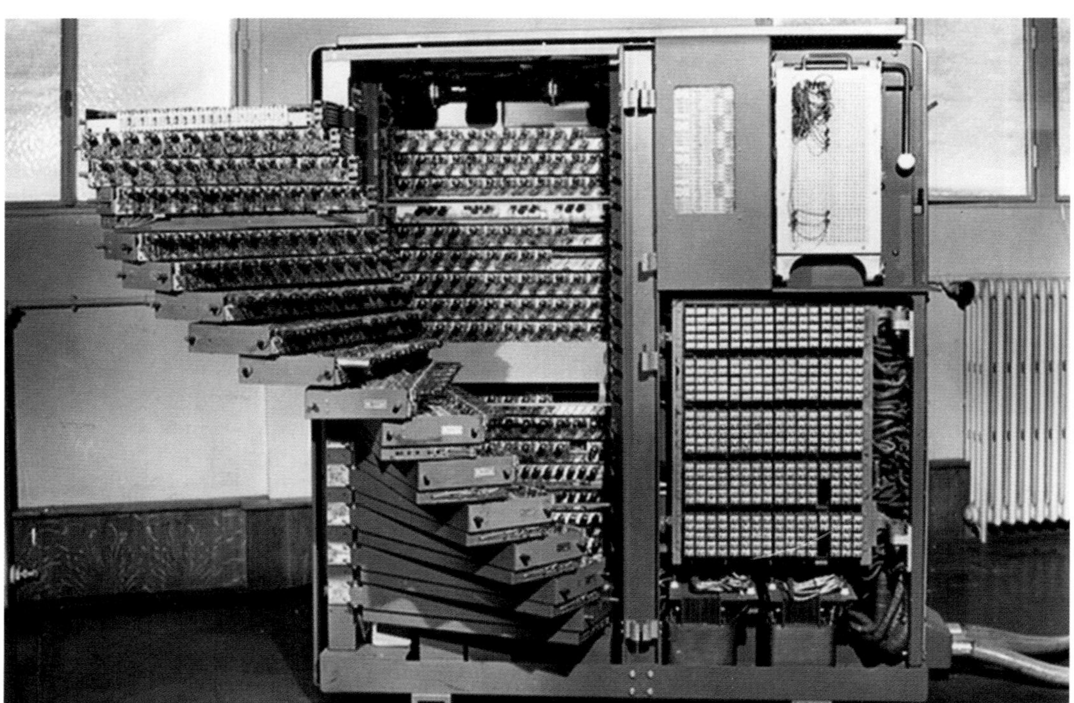

The first Gamma 3 electronic calculator by Bull.

Mechanical Compensation and Optical Compensation

In the 1940's, two main optical designers developed zoom lenses for cinema and television markets. The Zoomar was built in New-York after World War II by Franck Back, the Pan Cinor was designed by the French engineer Roger Cuvillier from SOM Berthiot company. Both systems used the principle of optical compensation.
A zoom of this kind has 5 groups of lens elements: 3 convergent groups (in green) separated by 2 divergent groups (in red) (fig. 1). The 2 divergent groups are connected with one another and able to move between the 3 convergent groups. The powers of the 5 groups are optimized so as to reduce displacement of the focal plane to the minimum. These systems do not, however, permit high levels of magnification, at which the loss of focus is unacceptable.
In 1955, Pierre Angénieux finds a way to improve the principle of mechanical compensation developed by Taylor Hobson in the mid 1930's. His new simplified design allowed to achieve a perfect sharpness while zooming and bigger zoom ratios. This system consists of 4 groups (3 convergent and 1 divergent), 2 of which are independently mobile but obey non-linear laws in such a way as to obtain total immobilization of the focal plane (fig. 2). This is made possible by the use of mechanical cams of very high precision, which appeared impossible at the time until Pierre Angénieux succeeded in producing them industrially.

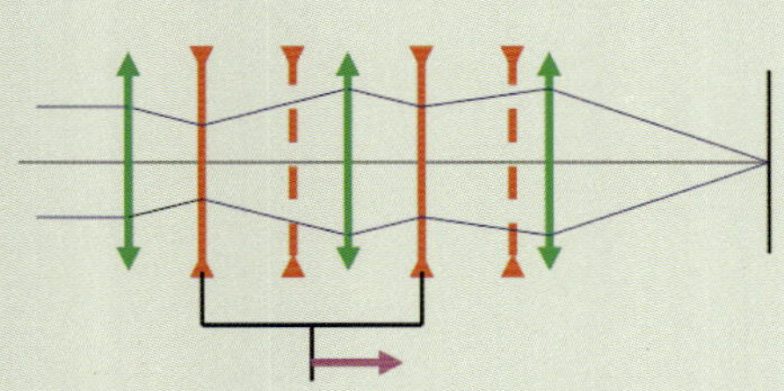

fig. 1

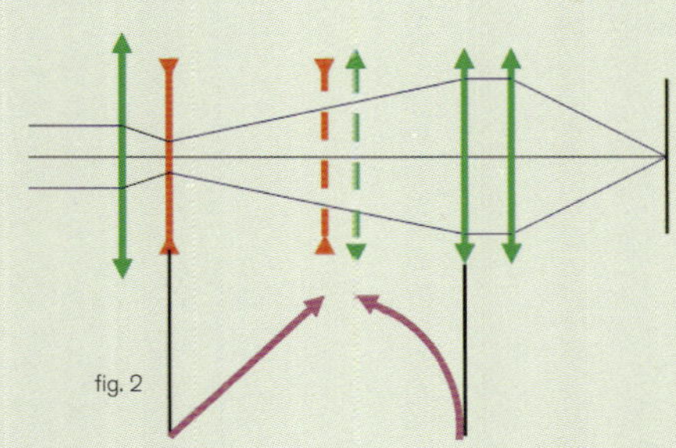

fig. 2

guidance of the optical group that uses sliding steel rods in a casing of aluminum alloy. The mechanical device (cam shaft) into which the optical groups are fitted is equipped with a spiral thread and grooves in a complex shape of the highest precision (in the order of ten microns) making it possible to transform the rotation motion of the rings into two motions of translation. The camera operator no longer needs to refocus when zooming. Slow zooming becomes possible.
This perfect mastery of the principle of mechanical compensation enabled Pierre Angénieux to ensure the reputation of his zooms and develop high degrees of image magnification. While the optically-compensated Pan-Cinor of SOM-Berthiot could attain no more than 5x, Angénieux zooms were able in time to arrive at 10x (1960), 18x (1967), 42x (1976), 72x (1994), and even 100x.

All of the present-day lenses with variable focal length, whatever their make, are based on the principle of mechanical compensation developed by Pierre Angénieux.

The Early Use of Angénieux Zooms in Cinema

Angénieux 16mm Zooms for Documentary and Artistic Cinema

The 16mm Format sets the Standard for News and Documentaries
Angénieux signed an agreement with the American manufacturers Bell & Howell for the distribution of its 4x zoom for 16mm film cameras in 1957. It was, however, precisely in this period that the amateur film market underwent radical change. As the Angénieux chief executive Jean Moret later explained in an interview with Herb Lightman, "The first zoom we designed was a 4 to 1 and the idea was to make a zoom lens for Bell & Howell 16mm amateur. They were selling an amazing number of 16mm cameras each month to amateurs, but it was just at that time they came out with the electric eye, so 8mm became the thing for amateurs (there was no Super 8 yet) and within few years 16mm was dead as an amateur format. The result was that our plan to make a great many zoom lenses for the 16mm amateur market was never materialized. However, that was probably the best thing that could have happened to us, because now we are essentially involved in supplying optics for professional" (*American Cinematographer*, March 1975) .
The range of zooms produced by Angénieux as from 1957 to meet the demands of this new amateur market comprised of a 4x, a 9-36 f/1.4 and 1.8, a 7.5-35 f/1.8, and a 6.5-52 f/1.8 for 8mm cameras followed by an 8-64 and a 6-90 for Super 8.

Bell & Howell then promoted Angénieux 16mm zooms for professional filmmakers. As Joseph Henry wrote in an article on the Bell & Howell/Angenieux 17-68 in *American Cinematographer* in March 1958, "The new lens is ideal for achieving special pictorial or compositional effects and for giving impact to the action scenes. In filming baseball game, for example, the wide angle setting can provide overall coverage of most of the playing field enabling the camera to see the plays develop. When the action becomes localized, you instantly zoom in for a dramatic closeup... For the 16mm travel film photographer the zoom lens can be particularly useful. In filming a market place, for example, he can make a wide establishing shot, then quickly zoom to telephoto position for revealing close-ups of people and interesting details within the scene."

1961, the First 10x Zoom for 16mm Cameras, the 12-120 f/2.2
A new revolution came in 1961, when Angénieux presented the first 10x zoom, an F=12-120mm f/2.2 for 16mm film cameras designed over the period 1958–1960. The launch of the 12-120 coincided with the appearance on the market of lighter 16mm cameras offering more freedom for hand-held shooting. This new equipment made new forms of filming possible. The 16mm zooms were used by TV channels all over the world for news and documentaries. They were also to triumph in the world of cinema.
The documentary genre called Direct Cinema originated in the United States and Canada in the early 1960s. The Canadian Michel Brault, one of its pioneers, stated in an interview with the BBC that 1960 was the year when it all began. In the new world where TV had come

Crown Princess Beatrix of the Netherlands with her 17-68mm zoom Beaulieu camera filming a cycling event during the Olympic Games in Tokyo, October 18, 1964. (photo Bride Lane Library/ Popperfoto/Getty Images).

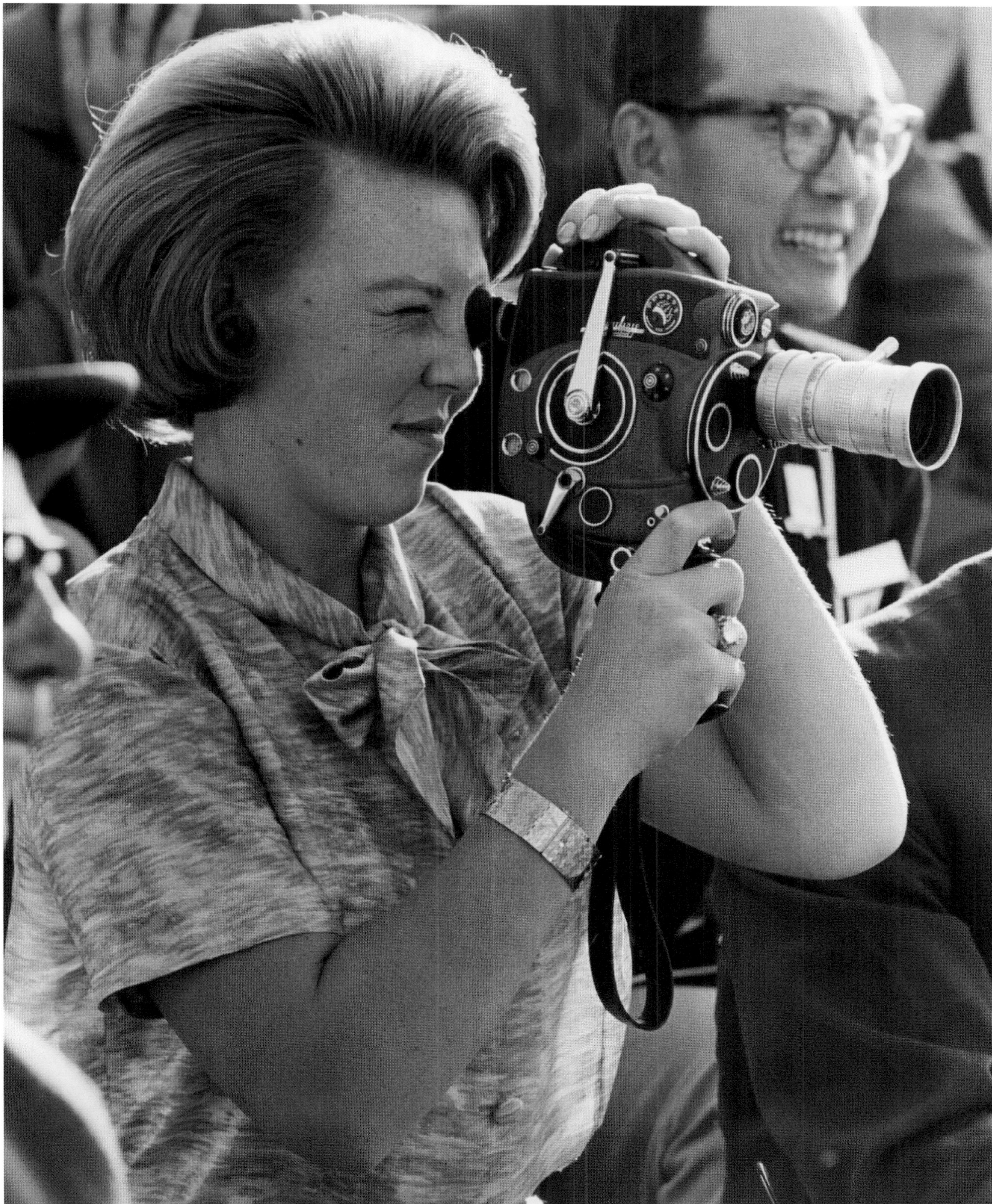

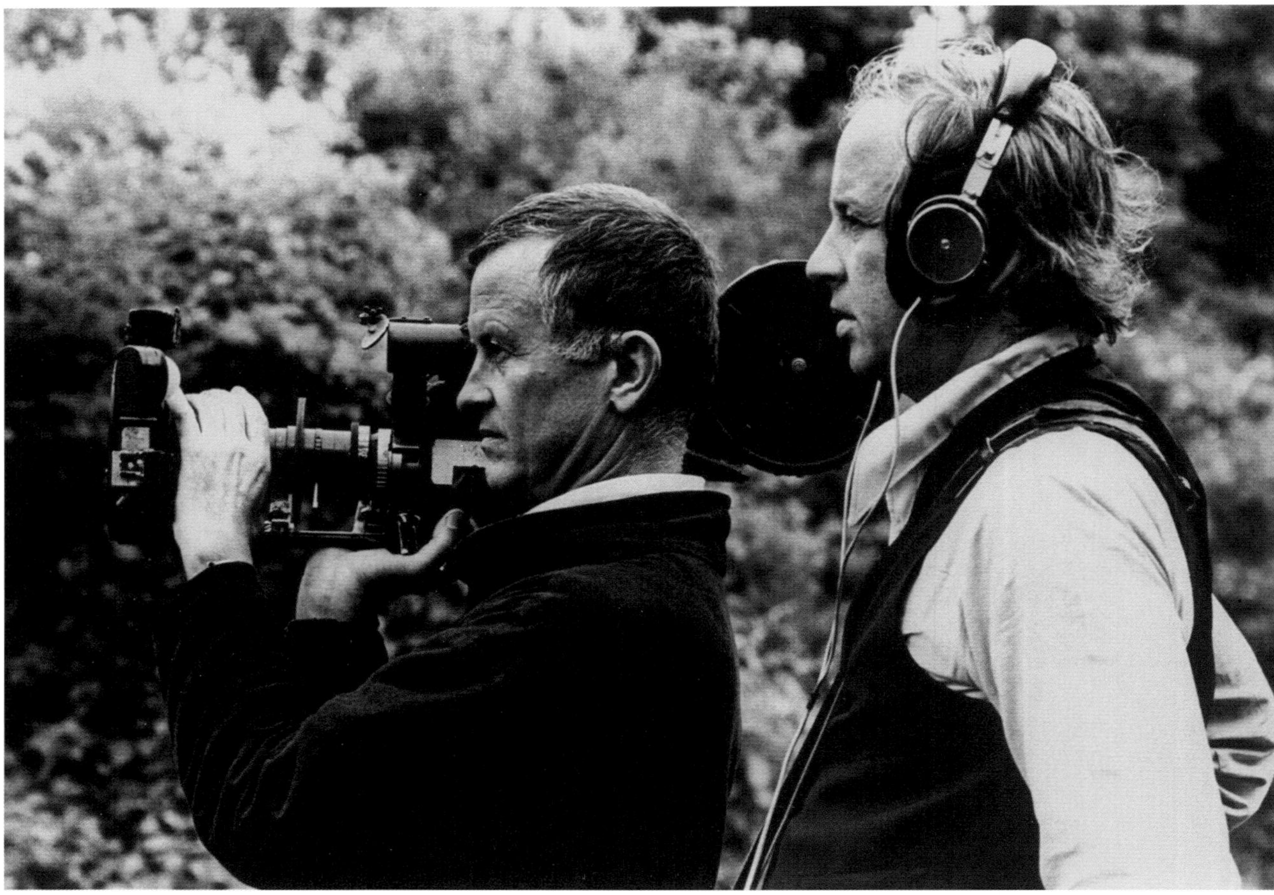

Filmmaking brothers Albert and David Maysles with a 12-120mm zoom (photo Marianne Barcellona / The LIFE Images Collection / Getty Images).

to compete with cinema, the practitioners of Direct Cinema took up the zoom to film more quickly and capture the new reality.
In the United States, these filmmakers documented the great events in American history, like the march on Washington and Woodstock. To follow John F. Kennedy in 1960 during his campaign for the Wisconsin primary, Richard Leacock and Albert Maysles mounted the 17-68 on an Auricon camera modified to facilitate filming. They lightened it, equipped it with a 400 feet roll for longer shooting time and a sync-sound system. *Primary*, the end result, produced by Richard Drew, is a milestone in the history of documentary film. The equipment was then used to shoot all the films produced by Drew Associates from 1960 to 1966. Donn Alan Pennebaker, the renowned maker of musical documentaries, used the same camera with a 12-120 zoom in 1967 to shoot his film *Don't Look Back* on Bob Dylan.

As Max Winter wrote in 'How D.A. Pennebaker Revolutionized the Rock Doc' (nofilmschool.com, November 2016), "Pennebaker was shrewd enough to notice that the Hollywood close-up, in which the camera zooms in on a character's face with a velocity and degree of magnification meant to produce the maximum emotional response from a viewer, could be used in documentary films to winning effect. And use it he did. We see here that zooming in on Dylan's face as he sings brings the 'character' Pennebaker is filming to life: the hawkish features, the will to perform, and the love of music itself combine to create a captivating figure who transcends what we call the documentary genre".

The French directors of the Nouvelle Vague or New Wave also took up these new lenses. The 12-120mm was the ideal companion for the brand-new Éclair NPR camera designed

by the French engineer André Coutant. François Truffaut, Éric Rohmer, Claude Chabrol, Agnès Varda, Jean Eustache, Jacques Rivette and many others found the tools they needed in Angénieux lenses. The cinematographers Michel Brault, Raoul Coutard and Nestor Almendros all made great use of the 12-120.
Angénieux 16mm zooms also became tools of creation for contemporary artists. The American Andy Warhol used a 12-120 to shoot performances in the Factory. The Canadian Michael Snow also used an Angénieux 12-120 to shoot his experimental films *Wavelength* (1967) and *La Région centrale* (1971).

The zooms developed by Angénieux after the 12-120 for 16mm cameras included the 15-150mm f/1.9–2.8 (1963), 12-240mm f/3.5–4.8 (1964), 9.5-95mm f/2.2 (1965), 12.5-75mm f/2.2 (1966), 10-120mm f/1.8 and 20-240mm f/2.2 (1967), 10-150mm f/2-2.8 (1977), and above all the 9.5-57mm f/1.6 (1971). The latter enjoyed particular success and was often used for documentaries, including Werner Herzog's *La Soufrière* (1977), where the team, including above all the directors of photography Ed Lachman and Jorg Schmidt-Reitwein, spent several days in Guadeloupe in the evacuation zone around a volcano threatening to erupt.

The filmmaker Alain Levent made this observation in an interview with Priska Morissey published by the AFC in March 2008, "The zoom was a kind of school for television, not least for its simplicity, as you work on a smaller surface in 16mm. The definition was good and it was necessary to go very quickly. This made it possible to replace tracking shots with zooms and above all not to change lenses. Moreover, the first successful zooms were for 16mm, because of their smaller optical size. There was above all the Angénieux zoom, the Zeiss 10-100mm, also outstanding, which appeared later, and small zooms like the Angénieux 9.5-57mm, a real gem. Very little use was made of lenses with fixed focal length in television, except for night shots, for example, which require lenses with a large aperture. The cameramen of the SFP company were trained for that. They had their own equipment, usually two zooms, a 9.5-57mm and a 12-120mm or 10-100mm. They also had a few boxes of fixed lenses but those were seldom used. More than other techniques, the zoom aroused a kind of passion, loved by some and hated by others. I think that its use is legitimate if the purpose is to generate emotion. The zoom can generate this emotion, like a tracking shot or a crane shot, even though it is undeniably an easier option".

The camcorders used for television became lighter and improved in quality from the end of the 1970s and through the 1980s. The 16mm camera gradually disappeared from the world of television, where the production was now done with specific TV cameras. The use of 16mm film was confined essentially to documentaries and independent movies, even though Super 16mm did become popular because of its high-quality image. Angénieux developped the 11.5-138 HR in the early 1990s and its wide-angle version, the 7-81 HR, in 1994.

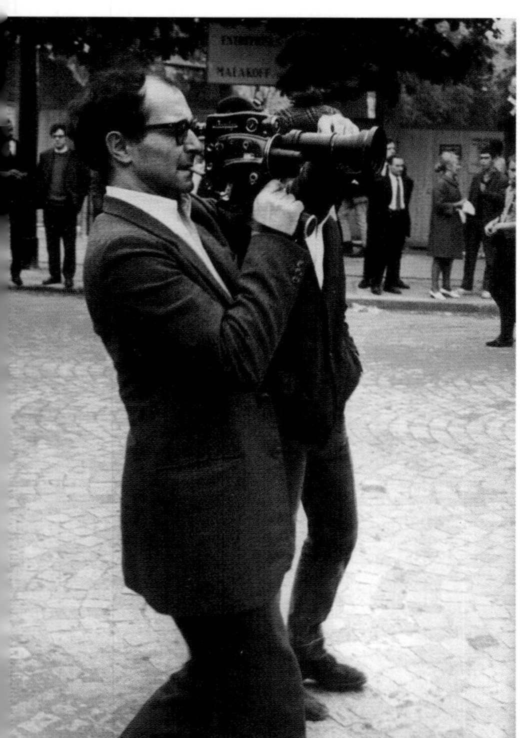

French director Jean-Luc Godard, a small 12-120mm zoom camera in hand, filming the events of May 1968 in France (photo Keystone-France /Gamma-Keystone, Getty Images.)

Angénieux 35mm Zooms and the European Cinema of the 1960s

1958, the First 35mm Zoom for Cinema, the 35-140 f/3.5

In the late 1950s, Pierre Angénieux designed a version of his 4x zoom adapted for the 35 mm movie camera. The prototype of the 35-140 mm was first used in 1958 to shoot Claude Boissol's *Julie la Rousse* and then Marc Allégret's *Les Affreux*, both photographed by Roger Fellous (AFC).

below
On the set of *Les Affreux* (1959), directed by Marc Allégret: Pierre Fresnay, Darry Cowl holding the prototype of the 35-140 f/3.5 zoom, and Roger Fellous, director of photography (photo provided by Maurice Fellous).

opposite
35-140 f/3.5 Angénieux zoom, type LA 2 CNC AP-15-1268, provenance Fellous (Cinémathèque Française collection, photo by Stéphane Dabrowski).

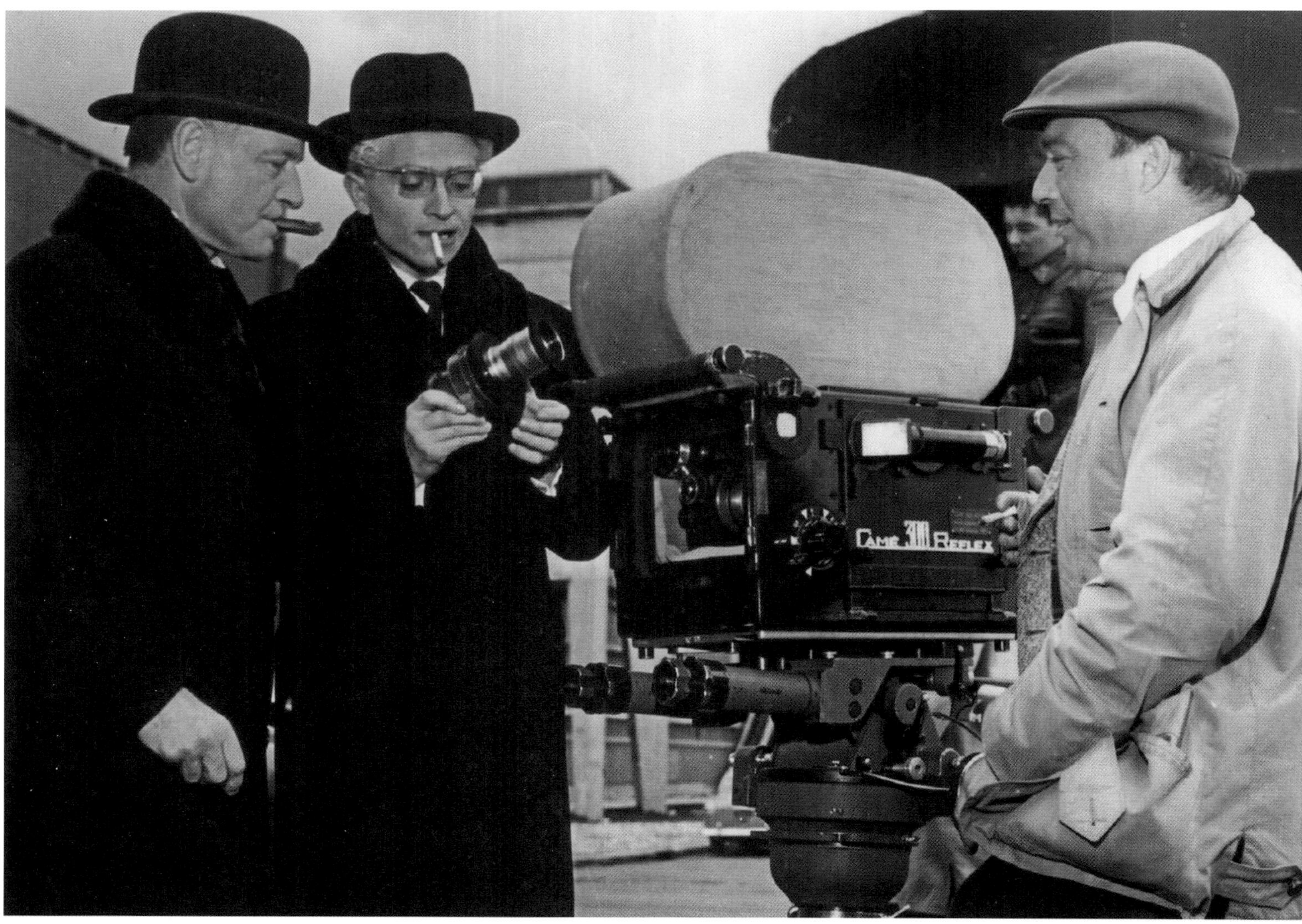

The Advantages of the Zoom over its Predecessors

Extract from an interview with Roger and Maurice Fellous by Henriette Dujarric ("Au banc d'essai des techniques nouvelles, Roger Fellous nous parle du 'Zoom,'" *Le Technicien du Film*, no. 50, May 15 – June 15 1959).

Photographers are very familiar with varifocal lenses. What are the advantages of the zoom, this newcomer, over its predecessors?
RF: It is much lighter and less cumbersome, details that offer a rapidity to which we were not previously accustomed with varifocal lenses. During the trials, we designed an intermediate mount so that it can be fitted quickly and simultaneously onto a Caméflex or a Camé 300.

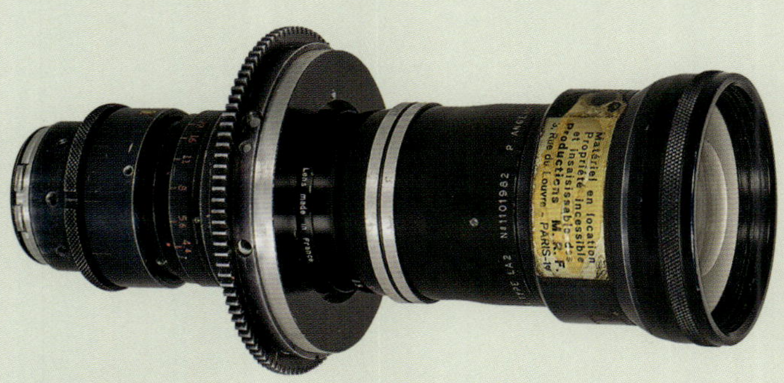

MF: A zoom offers the same definition and contrast as fixed focal lenses and can be used with the same diaphragms. The variation of focal lengths goes from 35 to 140 for focusing from 1.4 m to infinity and from 25 to 100 with the provision of an additional lens element for this purpose. It can be used at a fixed focal length, thus avoiding the need to change lenses. It thus has the advantage of instantly providing the director's desired focal length. The focal length is changed by rotating the lens.

Do the new Moviola Crab Dolly cranes not limit the use of these lenses?
RF: On the contrary, the combination of the two offers yet another advantage. I would really like to use them both together. In *Les Pétroleurs* [the film released under the title *Les Affreux*], the zoom enabled us to save a lot of time in a shot of Pierre Fresnay walking down the steps in Montmartre. In order to maintain the normal distance between the camera and the actor so as to frame the shot, it was necessary to place the camera at the bottom of the steps and use the zoom to simulate the back-tracking shot. I would have liked a Crab Dolly to complete the illusion by framing the height of the figure in the movement of normal descent, in other words, combining a simulated back-tracking shot and a vertical tracking shot.

« In some way, the concept of the 35-140 was revolutionary.
First, the focusing group was an inverted cemented doublet in order to correct the field aberrations at short focal lengths. Each element of variator and compensator is composed of three lenses: one single lens and a cemented doublet in order to compensate field aberrations at short focal lengths and spherical and chromatic aberrations at long focal lengths. The rear group begins with a thick lens in order to compensate remaining astigmatism and also allowed to reach the necessary back focal distance of the camera. Today we can calculate the MTF of this zoom and we can appreciate the performances reached by Pierre Angénieux. It's remarkable that this zoom was designed with manual calculations and logarithm tables, not with computers. »

Jacques Debize

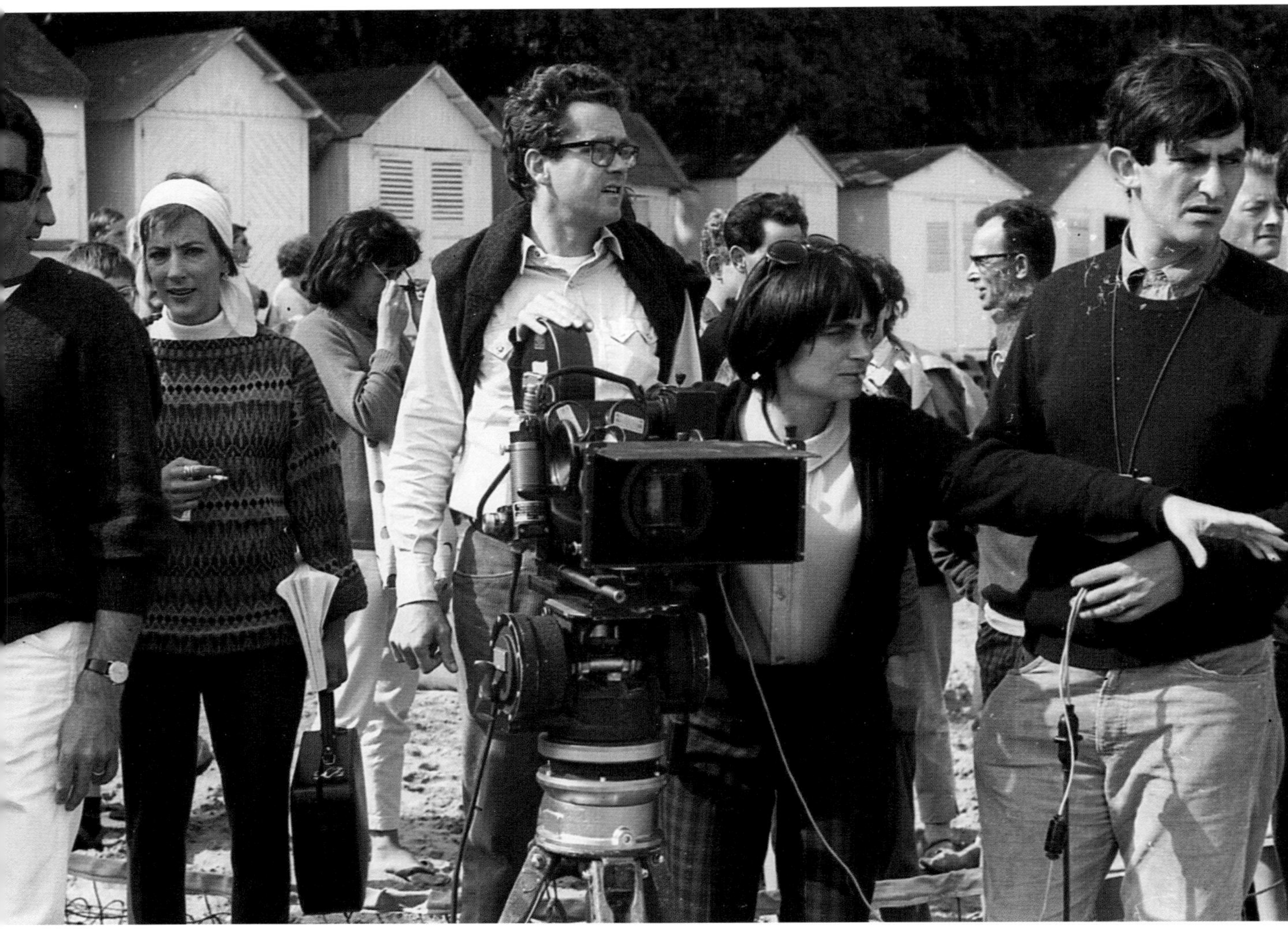

The filmmaker Agnès Varda on the movie set of *Creatures* (© Agnès Varda. All rights reserved).

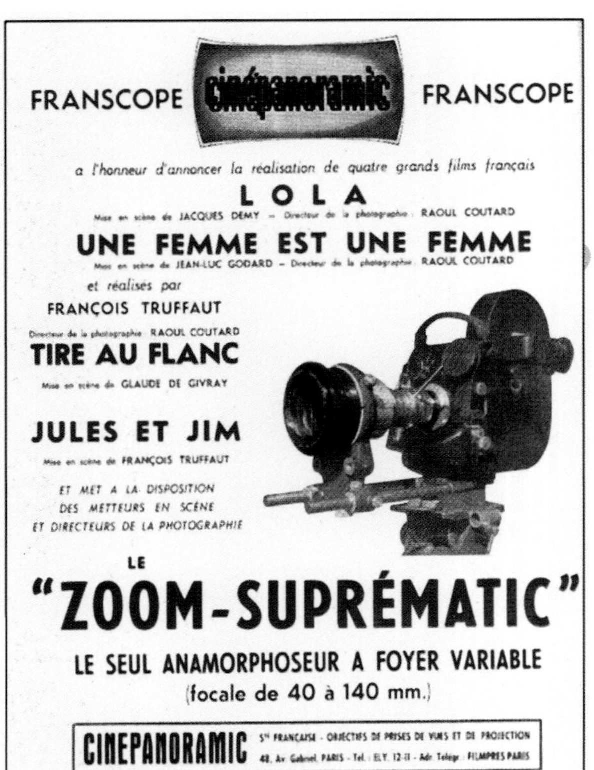

The 35-140 came into general use in the early 1960s among French directors of the Nouvelle Vague or New Wave. It was used in 1963 for practically all the images of Georges Lautner's *Les Tontons Flingueurs*, photographed by Maurice Fellous, Roger's brother.
This was also the great era of Cinemascope. The brothers Roger and Maurice Fellous worked with the camera operator Jean Dicop to design an anamorphic lens for the 35-140. The Franscope 40-140 was used on various French films, including Agnès Varda's *Les Créatures* (1966), Jacques Demy's *Lola* (1961), *Une femme est une femme* (1961) and *Le Mépris* (1963) by Jean-Luc Godard, and François Truffaut's *Jules et Jim* (1962). Raoul Coutard was the cinematographer of the last four films.

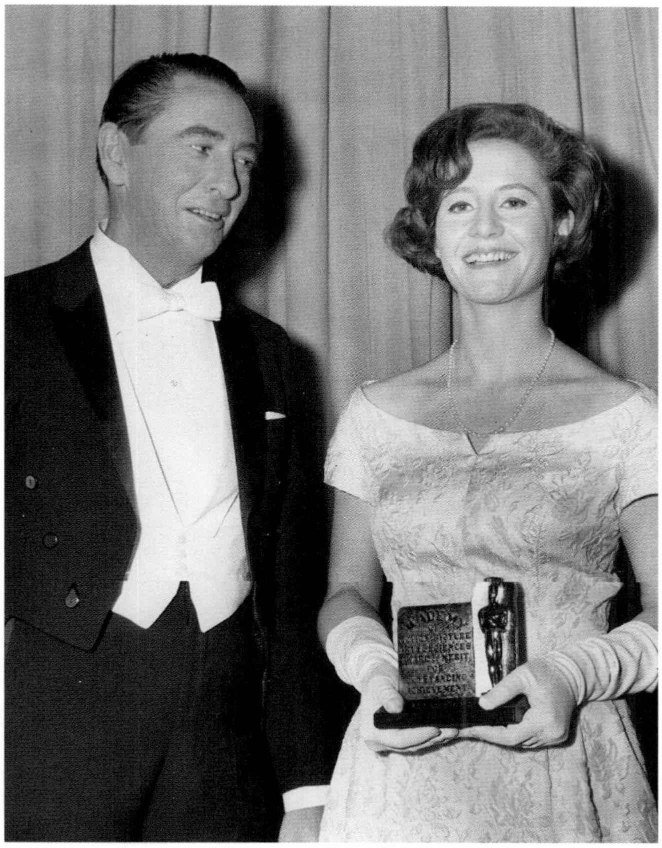

Pierre Angénieux's daughter Martine, then 25, receives the Scientific and Engineering Award from the American actor Macdonald Carey. 1964 was a year of triumph in the United States for Angénieux (collection of the Moret family).

1962, the First 10x Zoom for 35mm Movie Cameras, the 25-250mm f/3.2

The 10x version for 35 mm appeared in 1962, a 25-250mm f/3.2 T3.9[1] that was to become the benchmark for shooting with 35mm film for nearly two decades. Angénieux also brought out a 24-240mm and a 20-120mm.

This 10x zoom, a key invention in the history of the cinema, was to have a great influence on the cinematographic style of an entire generation. It was widely used on full-length films and documentaries for nearly 23 years until the improved 25-250 HP version appeared in 1985.

This outstanding technological achievement earned Pierre Angénieux a Scientific and Technical Award from the Academy of Motion Pictures Arts and Sciences (AMPAS) in 1964. Unable to attend the Academy Awards ceremony, he sent his daughter Martine, then aged 25, who later married Jean Moret, to receive his Oscar.

European filmmakers were quick to adopt the 25-250. In France, it was one of the favorite lenses of Raoul Coutard AFC, Denys Clerval and Nestor Almendros AFC ASC, whose work with the directors of the Nouvelle Vague enabled them to create a new cinematographic esthetic. As John Belton wrote, "The zoom is symptomatic of the evolution of the language of the cinema since the New Wave. Spatially distorting and inherently self-conscious, the zoom reflects the disintegration of cinematics codes developed before the Second World War" ("The Bionic Eye Zoom Esthetics," *Cineaste*, vol. 11, 1980). Firmly established in the French film vocabulary, the word *zoom* also entered everyday language despite the efforts of the Organisme de Vocabulaire Français to introduce the French spelling *zoume* and the term *baladage* for a tracking shot. It was widely used in Italy by Massimo Dallamano and Tonino Delli Colli, Sergio Leone's cinematographers, on the *Dollars Trilogy* (1964–1966) and *Once Upon a Time in the West* (1968). Massimo Dallamano combined a lateral tracking shot with a zoom in the famous train scene of *Bandidos* (1967), as did Tonino Delli Colli in *Once Upon a Time in the West*. Their technique in these films radically changed the way of shooting westerns. For Michelangelo Antonioni's masterpiece *Blow-Up* (1966), Carlo Di Palma AIC harnessed the power of different lenses and especially the zoom. As he stated in *American Cinematographer* (January 1993), the 25-250 made a great improvement in terms of quality and beauty, changing lens meant changing quality. The Italian cinematographer Pasqualino de Santis AIC used the Angénieux 25-250 on Franco Zeffirelli's *Romeo and Juliet* (1968), for which he won the Oscar for best photography. He also made great use of the panavised 50-500 APZA (Angénieux Panavision Zoom Anamorphique) to film Luchino Visconti's masterpiece *Death in Venice* (1971), winner of the Palme d'Or at the Cannes Film Festival. Nearly every scene of this film was shot with a zoom. As Michael Wilson wrote, "In Visconti's adaptation, *Morte a Venezia*, the zoom lens becomes the cinematic equivalent to Thomas Mann's free indirect style. The zoom allows Visconti to probe sympathetically the mind of Aschenbach while at

[1] The f-number (f/) corresponds to the geometric aperture of a lens (the ratio of its focal length to the diameter of the entrance pupil). It serves to quantify the opening of the diaphragm with respect to the frontal lens element and f/ is conventionally used as a prefix for the aperture number. The standard apertures are f/1, f/1.4, f/2, f/2.8, f/4, f/5.6, f/8, f/11, f/16 and f/22 (another possible notation is 1:2.8 for f/2.8). The T-number or photometric aperture takes into account the losses in transmission of light due, among other things, to refraction and reflection through the successive lens elements. The T-number is therefore a more pertinent characteristic as regards the real power of the lens. The geometric aperture still serves, however, to estimate depth of field or the extent of a bokeh. This is why both numbers are often retained in the specifications of a cinema zoom.

[2] Panavision is a major American motion picture equipment company. Both a client and a competitor, Panavision rented its own Panavision lenses as well as lenses from other manufacturers, "Panavised" to make them compatible with its cameras. The Panavision catalogue of equipement also includes other makes of lenses, including Angénieux zooms.

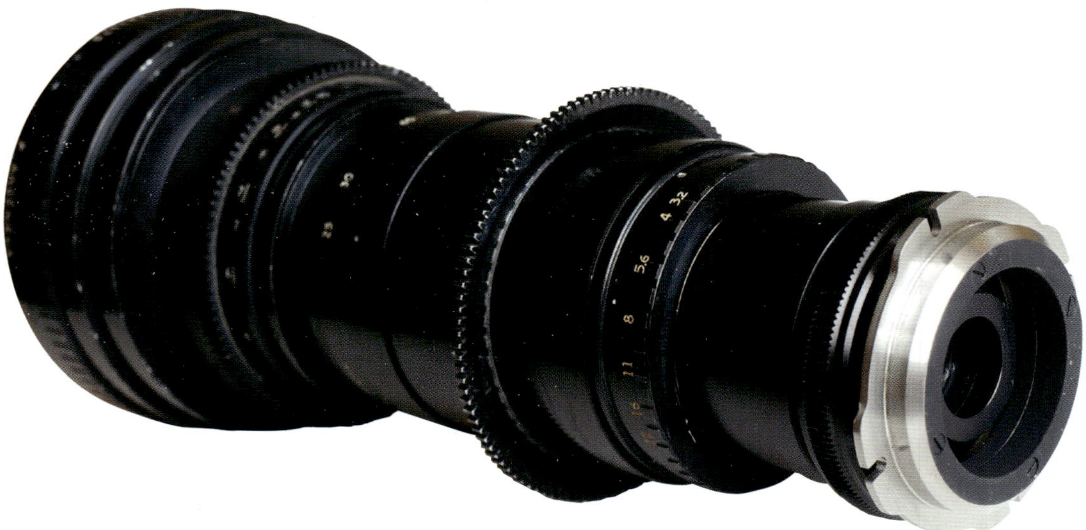

The first version of the Angénieux 25-250mm f/3.2 10x zoom for 35mm cinema (1962).

« I've been shooting with Angénieux zoom lenses for 35 years. I started in 16mm and owned several. Now I own most of them in 35mm. I've always been very fond of Angénieux lenses because of the look and the very useful focal lengths that they have. For example, I used the 9.5-57mm for many years in documentaries, and I thought it was an excellent lens. It was very high speed. The wide angle coverage and the length was perfect for documentary work. I also liked the 12-120mm. It has become a classic and worked on many productions. When I started to do features, the 25-250 was the only practical zoom lens around. Then it was improved with the HR version, it became one of my all-time favorite lenses, and I still have it 20 or more years later. »

Kees Von Oostrum

President of ASC – American Society of Cinematographers
Extract from an interview with Jon Fauer that appeared in *FDTimes*, April-June 2017

the same time ironically judging him. Moreover, the technical effects of the zoom, such as its manipulation of depth of field and its easy mobility, illustrate aspects of Gustav's inner character, most often those which he explicitly denies in a series of arguments which Visconti wrote for the film. Close attention to the film's zooms, then, can lead us to a greater understanding of both the main character and Visconti's ambivalent attitude towards him." ("Art is Ambiguous: the zoom in *Death in Venice*," *Literature / Film Quarterly*, vol. 26, 2, April 1998) Visconti also used the zoom in his next film, *Ludwig*, *La caduta degli dei* (*Ludwig,* 1972). Among the great Italian films of this period whose form depends on the zoom, David A. Cook (in *Lost Illusions*, University of California Press, 2002, p. 366) lists Vittorio de Sica's *Il giardino dei Finzi Contini* (*Garden of the Finzi-Continis,* 1970), Bernardo Bertolucci's *Il conformista* (*The Conformist*, 1970), Lina Wertmüller's *Mimì Metallurgico ferito nell'onore*

« The Angénieux zoom is wonderful in that the quality and image sharpness of each focal length is as good as a prime lens. For us, this was enormous progress. At that time there were not a lot of lenses that were as advanced and as high performing. The use of the zoom by Claude Lelouch almost became a symbol of the New Wave. »

Pierre-William Glenn

Director of photography and filmmaker, founding member of the AFC and chair of the CST (Commission Supérieure Technique de l'image et du son) from 2002 to 2018
Extract from an interview with Jon Fauer published in *FDTimes*, "A History of Angénieux", September 2013

(*The Seduction of Mimi*, 1972), *Tutto a posto e niente in ordine* (*All Screwed Up*, 1974), *Travolti da un insolito destino nell'azzurro mare d'agosto* (*Swept Away by an Unusual Destiny in the Blue Sea of August*, 1974), Liliana Cavani's *Il portiere di notte* (*The Night Porter*, 1974), Federico Fellini's *Amarcord* (1974), Pier Paolo Pasolini's *Salò o le 120 giornate di Sodoma* (*Salò, or the 120 Days of Sodom*, 1975), and Michelangelo Antonioni's *The Passenger* (1975).
Roberto Rossellini, an enthusiastic user of the zoom, filmed the opening of the Centre Pompidou to the public in 1977. These images bear unique witness to a key moment in modern culture. For the first time, we see the public face to face with contempory art. For this, he insisted on the use of just one lens unique, an Angénieux 25-250 fitted with an innovative device. Remote control of the zoom (with no image display) enabled Rossellini to decide when to zoom while filming. Jacques Grandclaude, the producer of this documentary, decided to film the director to work. The cinematographer Nestor Almendros has this to say, "This simply descriptive and educational film is in line with Rossellini's last work. Our camera literally walks from floor to floor, department to departement, showing things like how the

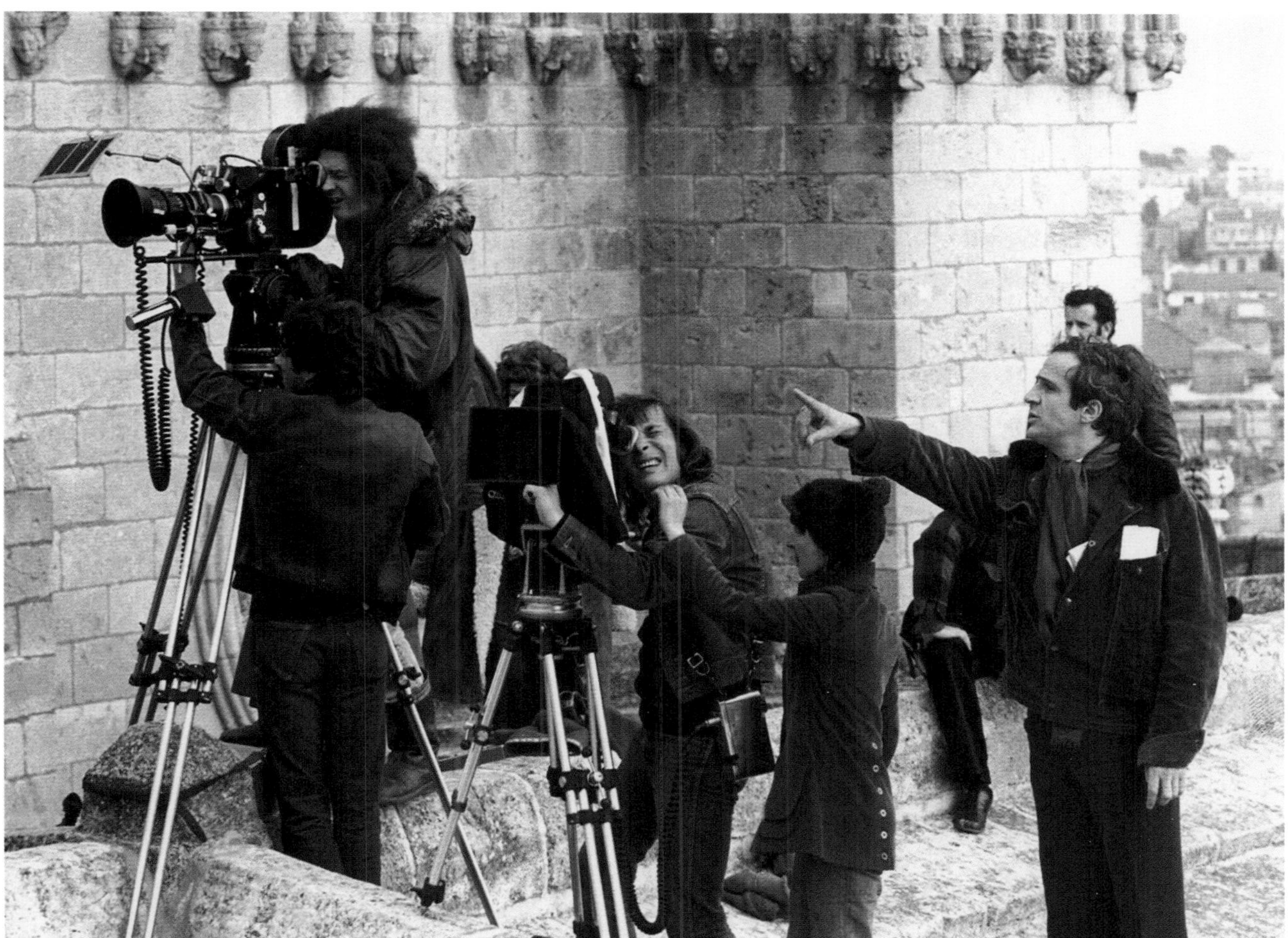

Pierre-William Glenn with a 25-250mm zoom camera filming *Such a Gorgeous Kid Like Me* (*Une belle fille comme moi*, 1972) directed by François Truffaut (on the right).
Production: Les Films du Carrosse.
Image provided by Pierre-William Glenn.

library operates and how the paintings are arranged on the museum's panels. The camera goes up the escalators with the public and proceeds through the rooms and galleries. Like a pointing finger, a zoom remote-controlled by Rossellini indicates a particular element of the building's structure or a painting on which he wants to focus. Environmental noise and snatches of the visitors' conversation accompany the images with no music or commentary. You could say that Rossellini himself guides the spectator by the hand in the discovery of the building." The combination of the film *Beaubourg, centre d'art et de culture Georges Pompidou* in its final version and the documentary on its making by Jacques Grandclaude constitutes one of the most educational works on the complexity of the creation of a film and the choice of shots. *Beaubourg* is the great director's last work.

The German director Werner Herzog also used the 25-250 for his two epic films set in the Amazonian forest, namely *Aguirre, der Zorn Gottes* (*Aguirre, the Wrath of God*, 1972) and *Fitzcarraldo* (1982). In the magnificent opening scene of *Aguirre*, the camera equipped with a 25-250 is placed far away on a peak in the Andes to film the conquistadors and their followers as they move down the mountain like a procession of ants lost in the mist, zooming in gradually to show that they are in fact human.
The Canadian director Sidney Furie used it for his musical *Wonderful Life* (1964) featuring Cliff Richard and the Shadows: "We went a little crazy with using the zoom lens on that picture," says Furie, grinning with the fond memory of how the novelty appeal of this 'new toy' inebriated his muses and induced using it to excess. Almost the entire opening twenty minutes of the movie are elaborate zoom-outs. They oohed and ahhed over it at the time though, believe me" (Daniel Kremer, *Sidney J. Furie: Life and Films*, University Press of Kentucky, 2015). Furie went on using the 10x zoom in other films, including *The Ipcress File* (1965), shown in Cannes in 1965, the winner of the BAFTA award for Best British Film that year.
Greatly influenced by the Nouvelle Vague or New Wave, the British director Richard Lester made intensive use of the 25-250 zoom for his Beatles musicals *A Hard Day's Night* (1964) and *Help!* (1965). As Chris Pizello wrote in *American Cinematographer* in February 2003, "*A Hard Day's Night* shattered movie conventions, much as the band would later redefine the parameters of pop music. Lester gleefully discarded the standard visual ritual of an over-the-shoulder shot, a wide shot and a close-up. Instead, the director and cinematographer Gil Taylor, BSC, adopted a roving, multiple-camera technique (aided by new, versatile 10:1 zoom lenses) so that the Beatles could move about freely and not worry about technicalities like hitting marks. This fast, fresh brand of filmmaking was a perfect fit for the film's tiny budget, tight schedule and simple black-and-white aesthetic." The Beatles film *Magical Mystery Tour* (1967) was shot with a 25-250 combined with a 12-120 on a shoulder-held Éclair NPR.

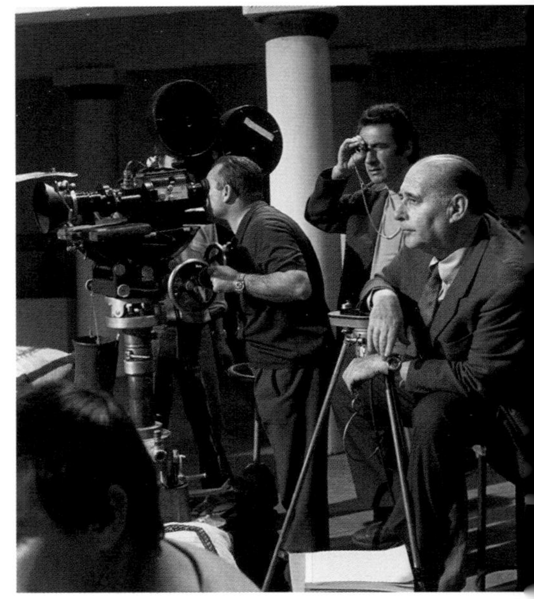

Italian director Roberto Rossellini on the set of the film *Socrates*. 1970. On the camera, a 25-250mm zoom (photo by Reporters Associati & Archivi\ Mondadori Portfolio, Getty Images).

opposite
John Lennon (far right) with three crew members during the filming of *Magical Mystery Tour*, the film The Beatles shot for the BBC, in Newquay (Cornwall), Great Britain.
On the camera, a 12-120mm zoom (photo by David Redfern / Redferns / Getty Images).

Angénieux 35mm Zooms and the New Hollywood

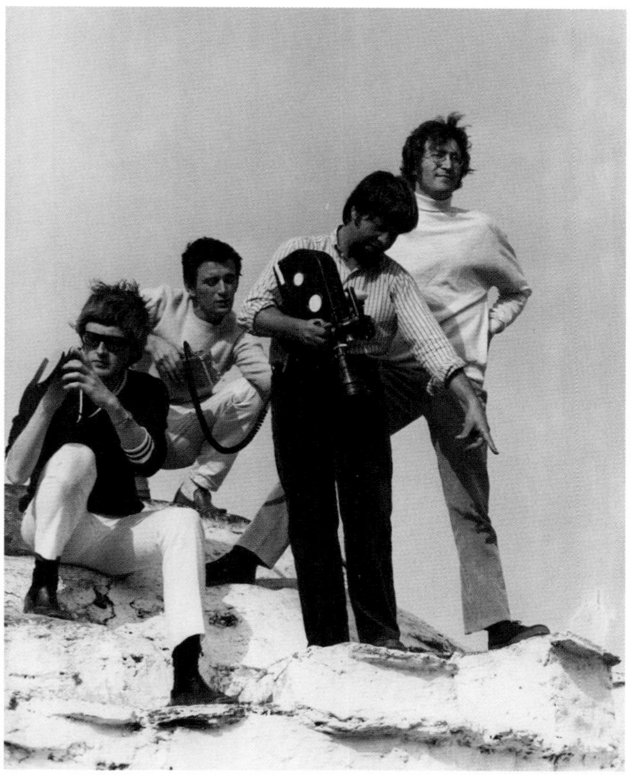

« When we started shooting what was called Cinéma Vérité, the Angénieux 10:1 zooms, the 25-250 for 35mm and the 12-120 for 16mm were the lenses of choice at that time. In the beginning there were maybe four or five shooters in America who had them. After that, everyone did. »

Haskell Wexler ASC

Extract from the interview with Jon Fauer that appeared in *FDTimes*, "A History of Angénieux", September 2013

Despite the enthusiasm in Europe, Hollywood directors were reluctant to use these zooms on major productions. The quality of Angénieux zooms received, however, the unconditional backing of the legendary US movie equipment company Panavision, which produced "Panavised" versions of the 25-250 and then the 20-120 (labelled APZ for Angénieux Panavision Zoom) for its cameras. Panavision also produced an anamorphic version of the 10x zoom called the 50-500 APZA (APZ Anamorphic). It was not until the late 1960s that the Angénieux 25-250 began to enjoy a certain degree of success in Hollywood, initially with directors who had worked in TV, including Robert Altman, John Frankenheimer, Sydney Pollack, Sydney Lumet, Stuart Rosenberg, and Arthur Penn. Sydney Lumet was one of the first to use the 25-250 in *Fail Safe* (1964). John Frankenheimer used a 25-250 on his 35mm cameras to shoot the film *Grand Prix* (1966). The "Panavised" anamorphic version was widely used in *Le Mans* (1971) starring Steve McQueen. Haskell Wexler ASC got to know Angénieux zooms on the set of *Who's Afraid of Virginia Woolf?* (1966), the closing scene of which was filmed with the brand-new 24-240. This work earned him the Oscar for best black and white photography. For his first movie in color, *In the Heat of the Night* (1967), Wexler followed the example of Raoul Coutard and used the 25-250 zoom for numerous scenes, including the one where Wilson tries to flee across the bridge over the Mississippi, where the long zoom produces a documentary-like effect. This was one of the first American movies to make large-scale use of the zoom. He continued his experience with the 25-250 in *Medium Cool* (1969), a cult movie of fictional reality in which Robert Forster plays John Casselis, a TV news cameraman covering the Democratic Convention of 1968 in Chicago with an Éclair NPR and a 12-120 mm zoom. The *New York Times* film critic Vincent Canby underscores the movie's historical importance: "The result is a film of tremendous visual impact, a kind of cinematic *Guernica*, a picture of America in the process of exploding into fragmented bits of hostility, suspicion, fear and violence." Haskell Wexler also used the 25-250 on *Bound for Glory* (1976), which earned him his third Oscar.

The cinematographer Robert Surtees used a 50-500 APZA in his work with Mike Nichols on *The Graduate* (1967). The movie received seven Oscar nominations, including best film and best photography. Mike Nichols won the Oscar and the Golden Globe for best director.

Another point of reference for American films shot with the 25-250 is *Easy Rider* (1969). The movie started as a low-budget production.

Dennis Hopper hired László Kóvacs as cinematographer. László had worked before with Richard Rush with whom he "perfected the zoom-lens technique Rush dubbed 'critical focus,' in which, in the same shot, you can change the focus from the foreground to a face in the background, and have one element crystal-clear and the rest of the elements disappear. It's very smooth and makes for an entirely different poetic effect than the cut." (Michael Sragow, *Picture Perfect* www.baltimoresun.com). Kóvacs used the Angénieux 25-250 on the Arriflex camera of a friend of himself Vilmos Zsigmond HSC ASC for most of the movie. *Easy Rider* won the prize for best first film at Cannes in 1969 and found favor with the general public as the third biggest box-office hit of the year.

« We panned and tracked to get him to the bed. We're still fairly far back on the set with the zoom. Basically the look of the picture is that zoom. I think we used it on almost everything. It's constantly moving in and out. It also helped the match enormously because we didn't have to switch lenses. We were just setting on the zoom that worked. I wanted to express the despair of an affair that is just below the waist and he has no real connection to it, so the montage is how it goes on once conversation fails. It's quite extreme but it needs to be. There is no interpersonal interaction. It's not just for good taste that we skipped the sex. »

Mike Nichols

(cited in Rob Felds', *Trouble in Mind. Ce plaisir qu'on dit charnel*, 1971)

Robert Altman was one of the first American directors to make regular use of zooms in his films. Various scenes of *The Long Lost life of Edward Smalley* (1963) were shot with an Angénieux 25-250, which plays a key part in this TV series and was also used in others of the late 1960s. In *I Spy*, shown on NBC between 1965 and 1968, the 25-250 was equipped with three motors (zoom, focus and diaphragm) so that it could be controlled by a single camera operator. Frank Phillips ASC used the 10:1 on season two of *Hawaii, Five-O* (1969), for which he was awarded the Emmy for best photography. For *Owen Marshall, Counselor at Law* (1971–1974), the director of photography Walter Strenge "was on the 21st day of a 23-day schedule. He said that so far, he had photographed every scene in the picture with the Angenieux 25-250mm lens. 'It has not been off the camera except to be cleaned and checked each evening,' he added." (quoted in Nick Hall, *The Development and Use of the Zoom Lens in American Film and Television: 1946- 1974*, PhD thesis, University of Exeter, 2012)

Robert Altman continued to work with zooms when he began making full-length feature films in the late 1960s. An anamorphic 50-500 is widely used in *MASH* (Palme d'Or at Cannes in 1970), a satirical comedy set in a military hospital in Korea. Robin Wood suggests in *Lost Illusions* (p. 362) that while Altman used the zoom both to guide the viewer's awareness and to assert his own presence in the film, he also realized the possibilities it offered of dissolving space and disturbing our sense of physical reality. The cinematographer Vilmos Zsigmond learned to use the zoom while working with Altman on *John McCabe* (1971). The Angénieux 25-250 was also a tool of creation for young American directors of the period like Francis Ford Coppola, Steven Spielberg and Martin Scorsese. Scorsese, who was part of the team that filmed Woodstock, used it on a regular basis in the 1970s. Gordon Willis shot the opening scene of Coppola's *The Godfather* (1972) with a 25-250 controlled by software developed by Tony Karp to obtain very slow, fluid movement. The director of photography Owen Roizman ASC used on on William Friedkin's *French Connection* (1971), the first R-rated movie (for spectators aged 17 and above) to win an Oscar, in this case for best film in 1972. Roizan also used it for *The Exorcist* (1973), again directed by Friedkin, which was nominated for ten Oscars including best film. The director of photography Bill Butler used “Panavised” 50-500 zooms on a tripod and hand-held camera for Steven Spielberg's *Jaws* (1975). The combination of a tracking shot and a zoom on Sheriff Brody, witness to an attack by the shark, made this a cult scene in the world of cinema. *Jaws* was an even bigger box-office hit in North America than *The Godfather*.

In 1964, Angénieux developed a 20X zoom for 16mm cameras, a 12-240. The lens remains compact and precise even at the maximum aperture. The British firm Samuelson Film Service adapted this for 35mm in the early 1970s. Transformed into a 25-500mm with an aperture of T9, the modified Samuelson-Angénieux zoom was used by director of photography Denys Coop BSC on the thriller *10 Rillington Place* (1971) directed by Richard Fleischer. The American director Stanley Kubrick, resident in the United Kingdom, made movie history the same year with the famous opening scene of *A Clockwork Orange* (1971), shot with a Samuelson-Angénieux 25-500. Kubrick, who preferred to own his equipment, insisted on having a 20x zoom and so the Cinema Products Corporation contacted Angénieux to have a 20x lens based on the 12-240 zoom made for him at Saint-Héand. The 24-480 T9 Cine Pro appeared when Kubrick was making *Barry Lyndon* (1975) and he used it together with the 25-250. As journalist J.S. Bernstein wrote:e: “For most of Hollywood history, a zoom in progress, whether a zoom-in or a zoom-out, was an underused technique. In Barry Lyndon the zoom movement takes center stage and performs in an elaborate way. Kubrick elevates the ‘lowly’ zoom movement into a Grand Technique.” The cinematographer John Alcott ASC BSC won the Oscar for best photography. This experience prompted Angénieux to bring out the 25-625 T4, a 25x zoom for 35mm cameras, in the early 1980s.

Actor and director Dennis Hopper and cameraman Lásló Kovács prepare to shoot a scene from *Easy Rider* in Taos, New Mexico, in 1969. On the camera, the 25-250mm zoom (photo by Michael Ochs Archives / Getty Images).

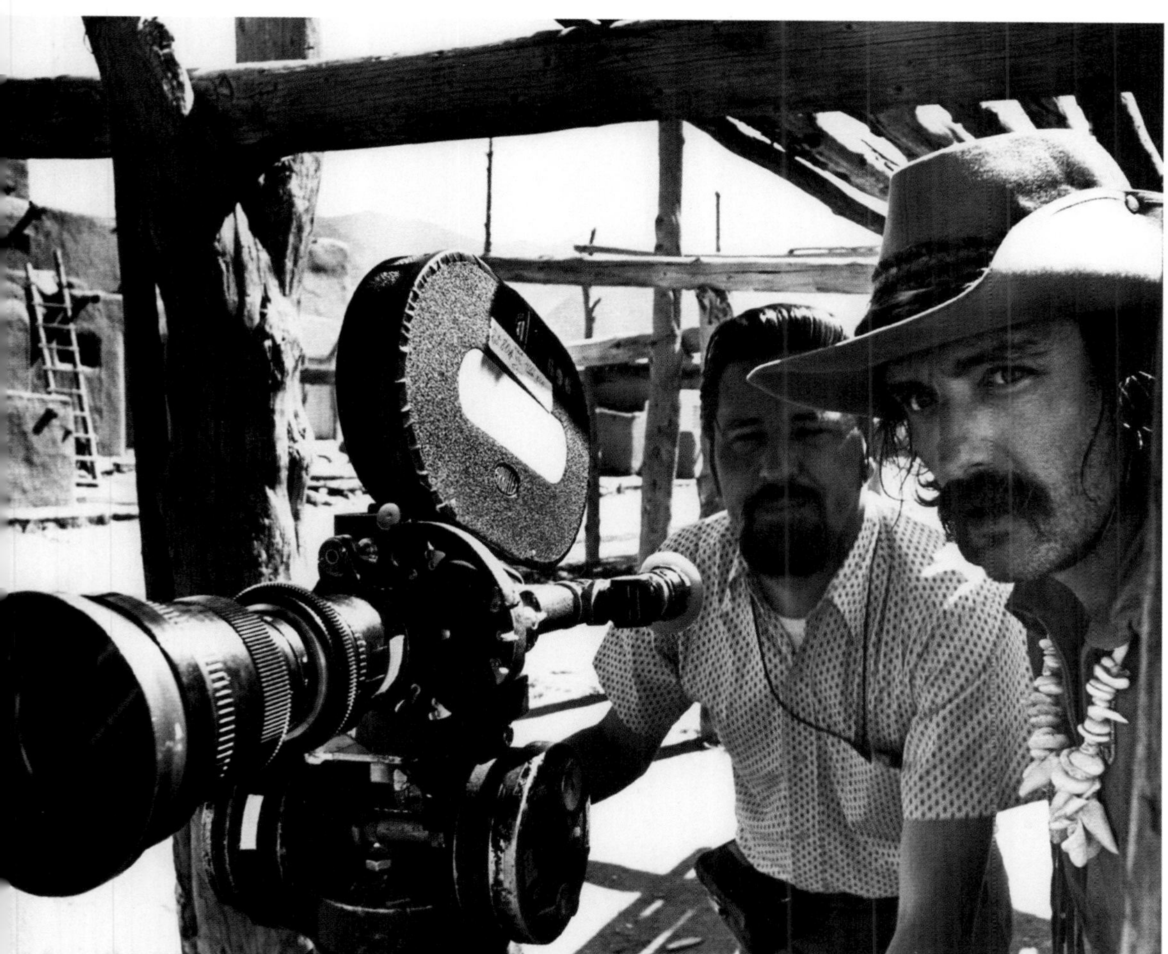

Angénieux 35mm Zooms and Asian Cinema

« Because you can do very small zooms that are not noticeable, you can get your emphasis all the time with a zoom, and it's lovely with that. And sometimes you don't even notice that. You are not supposed to, most of the time. It's not *zoom-zoom*, like that, it's just a little bit. Sometimes combining with a tracking shot you can zoom in. I love the zoom. I think it's wonderful, particularly now. »

Satyajit Ray

Extract from the interview that appeared in *Film Comment Summer*, 1968

On the other side of the planet, some Japanese cinematographers took up the zoom in the early 1960s. Don Ichikawa used several 25-250 lenses on Arriflex cameras in filming *Tokyo Olympiades*, his documentary on the Tokyo Olympics. Akira Kurosawa, one of the masters of movement in cinema, used the 25-250 for the first time in *Dodes'ka-den* (1970), largely shot in a shantytown, and subsequently continued to do so.
Indian filmmakers make regular use of Angénieux zooms. The director Satayajit Ray, winner of an Oscar for lifetime achievement, began to employ them in shooting *Charulata* (1964), one of the finest films in the history of Indian cinema (Ray was also awarded the Silver Bear for best director at the Berlin Film Festival in 1965).

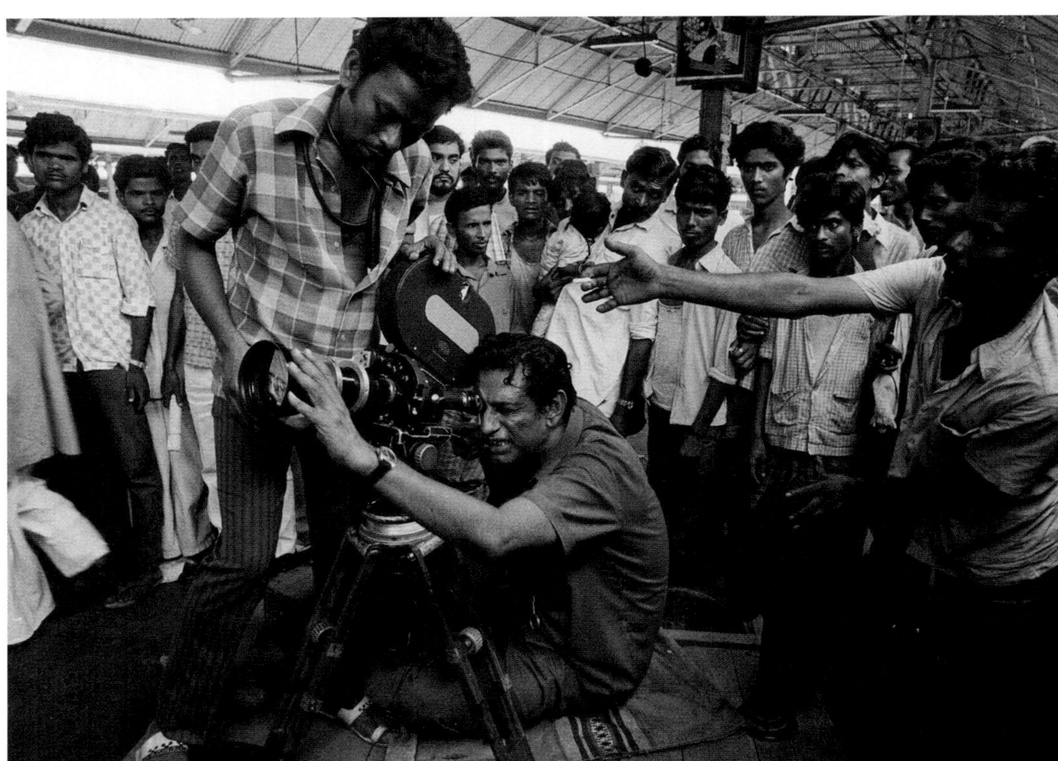

Bengali filmmaker Satyajit Ray filming a scene of *Sonar Kella* (*The Golden Fortress*) in Calcutta (photo Nik Wheeler /Sygma / Sygma, Getty Images).

The 1980s, Improvement in the Quality of Angénieux 35mm Zooms

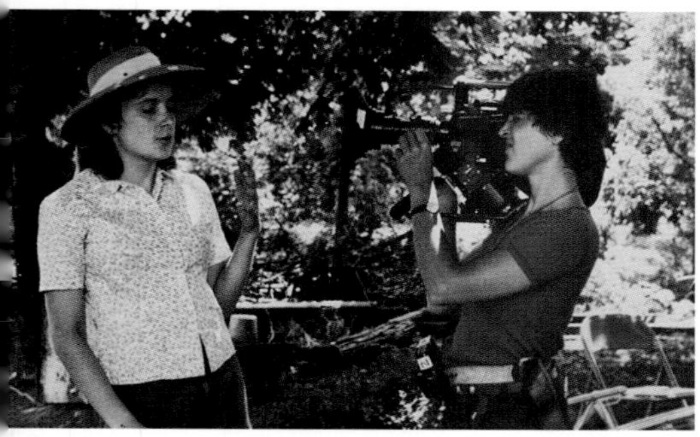

Cinematographer Nancy Schreiber ASC with a 10-150mm zoom camera (photo provided by Nancy Schreiber).

The first Cooke 25-250 f/2.8, named the Cooke Super Cine Varotal, had appeared in 1978, over fifteen years after Angénieux, and was soon followed by an improved version, the Cine Varotal 25-250 Mark II T3.9.
Angénieux's response was the 25-250 HP (High Performance) T3.7, an enhanced version of its first 10x zoom with a larger aperture and a sturdier mechanical system than its predecessor. Secondary chromatism was considerably reduced by the use of fluorophosphate glass and the addition of supplementary glass element to the focusing group helped to reduce chromatic aberration.
The third generation of the 25-250 was born in 1990 with the 25-250 HR (High Resolution). As Jacques Debize stated, "Major improvements were made to both the optical and the mechanical aspects. By adding a UV filter in front of the lens, internal focusing was possible, providing a constant volume lens in order to keep the inside body free of dust or condensation when focusing." At the same time, Angénieux also improved its range of wide-angle zooms with the 17-102 HP, which came out in 1988.

These two zooms were to be used in countless productions all over the world. The 25-250 HR is a legendary milestone in the history of Angénieux.

« I started my journey in the film industry as an electrician in New York, eventually becoming a gaffer, mostly on movies and commercials. When I started to move up to cinematography, the work at that time was in documentaries. I taught myself how to shoot handheld, first with an Éclair NPR then with a new Aaton which the Abel brothers of Abel Ciné maintained and repaired in their basement in Staten Island. I had been gaffing a show in Nigeria after buying my first 16mm camera, and arranged to stop in France to visit the Angénieux factory. There I bought my first lens, a 9.5-57 zoom, handpicked at the factory.
Later I supplemented my package with a 10-150 Angénieux zoom both outfitted with Chroiszel fluid zoom controls. As my cinematography career progressed and I began shooting features and television shows, in 35mm, and later digital, I would often use the Angénieux 25-250 HR. Great color and contrast. »

Nancy Schreiber ASC

Director of Photography
The Birth and Evolution of a Profession

Marc Salomon
Historian of directors of photography

The Pioneers

The cinematographer's profession is the oldest in the world of cinema.[1] As the French director of photography Louis Page wrote: "He was cinema all by himself for a long time." The first amongst them, the cameramen of the Lumière brothers, were globetrotters in pursuit of images as early as 1895, travelling the world to film newsworthy events and record scenes of war. In addition to filming, they handled the developing, printing and projection of the material. The Lumière camera, a simple wooden box with a jerky cranking system and a single lens, also worked as a projector when placed before a lamp. This was soon followed by other models, including the first Pathé camera (with a Voigtlander Heliar 51mm f/4.5 lens) in 1897 and the British Prestwich in 1898.

It was not until a decade later that the foundations of a cinema industry were laid with companies like Pathé, Gaumont, and Éclair in France as well as the first production companies and their studios on the East Coast of the United States but also in Denmark, Germany and elsewhere, where the different professions were to take shape gradually (directors, set designers and decorators, electricians and so on). The basic elements remained unchanged for years, however: a simple set, sometimes no more than a painted backdrop, receiving daylight through windows, sometimes shaded by awnings or blocked by black curtains, and the camera frontally filming a medium shot. According to Louis Forestier, who started at Gaumont in 1908, a cameraman was asked to do little more than produce a light, clear image and keep the actors in the frame. Operating a camera was regarded for a long time as just turning a handle and the first articles to appear in the specialized press were confined to anecdotes about accidents while filming and cunning tricks.

« The cameraman stands at the natural confluence of the two main streams of activity in the production of a film: where the imagination meets the reality of the film process. »

Freddie A. Young

Director and cinematographer, born in London in 1902

The equipment slowly evolved. The renowned Pathé Professionnelle camera appeared on the market in 1910 and enjoyed great success all through the silent era. It is regarded as the most used in the world until 1918. Generally equipped with a Zeiss Tessar 50mm f/3.5 lens, it was produced by the Krauss factory in Paris. It was used, for example by Billy Bitzer with a Goerz Kino-Hypar 50mm lens to shoot the films of D.W. Griffith. Other, more sophisticated cameras appeared on the market, however, like the French Parvo Debrie in 1909 and the famous Bell & Howell Model 2709 in 1912, made in Chicago and equipped with

The American cinematographer Billy Bitzer and the director D.W. Griffith (photo © Wisconsin Center for Film and Theater Research).

[1] The director of photography has not always been called this, being referred to originally as a camera operator or cameraman and listed in the credits for a long time under "Camera Operator" or "Photographed by...". It was not until much later, with the development of the movie industry and concern with professional recognition, that the term "Director of photography" appeared. Terms like "Lighting cameraman" and "Cinematographer" are also to be found. Vittorio Storaro referred to himself as an "Autore della fotografia" before opting for "Cinematografo".

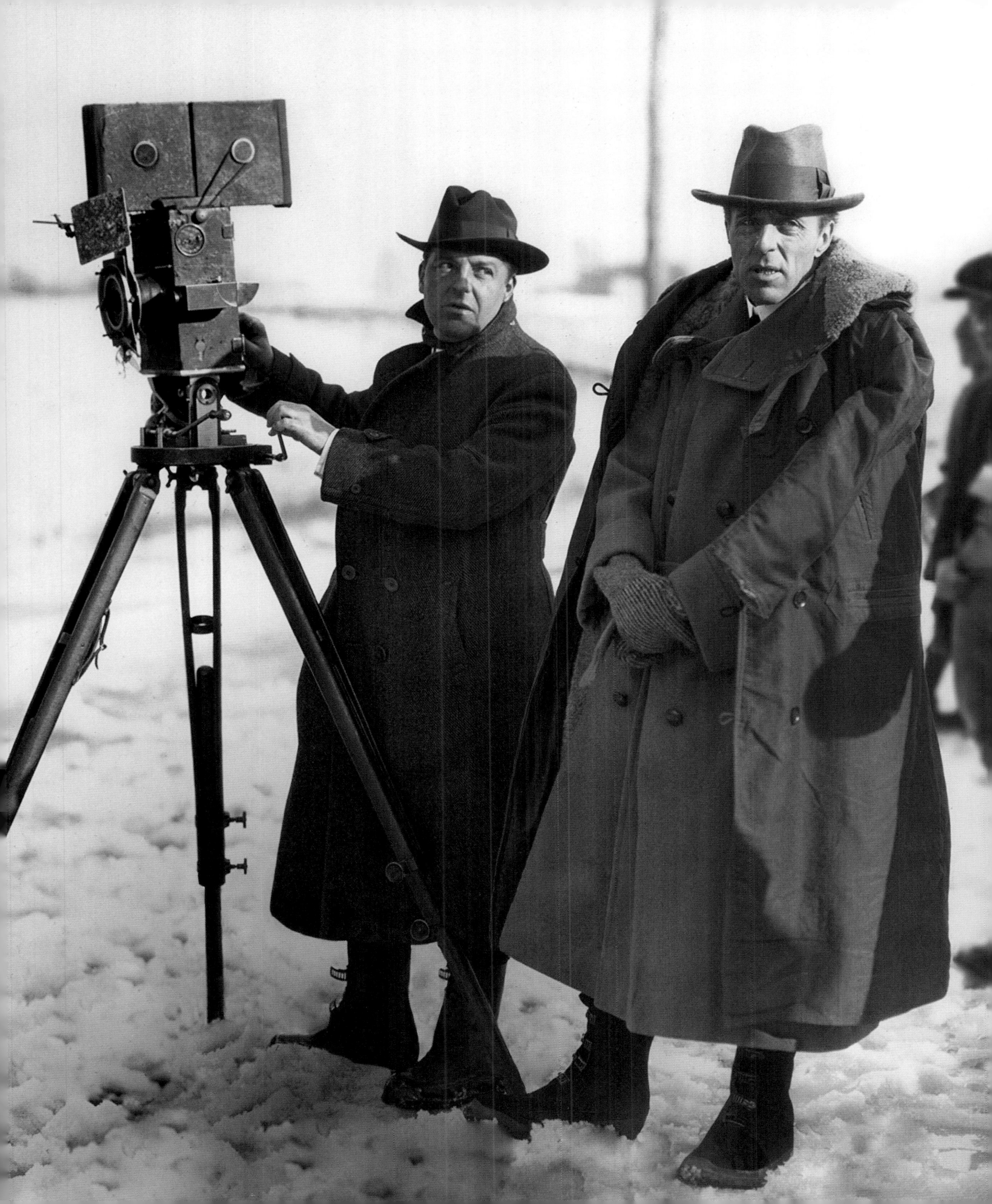

a four-lens turret. This remained one of the most highly respected cameras even after the appearance of the renowned Mitchell in 1921, again with a four-lens turret. The turret was also a feature of the Caméréclair launched by Éclair with four lenses in 1920 and followed by a new six-lens model in 1926. It was in the same year that Debrie presented the Parvo L, which was to remain the most popular camera (immortalized by Dziga Vertov in *Man with a Movie Camera*) long after the introduction of sound. It should be recalled that it was normal during the silent era for cameras and lenses to be owned by their operators. The Caméréclair and Parvo were the most widely used in France, where René Guissart, Léonce-Henry Burel, and Marc Bujard were among the few cameramen able to afford the Bell & Howell, the most expensive of all.

A Few Words on the Director of Photography

Above and beyond the technical evolutions and changes that constitute milestones in the history of cinema, the essential function of the director of photography will remain always as stated by Billy Bitzer, the trusty sidekick of David Wark Griffith. As he put it in his memoirs with simplicity and the humility of a true pioneer, "What Mr. Griffith saw in his mind we put on the screen." Directors of photography are responsible for the image of a film all the way from conception to color grading. After reading the screenplay, they decide in agreement with the director on the photographic atmosphere and style required, and ensure their consistency all through the shooting. After finding the various sets and locations, they draw up the list of equipment, including camera, lenses, film and lighting elements. They can also carry out all the technical and artistic trials they deem necessary in order to decide—again in agreement with the director—on an esthetic approach from filming to postproduction (on film or digital). If they do not operate the camera themselves, they will in any case supervise all the shooting parameters with the aid of the assistant cameramen, electricians and grips. They work in close collaboration with the other heads of department (sets, costumes, make-up, etc.) and their work regards both the artistic aspect and the financial constraints.

The Exploration of Light

Artificial lighting with arc lamps and mercury vapor lamps compatible with orthochromatic film made its timid appearance before 1910 as a safeguard against the loss of days of shooting due to the vagaries of the weather. The initial aim was to obtain steady, homogenous light with no particular effects. Certain cinematographers then began to experiment directional light, contrast and even penumbra to give the effect of light from a table lamp, an oil lamp, a fireplace, a window and so on. The 1910s saw such experiments in the United States, but also in Denmark, France and elsewhere. Alvin Wyckoff in the USA, Georges Specht in France, Axel Graatkjaer in Denmark, and Boris Zaveliev in Russia were among those who began to break away from purely functional lighting in an effort to produce more expressive images of greater esthetic quality. It was the task of these early generations of cinematographers to innovate, to pave the way, and finally to invent the profession, defining

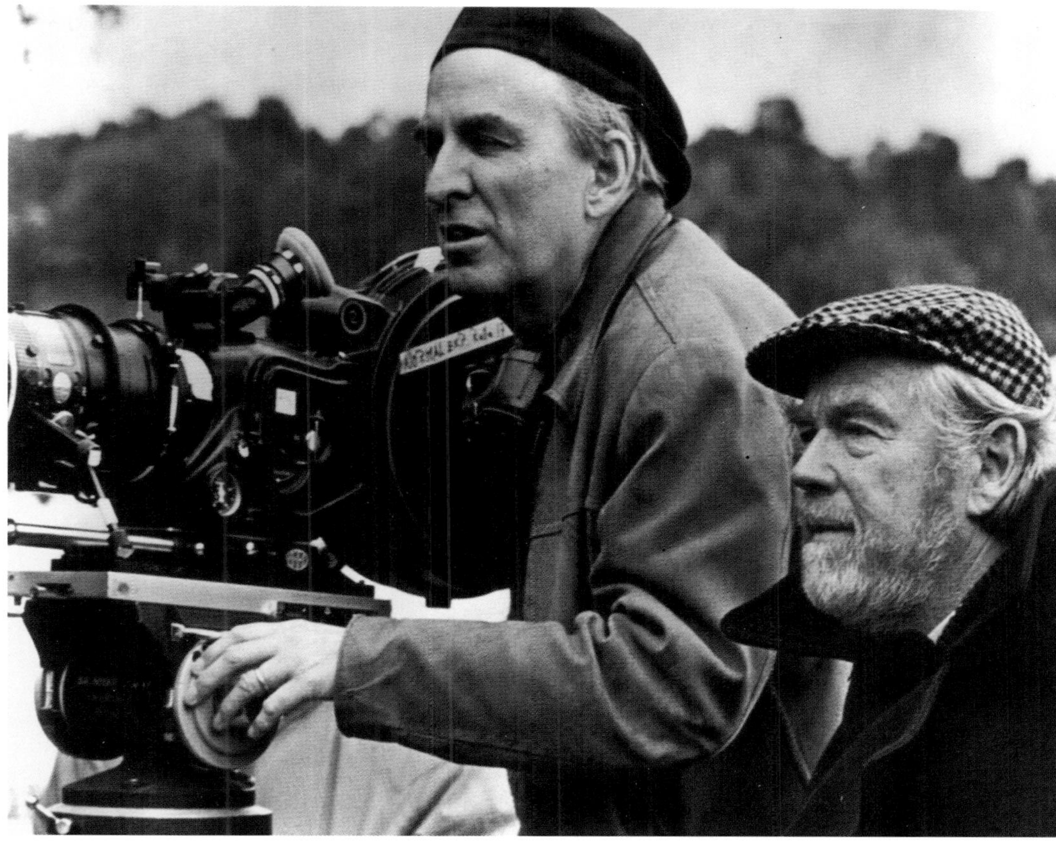

Swedish director and screenwriter Ingmar Bergman with his Academy Award winning cinematographer Sven Nykvist, on the set of his movie *Fanny och Alexander* (*Fanny and Alexander*) (photo by Sunset Boulevard / Corbis, Getty Images).

the field of their responsibilities through experience while mastering and improving a still rudimentary technique offering little in the way of good and reliable tools. Axel Graatkjaer and Johan Ankerstjerne in Denmark and the brothers Henrik and John Julius Jaenzon in Sweden blazed the trail followed soon after by the great cinematographers of German expressionism in the 1920s, including Karl Freund, Carl Hoffmann, and Fritz Arno Wagner. Most of these early practitioners learned on the job through practical experience. Some had a background in photography (Karl Strüss, Charles Rosher, and Arthur Edeson in the USA, Daniil Demoutski in the USSR, Giovanni Vitrotti and Anchise Brizzi in Italy, Fritz Arno Wagner in Germany), some were projectionists (Axel Graatkjaer in Denmark, Karl Freund in Germany, Otto Heller in Czechoslovakia, Victor Milner, Sol Polito, and Nicolas Musuraca in the USA, Georges Périnal in France, and Boris Voltchek in the USSR), and some were war correspondents (Édouard Tissé in Russia), but most of them had to learn how to develop and print before actually filming.

Early Recognition

Often working under contract for a production company or studio,[2] cinematographers were not known as the authors of their images and their names did not appear in the credits. Little by little, progress in lighting techniques and the development of the star system, above all in the USA, brought certains names into the spotlight, including Billy Bitzer with the films of D.W. Griffith starring Lilian Gish, Charles Rosher and the films with Mary Pickford, Léonce-Henry Burel in France and the films of Abel Gance, and John Julius with the director Victor Sjöström in Sweden.

The profession developed its own associations in the USA as from 1913 and these merged to form the American Society of Cinematographers (ASC) in 1919. It was in 1922 that the ASC began to publish the monthly magazine *American Cinematographer*, which still

[2] The status of operators and technicians in general has certain evolved over the years and differs from country to country. Usually under contract to a studio or producer in the West but state employees in the former Soviet Union and satellite countries, these professionals have gradually become more independent but also more precariously situated since the 1950s with the end of the studio system.

addresses the movie scene from the standpoint of directors of photography today.[3] It was not until 1927, however, with the creation of the Academy Awards, that directors of photography were fully recognized and rewarded for their work. Charles Rosher and Karl Strüss won the first Oscar for best photography for *Sunrise* (1927), directed by F.W. Murnau. This growing recognition was accompanied by increasing responsabilities all through the 1920s. The idea that not all action necessarily takes place in broad daylight, a taste for chiaroscuro, the deft interplay of light and shadow, the pursuit of an atmosphere other than the faithful reproduction of a supposedly naked reality with no effects all required at least as much education of the public taste as daring and determination on the part of a few directors and cinematographers. Karl Freund made this statement in 1926: "I want to show that the role of the camera has been wrongly subordinated so far to the other elements making up an image and that it is the mission of the camera to create a mode of expression specific to the cinematographic language. As far as possible, I want to get away from the traditions of the spoken scene and constantly discover new pictorial effects and new dramatic possibilities for film."

The Revolution of Sound

Cinema underwent its first great technological and esthetic mutation in the late 1920s with the transition from silent movies to talkies. The changes involved as regards film, lenses, and lighting swept away all the expertise built up over the years. It was necessary to master new equipment and its constraints (panchromatic film, incandescent lighting, heavier cameras, corrected lenses, and so on) as well as a completely new art of directing and photography (cutting, framing, etc.). According to James Wong Howe, even experienced cameramen had to prove themselves again with the new tools. It was also during this period that the new technical demands and the weight of the material led to the emergence of the camera crew as we know it today, comprising a director of photography, a cameraman, and various assistants. To a certain degree, the 1930s can be seen as a period of mastering the skills already possessed in the 1920s all over again on a different basis. The career of Gregg Toland and his pursuit of depth of field encapsulate all of the esthetic evolution that accompanied technical advances such as more sensitive film (from 20 to 160 ISO in a decade), antireflective coating for lenses, more powerful and effective lighting (with the whole range of arc lamps developed for Technicolor), and widespread use of exposure meters (Weston, Norwood). Well aware of what he owed to his cinematographer, Orson Welles made sure that the name Gregg Toland received equal billing beside his own in the credits of *Citizen Kane* (1941).

Publications & Associations

The early stages in the recognition of directors of photography involved the publication of books, the creation of new training schools, and the birth of professional associations. In the USSR, the cinematographer and theorist Vladimir Nilsen published two works in 1936, one of which was translated into English as *The Cinema as Graphic Art* and appeared in several editions, and Anatoli Golovnia's *Svet v iskusstve operatora* (*Light* in the Art of the *Cameraman*) appeared in 1946. In the USA, John Alton, the master of film noir, published the classic *Painting With Light* in 1949. Logically enough, training schools opened with courses for cinematography and other professions. While the two oldest were founded in the 1920s, namely the VGIK in Moscow and the École Louis Lumière in France, the IDHEC in Paris, the Łodz school in Poland and the FAMU in Prague opened after World War II.

As regards the professional associations, the British Society of Cinematographers (BSC), based on the model of the ASC, was born in 1949 and the Italian AIC in 1950. These were

[3] Mention should also be made of the monthly *International Photographer,* which first appeared in 1929. It was renamed *ICG Magazine* in 1999.

followed by the DDF in Denmark, the FSF in Sweden, the BVK in Germany, the RGC in Russia, the AFC in France, the HSC in Hungary, and the AEC in Spain. It is through the efforts of the Italian cinematographer Luciano Tovoli that these associations joined forces in 1992 to form the European federation IMAGO.

The Post-war Period

The post-war period also saw a radical renewal of filmmaking with directors addressing issues of everyday reality outside the confines of the studio. Emblematic and often atypical cinematographers like G.R. Aldo, Walter Lassally, and Raoul Coutard played their part in movements like Italian Neorealism, British Free Cinema, and French New Wave. The camera operator was no longer the "king of the studio" as Günther Krampf wrote in the 1920s, but at the service of the story even if it meant riding roughshod over the established canons of an outmoded esthetic. The new conception is perfectly encapsulated in Ghislain Cloquet's remark, "Cinematographers used to light their shots, I light the film." This simultaneously thematic and esthetic mutation, expressed essentially in black and white, affected the entire world of filmmaking, as demonstrated by the Bengali cinema of Satyajit Ray and Cinema Novo in Brazil. The history of the profession of director of photography is marked by a constant succession of more or less crucial technological innovations. The 1950s thus saw the introduction and large-scale adoption of color thanks to Eastmancolor (after the

International Recognition

Numerous awards have been created since the 1940s all over the world to highlight and reward the work of directors of photography, starting with the Oscar in the USA and continuing with the Condor de Plata in Argentina (1943), the Ariel in Mexico (1948), the BAFTA awards in the UK (created in 1948 but not including cinematography until 1964), the Coruja de Oro in Brazil (1969), the César in France (1976), the David di Donatello in Italy (1981), the Deutscher Filmpreis in Germany (1982), the Goya in Spain (1987), and the Nika in Russia (1988). In Japan, the Mainichi Film Award (Mainichi Eiga Konkūru) was created in 1946, the Blue Ribbon Award (Burū Ribon Shō) was presented from 1950 to 1965 by an association of film critics (and won three times by Kazuo Miyagawa), and the Japan Academy Award (Nippon Akademī-shō) in 1978. Mention should also be made of the Italian Nastro d'Argento, founded in 1954 and awarded seven times each to Giuseppe Rotunno to Luca Bigazzi, just ahead of Tonino Delli Colli, Vittorio Storaro, Gianni Di Venanzo, and Armando Nannuzzi. Teramo, Di Venanzo's hometown, has awarded the Esposimetro d'Oro every year since 1996 for lifetime achievement in this field.

The few festivals entirely devoted to cinématography include the Manaki Brothers Film Festival, founded in 1978 in Bitola (formerly Jugoslavia, now the Republic of Macedonia) and the Camerimage Festival in Poland. Founded in 1993 and devoted exclusively to the cinematographic image, this presents a yearly lifetime achievement award accompanied by the publication of a superb album, lavishly illustrated with stills, set photos and personal images, covering the career of the winning director of photography.

Finally, it should be noted that cinematography has had its own place in the Cannes Film Festival too since 2013, where the Pierre Angénieux ExcelLens in Cinematography award is presented every year to honor the exemplary and innovative work of director of photography. Previous winners include Philippe Rousselot, Vilmos Zsigmond, Roger Deakins, Peter Suschitzky, Christopher Doyle, and Ed Lachman.

almost undisputed reign of Technicolor), CinemaScope, and then the zoom. In each stage, the art of cinematography appears to mark time, both because the new tools require further honing and because they must be mastered and adapted to the needs of filmmaking. The initially clumsy, ingenuous or inappropriate use of color, CinemaScope, and zoom effects gave way in the 1960s to films of greater mastery in which technical and esthetic elements are combined with greater fluidity and discretion by directors of photography like Carlo di Palma, Conrad Hall, Sven Nykvist, Jaroslav Kucera, Ghislain Cloquet, Pierre Lhomme, David Watkin, Kaskell Wexler, and Luis Cuadrado. The renewal of color cinematography developed through the work of these figures was to characterize the following decade. Color and black and white did, however, continue to coexist until the mid-1960s, as demonstrated by the award of an Oscar for each category up to 1966.

Modernity in Color

The 1970s saw a number of major technological advances that were to steer the cinematographic image lastingly toward greater realism in accordance with the subjects addressed: lighter cameras equipped for sound (the American Panaflex and the German Arri BL), more powerful and effective lighting of the daylight-balanced kind (HMI), Kodak Eastmancolor 5247 film, and a series of fixed lenses with more light (apertures between T1.1 and T1.6) produced by Zeiss, Panavision, and Canon.[4] Vittorio Storaro, Nestor Almendros, Gordon Willis, Vilmos Zsigmond, Billy Williams, Bruno Nuytten, Robby Müller, and Pavel Lebechev were part of a new generation of directors of photography exploring a different use of color (desaturation or hyperrealism) and light. In their pusuit of a more or less stylized form of naturalism, light was no longer either a purely technical contingency or subordinated to the rendering of color or a more or less arbitrary and outmoded artistic construction.
The sensitivity of film began to increase with Fuji (250 ISO) and Kodak (320 ISO) in the early 1980s and went on to reach 800 ISO in the late 1990s. This made it easier to shoot scenes in available light or in urban spaces by night, as demonstrated by the work of William Lubtchansky in Juliet Berto's *Neige* and Ric Waite in Walter Hill's *48 Hours*. Nestor Almendros had of course already embarked in 1976 with Terrence Malick's *Days of Heaven* on a magnificent exploration of ways to capture the subtlest nuances of daylight all the way from dawn to dusk with Panavision Super-Speed lenses, testing the equipment of the period to its utmost limits.

The Digital Revolution

The advent of digital technology at the turn of the century marked the last revolution to date with first-generation digital cameras (HDCAM) like the Sony CineAlta F900 and F950, the first Panasonic Varicam, and a mini-DV range (Sony and Panasonic).
The directors George Lucas (*Star Wars II* and *III*), Michael Mann, and David Fincher, and the directors of photography David Tattersall, Dion Beebe, Harris Savides, and Claudio Miranda were the pioneers in this field, exploring—not always without difficulty—the new possibilities opened up by the cameras of Sony and Thomson Viper (Michael Mann's *Collateral* in 2004). In France, Pitof's *Vidocq* (2001) was the first full-length film shot entirely with a digital camera, the Sony F900, which was also used for Alexander Sokurov's *Russian Ark* and Lars Von Trier's *Dogville*.
In Denmark, directors Thomas Vinterbergh and Lars Von Trier, founders of the Dogme 95 movement, used small Sony cameras (PC3 or VX-1000) to make films like *The Celebration* (*Festen*, 1998) and *The Idiots* (*Idioterne*, 1999).
Digital cinema gradually came to predominate in the space of about fifteen years but without completely eliminating the use of film. Agfa and Fuji disappeared, however, leaving Kodak

[4] It should be recalled that in the 1960s Angénieux sold a 50mm M1 lens with an aperture of f/0.95 (T1.1), initially reserved for specific applications (video surveillance) before apparently being modified and adapted for filming and photography. It still featured in the Angénieux catalogue in 1976.

alone on the market. The real transition to digital came around 2006 with the appearance of new cameras with larger sensors like the HD Super 35.[5] Dean Semler used a Panavision Genesis to shoot Mel Gibson's *Apocalypto*.

Then came RED (RED One and all the range of Epic, Dragon, Weapon, Helium, Gemini, etc.), ARRI D20 and D21 (followed by Alexa and Amira), Aaton (the sadly missed Penelope Delta), Sony F35 (then F55 and F65), and Panasonic (VariCam 35, LT, Pure and EVA-1).[6]

The arrival of the ARRI Alexa in 2010 radically accelerated the shift to digital cinematography and the production of film cameras was drastically reduced in just two years before ceasing definitively (with the exception of ARRI and Panavision 65mm cameras). By now, in addition to fine detail, improved dynamics in highlights, and better color, directors of photography are also offered the depth-of-field effects that were so sadly lacking in first-generation digital cameras.

The most recent evolution to date (2018) regards full-frame cameras (Sony Venice, Arri Alexa LF) and camera with large sensors (Arri Alexa 65, RED DSCM2 Monstro, Panavision DXL et DXL 2), which are broadening the horizons of digital cinematography still further.

[5] Super 35 sensors approach the ratio of 16/9 with a width of more or less 24mm. Full-frame sensors approach the 24x36mm standard and large-format sensors have a base of about 41mm or even 54mm in the case of the Alexa 65.

[6] More affordable cameras equipped with an S35 sensor, like the Canon C300 and Sony FS7, enjoyed a certain amount of success for low-budget films and documentaries.

The Testimony of Cinematographers

«The function of the cinematographer is to be the bridge, the conduit between the audience and what's in front of the camera.
To me there are only three people in cinema: the person in front of the camera, the audience member, and the person who is the real passage between them—the cinematographer. Of course director and producer facilitate the interface. But for the engagement to be direct and compelling, I really believe we have to be transparent and remove ourselves enough so that the passage between the actors and audience is direct. In Asia, we move fast, we are in very tight spaces, we do not have a big crew. So, the organic nature of the camera and the lens is very important to me, the balance and how it feels.»

Christopher Doyle HKSC

Winner of the 5th Pierre Angénieux ExcelLens in Cinematography award at the Cannes Film Festival in 2017

«Well ... What do we do as cinematographers? I hope this doesn't sound like an obvious answer ... we of course translate stories into images with the director's vision, the production designer's shape and form, the crew's skills and artistry and the emotional sensitivity of the actors ... Literature can enter the interior world of the character, but it's much more difficult in writing to show where you are – authors spend paragraphs and pages illustrating place. In film, you can show the exterior world with one shot, but it is much more difficult to enter the interior world of a character with a camera. That's what I'm always trying to explore with a director: how you enter the interior world of the character and his emotions. Images are metaphors to uncover something not hidden on the surface that create an emotional landscape for your characters and story. Images are the subtext for the psychological world that you create for the characters. When I research a period for a film, I might not only look at other films of that time, but also writers, music, and painters, or whatever visual cues the script gives me. I go to directors when I've read a script and present them with a kind of visual notebook. Some directors embrace that – even if it isn't their ideas, at least it starts a forum for discussion. So you have to find your footing with the director for how you create that language, because for me the language of cinema is images.»

Ed Lachman ASC

Winner of the 6th Pierre Angénieux ExcelLens in Cinematography award at the Cannes Film Festival in 2018

« The role of the director of photography in a film is quite simply to make the screenplay visible in an interesting way and in an appropriate style. When I work, I think I can introduce surprises for the director, pleasant surprises of course.
There are directors who do not know exactly what they want visually, and they rely on the director of photography for that. It is always my goal to bring my personal vision while creating the images of a film. I am happy when the director brings his personal vision but I have very rarely met a director who told me what I needed to do.
A wise director knows how to leave room for his main collaborators and let them contribute what their tastes and professional skills make possible, to work as well as possible. That's why they were chosen.
I had the good fortune in my life to meet David Cronenberg, with whom I've made eleven films. Working together is a new joy every time, a different artistic and intellectual challenge for each movie.
We work in a relationship of trust.
We often decide together on the spot about the rhythm to give the film, the number of shots. It's the way I prefer to work. »

Peter Suschitzky ASC

Winner of the 4th Pierre Angénieux ExcelLens in Cinematography award at the Cannes Film Festival in 2016

« Working as director of photography is like entering the imagination of another person, namely the director, and looking at things together in search of a rythm, a vision to be shared through the formation of images. It is a commitment that goes beyond a vision, however, a whole perception that you try to create together and share, developed also through words, sounds, music, and all kinds of inspirations. For me, this means being totally committed and responsive towards the projects and the people I work with ... The sources of inspiration are endless, especially outside cinema ... It seems essential to me to go and look elsewhere in order to make films or any form of art. There is not only cinema or painting or music. All these forms coexist with one another and draw on one another. This is where their richness comes from ... »

Eponine Momenceau

Born in 1985, Eponine Momenceau worked as director of photography in 2015 on Jacques Audiard's film *Dheepan*, winner of the Palme d'Or at Cannes the same year

The Zoom, a Multipurpose Lens, a Semantic and an Esthetic Tool

Pascal Martin
A university professor, Pascal Martin teaches at the École Nationale Supérieure Louis-Lumière and has been a lecturer at the FEMIS since 2004.

As a young teacher at the École Louis-Lumière in the 1980s, I had the good fortune to discover the Angénieux factory, which had already developed unique know-how through a deft combination of artisanal experience and cutting-edge industrial skills. Jacques Angénieux[1] showed me round and mentioned that he associated the pratice of optical testing for quality control with the need to create images. I included this remark at the end of the second of two articles in *Le Photographe*,[2] one on aberrations and the other on tests for lenses, and was astonished to come under criticism from certain quarters on the grounds of disrespect for science. Some thirty years later, having been involved in the training of several generations of directors of photography and remained in contact with many of them, my opinion remains unchanged and is indeed more deeply held. The function of a lens[3] is to create images. In actual fact, however, the question is far more complex than it may appear, as this can only be done through great technical expertise. Let us therefore begin by simply asking what a lens is.

A lens is an optical system that creates the image of an object by means of a phenomenon well-known to pupils of secondary schools. It can be described as an interface that transforms a scene, a space with three dimensions, into a surface (of the film or sensor), which has only two. As an initial approximation, it is easy to demonstrate that this transformation depends on a set of quantitative characteristics (focal length, distance of focusing, aperture, coverage circle, and range in the case of a zoom), to which a number of qualitative characteristics should be added, the most evident being the degree of correction of aberrations, optimization, and the levels of resolution sought at certain values of contrast as expressed by the modulation transfer function (MTF). The combination of all these elements gives a lens its own particular identity. This is why it is not inappropriate to regard it as a tool in both the semantic and the esthetic fields. These two concepts are connected primarily with the way in which it is seen by users. This is the question we shall address in this brief text, first drawing a distinction between fixed and variable focal length.

The Lens as a Semantic Tool

Meaning is imparted above all by the way in which the lens is used, how it is positioned in space, near or away from the scene to be filmed, its height above the ground, and its inclination. By correlating these parameters with the choice of focal length and distance of filming, the camera operator established the value of the shot and determines both the magnification of the image and its rendering in perspective. Several magnitudes are connected in optics as applied to the sphere of images and focal length is a very relative element that depends on the coverage required of the lens and determines the angle of field. The latter concept is so implicit that the terms *short focal length* and *wide angle* are often confused. While this set of values defines the frame, an image is the retranscription of a volume that is only made possible by including the third dimension obtained from depth of field. One of the parameters making it possible to modulate this is the aperture of the diaphragm,

[1] Pierre's youngest son, a graduate of the École Supérieure d'Optique, joined the company for a time.

[2] 'C'est aberrant,' *Le Photographe*, March 1990, pp. 32–46, and 'Si on testait,' *Le Photographe*, November 1990, pp. 52–56.

which has a threefold effect. It introduces the variable of exposure as a photometric parameter, it affects the importance of the depth of field, as stated above, and hence, intrinsically, the presence or absence of blurring in foreground and background, and it affects the quality of the image by minimizing residual aberrations and modulating their impact of diffraction.
Having briefly outlined the quantitative parameters, we must also recall one peculiarity of the moving as opposed to the fixed image, namely the fact that the camera operator can change focus during the shot, which is called a rack focus. This is an authentically semantic function in itself, made all the more evident by combination with camera movement.

The Lens as an Esthetic Tool

The calculation of optical combinations is crucial to the end quality of the system. Optical engineers have five parameters of the lens elements to work on: two radiuses of curvature (spherical or aspheric), thickness on the axis, index of refraction, and constringence or V-number, which indicates the behavior of the glass in relation to wavelength. They then determine the position of each of the elements with respect to one another within a mechanism requiring compliance with very strict position tolerances in the order of a few hundredths of a millimeter. Controlling the path of the light also entails eliminating everything that can impair the quality of the images, e.g. by causing unwanted reflection. Inside the lens, the mechanism is therefore anodized and sculpted to create light traps, the lens elements are vacuum-coated with thin layers, and their rims are varnished. The interplay of destructive interferences is used not only to increase transmission by minimizing the loss of light at the entrance of each dioptre but also to prevent the formation of unwanted images. If this description appears to have little to do with the esthetic dimension referred to above, it is because of failure to appreciate the effect of these factors on the image. The degree of correction of the different aberrations in different points of the field, both in width and in depth, will clearly have an effect on how the scene filmed appears, on its relief, contrast, and quality in terms of the rendering of details and values. The nature of the glass used and the antireflective coating of the lens elements also effect the color of the image as well as the presence and quality of flare.[4] While all these factors can be scientifically determined in advance by means of highly complex methods such as the calculation of wavefronts and PSF, and the selection of suitable Zernicke coefficients with respect to the chosen MTF, what they mean for the user of the appearance of the image and the way in which the lens renders a set of values referred to in more general terms like warm and cold, hard and soft. These are indeed not so much simple adjectives as a highly codified language that camera operators know very well. While warm and cold refer to the tonalities of the lens as regards the rendering respectively of reds and of blues and greens, hard and soft refer to sharpness and contrast: the harder the lens, the sharper the

[3] A lens can also be called an objective. The specific characteristics of lenses for photography and cinema are not addressed here.
[4] Lens flare is the internal scattering of light in such a way as to impair the quality of the image.

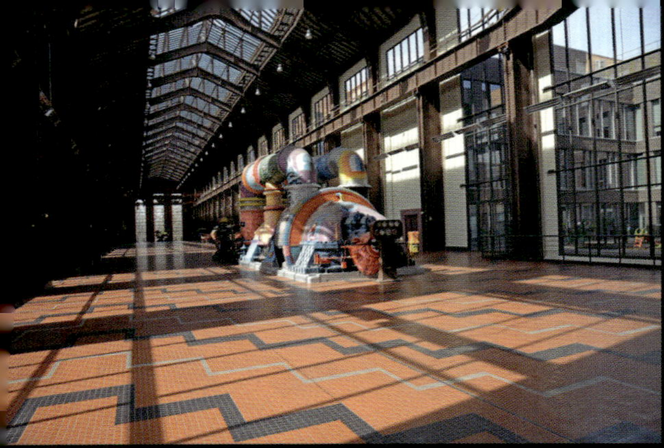
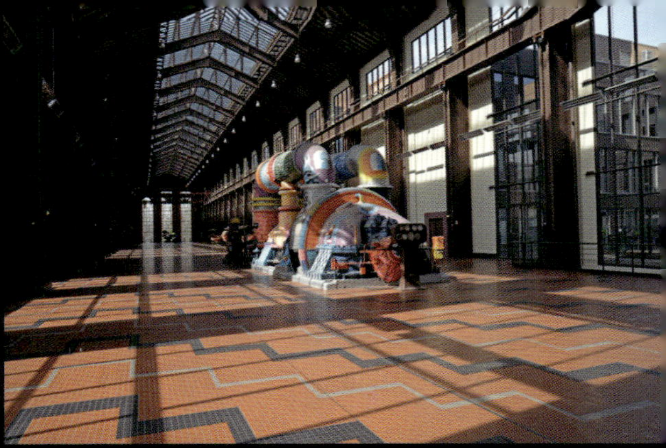
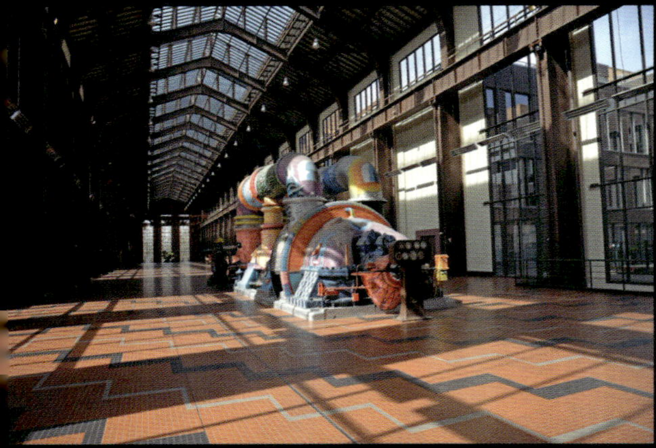
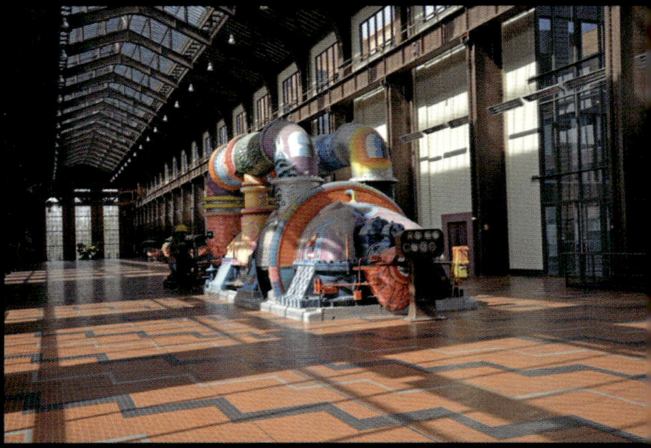
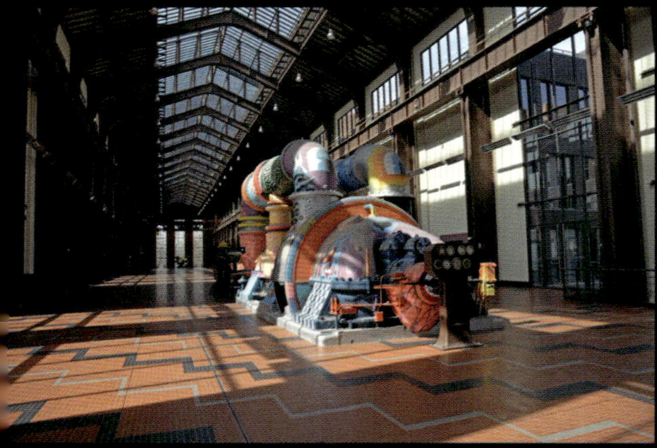
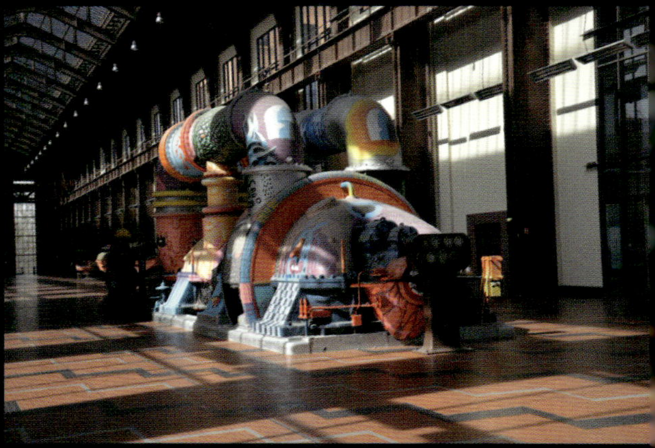
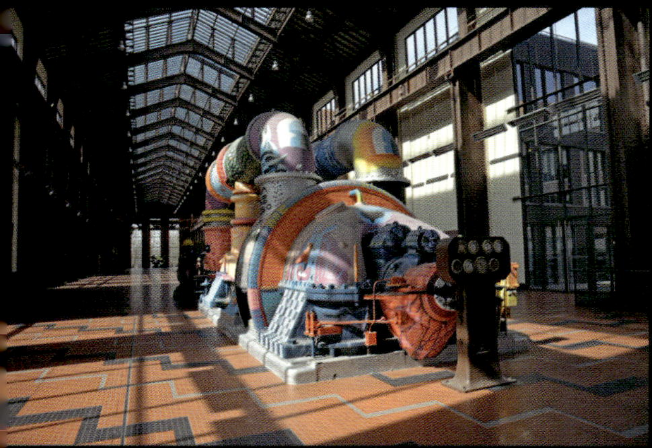
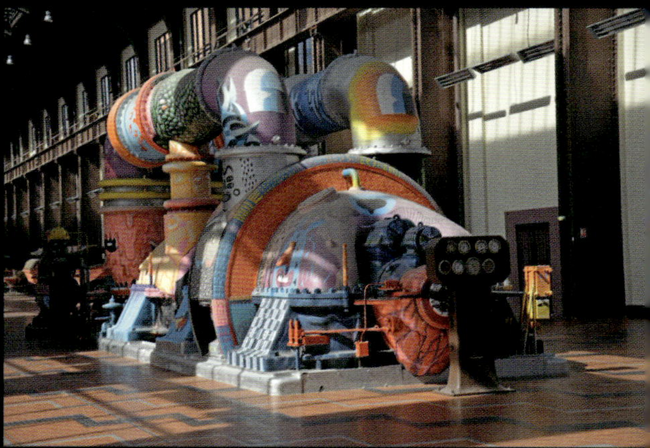
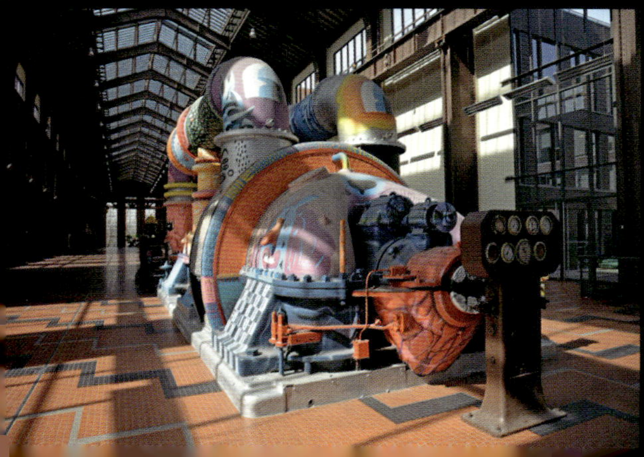
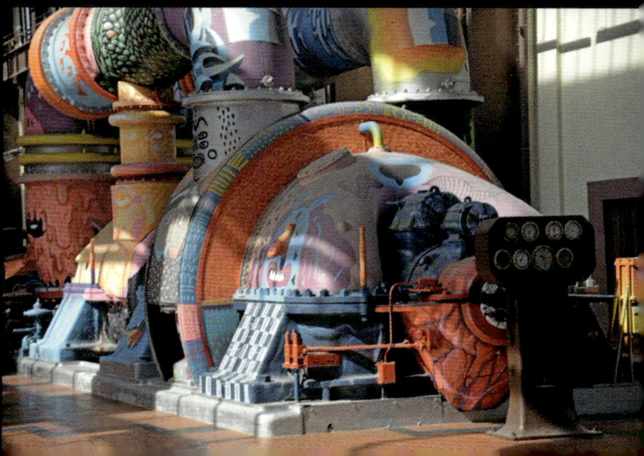

opposite
Comparison of a zoom and a tracking shot.
The perspective changes in the latter (left column) but not in the former (photos by Pascal Martin at Cité du Cinéma La Plaine, Saint-Denis).

image. It should be understood, however, that too much hardness can mean a stark image and be regarded as a flaw, whereas a soft-focus lens gives an element of glamour and is not necessarily incompatible with sharp edges.[5]

The Case of the Zoom[6]

While the above elements obviously apply to lenses of fixed focal length, they take on particular significance in the case of the zoom. The difficulties of conception increase exponentially with the extent of the variation in focal length. It is in fact necessary to preserve both the qualitative criteria, as far as possible, and the focus at every focal length. The technical term *compensation* derives from the need to compensate any loss of focus during the variation of focal length. The compensation can be optical or mechanical or a combination of both. Pierre Angénieux achieved an authentic feat of optical engineering by devising the mechanical compensation system that made it possible to attain a ratio of 10:1 as early as 1961 with his renowned 12-120. The first industrially produced varifocal lens with optical compensation was the Par-Cinor 20-60mm f/2.8, developed by Roger Cuvillier for SOM-Berthiot in 1950.
While it is not our intention here to tell the story of variable focal lengths, the pursuit of a lens that can be described as universal is concomitant with the evolution of optics as applied to the image. Photographers in the 19th century had sets of lenses, i.e. of tubes with an interchangeable lens at either end, offering different combinations of focal length. Lens turrets were then developed in cinema for numerous cameras, the best-known of which being the Caméflex 35mm and the Paillard Bolex 16mm. The goal in every case was a multifunctional system. When the real zoom made its appearance on the film set, it was precisely this idea of multifunctionality that constituted an impediment, as we shall now see. At that time, but certainly not now, the zoom was markedly inferior in qualitative terms to the equivalent primes, and many camera operators refused to use them for that reason. The images produced were often blurred and marred by aberrations such as degrees of distorsion visible even to non-experts. In actual fact, however, this was due to incorrect use combined with misunderstanding of a principle of image formation that is, however, easily explained. It is certainly not for nothing that critics of the zoom like Jean-Luc Godard called it an enemy of cinema. Filmmakers were described in the specialized press during the early 1960s as suffering from "zoomitis". This stems from a confusion between two things that are completely unconnected despite the appearace of a dialectical relation and the effect of drawing closer or farther away generated, namely the mechanical tracking shot and the optical tracking shot or zoom.
The former involves physical movement of the camera and the latter the simple operation of varying focal length. The essential difference between the two is connected with the intrinsic consequences of the effect produces, as one alters perspective and the other only magnification.

[5] I would describe Angénieux lenses as rather warm and midway between hard and soft.
[6] It should be recalled that *zoom* was the name registered by Angénieux for the 17-68/2.2 and entered the French language subsequently.

Perspective involves structural modification of the distancing of elements from one another. There is a kind of dilation of space that is a function of the distance at which the viewer is placed. A mechanical tracking shot corresponds to the visual experience of a spectator in motion. A zoom corresponds instead rather to reframing on the same axis. There is no change in perspective but only in visual terms. It corresponds to a refocusing of attention and it is by no means absurd to draw a parallel here with foveal and extra-foveal vision. The retina is made up of receptors, whose configuration in its center makes it possible to see details and corresponds to a narrow angle of vision. When we read a word in the middle of a text or look at something, the image is formed at this precise point of the eye called the fovea. This is foveal vision. Extra-foveal vision takes place outside this very small area of the visual field and becomes hazier the farther away it is from the center of the eye.

This is not the case with a cinematographic image, the entire surface of which should in theory be equally distinct. When a director wishes to draw the viewer's attention to a particular point, a zoom in appears the most obvious choice. When instead he wishes to do the opposite, he can zoom out. It is therefore easy to understand that these two shots have fundamentally different functions and that zooming in rather than moving the camera closer is not only much simpler but also means something different.

It was this incorrect use of the zoom that was attacked at the outset. While this is understandable, it was not the tool but the user that should have been criticized. It seemed easier to use the zoom than lay the rails for a tracking shot. A considerable amount of time was saved and the production companies may well have encouraged this practice, now fortunately a thing of the past. At the same time, the zoom immediately aroused interest in TV news reporting by enabling the operator to focus on a detail of a scene without actually moving closer, especially in dangerous situations.

A New Way of Working

The lens with variable focal length is today indispensable for the general public.[7] Most photographic cameras are sold with a transtandard zoom, i.e. with a range of focal lengths extending beyond the normal at both ends. This increasingly large range is also a characteristic of both amateur camcorders and professional video cameras. A 20x zoom, which would cause problems at larger focal lengths if image stabilization systems did not exist, is standard issue today. All TV cameras have obviously been equipped with such devices for decades now.

Cinema is still, however, lagging behind as regards the exclusive use of a zoom lens in filming. Few movies are shot entirely with a zoom even though there is nothing to prevent it, as shown recently by Jordan Peele's *Get Out*, nominated for four Oscars in 2018, on which the director of photography Toby Oliver used nothing but zooms, the Angénieux Optimo 15-40, 28-76, 45-120 and 24-290. It is therefore reasonable to wonder about

[7] This text only regards the optical zoom. The digital zoom of most smartphones is a device to enlarge the image and refocus.

[8] An advertising slogan of the 1980s for a 80-200 zoom aimed at photographers offered 121 different focal lengths. While this may have worked commercially to attract potential buyers, I defy anyone to distinguish visually between two lenses with just 1mm of difference in focal length.

the reasons for this reluctance. In the case of top-of-the-range products like those of Angénieux, qualitative comparison of primes and zooms gives no grounds for ruling the latter out, contrary to the idea still existing in the minds of many users. If one lens of variable focal length is recognized as equivalent to a whole series of primes,[8] its multifunctional and economical character is clearly evident. The zoom can thus replace a lot of different lenses. As stated above, this does not of course alter the need to change the position of the camera, as would be necessary with any lens. Is it then simply the fact that camera operators prefer to prolong the existence of an older method developed at a time when using a zoom could be construed a a lack of professional rigor?

Old habits die hard but digital tehnology has completely revolutionized working methods, the data provided by the lens (aperture, focusing distance, and focal length) can inform certain decisions to be taken in the postproduction stage, and the multifunctionality of the zoom can prove very useful, especially when combined with reduced size, that makes it the first choice for steadicam and now also for drones.

One last point that may still affect camera operators' choices is the use of large apertures. Even though the best zooms on the market, like the Angénieux Optimo range, offer exceptional stability of aperture at all focal lengths, they certainly have less luminosity that the equivalent large-aperture primes. Given that digital cameras now have a nominal sensitivity of hundreds of ISO, however, this is not necessarily a discriminating factor. And if very limited depth of field is required, a soft-focus effect is possible. The Angénieux engineers can cope perfectly well with all these requirements and may well have something new in store too.

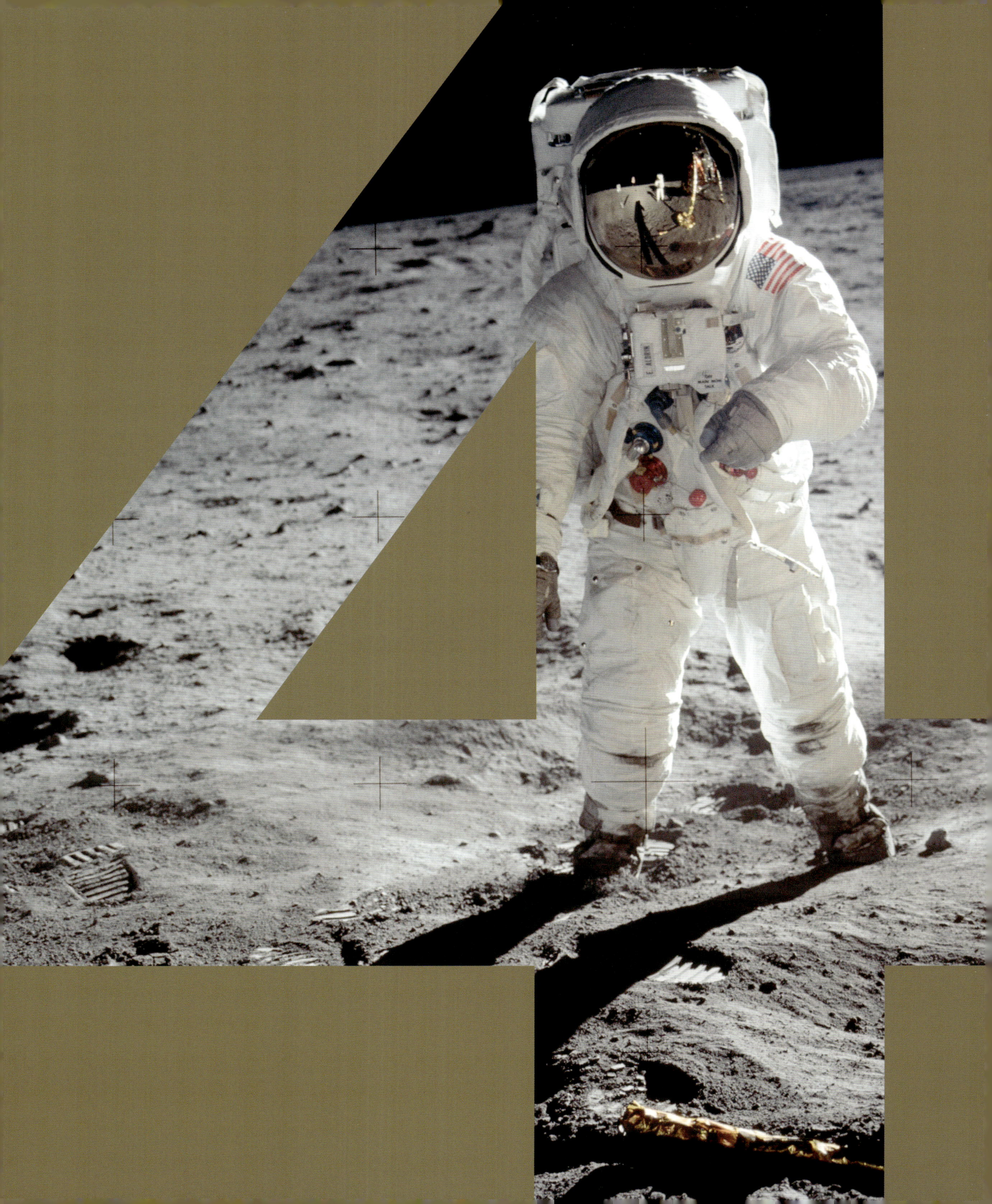

SUCCESS STORIES
AN INDUSTRIAL ADVENTURE LIKE NO OTHER

In addition to his work for motion pictures, Pierre Angénieux has played a **major role** in various industries, from photography to television, not to mention space, medicine, security and optronics. Sometimes a **pioneer**, sometimes a **creator** of solutions, sometimes a key player and sometimes a **leader**, he has always been able to make his contribution known and adapt to the evolution of different markets. His motivation, "to be at the service of...".
This is still the case today.
With his legacy, the company remains an example of **innovation** and expertise worldwide in a number of highly competitive sectors.

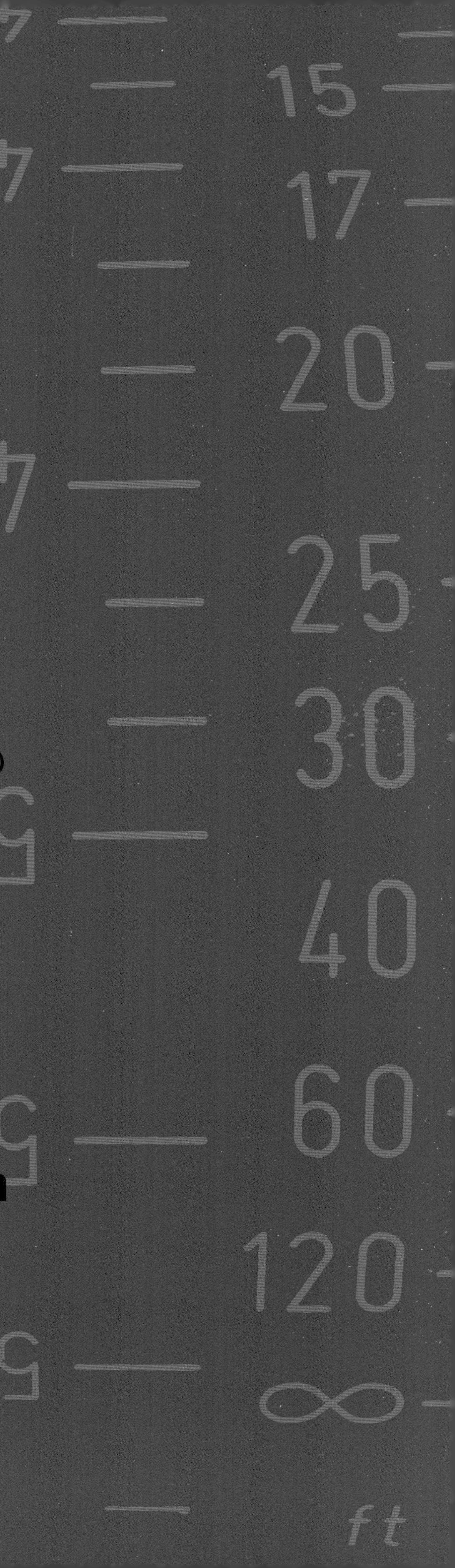

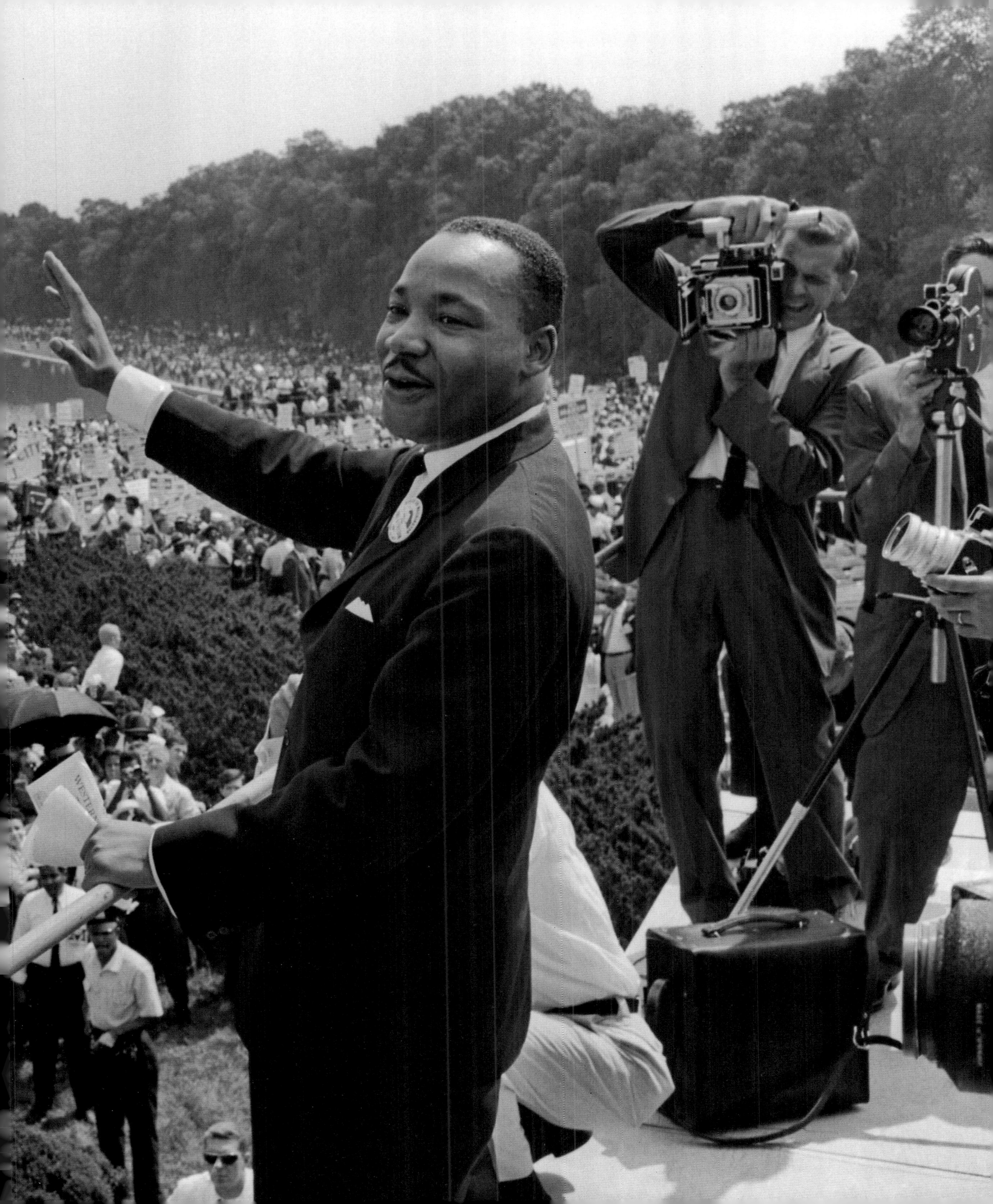

Photography

Angénieux Excellence, at the Service of Photographers

After the end of World War II, the first output from Établissements Angénieux turned toward photography, a sector that was, at that moment, flourishing more than motion pictures, of which the founder was particularly fond.
Pierre Angénieux left his mark on the history of photography when he invented the retrofocus in 1950. This major innovation would contribute to the universal use of 24x36 SLR (single lens reflex) cameras. With the explosion of zoom lenses in motion pictures, making photographic lenses became less important, but they returned in force in the 1980s, under the impetus of Pierre Angénieux's son-in-law Jean Moret, then head of the company.

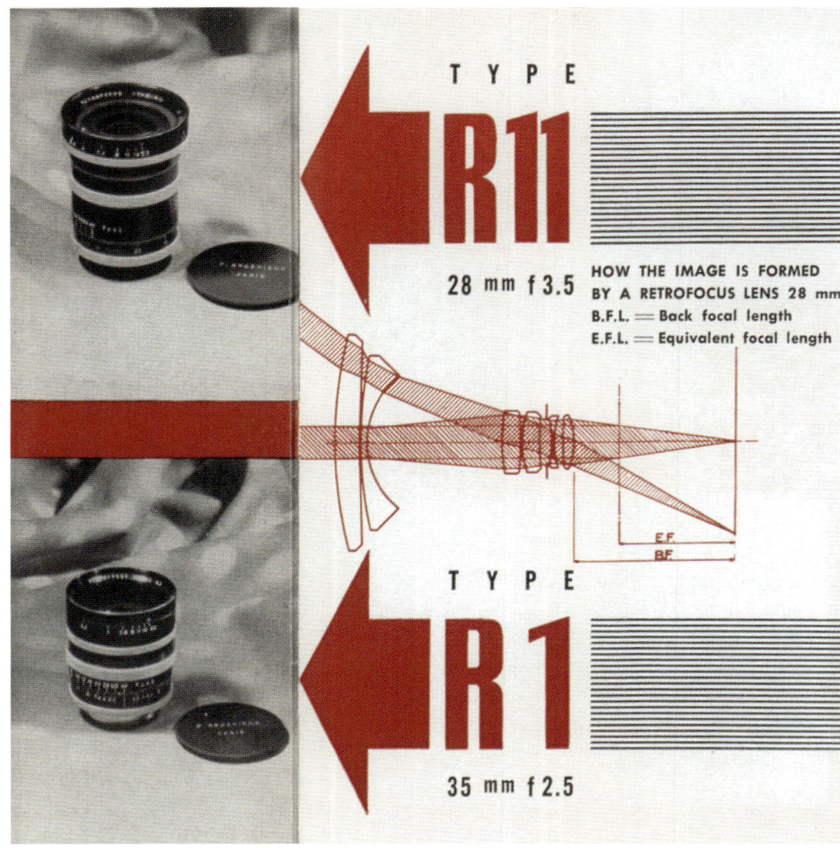

English-language brochure of Pierre Angénieux's lenses: Retrofocus for 16 and 35mm SLR cameras (Granet family collection).

The Invention of the Retrofocus Revolutionized the World of Photography

The retrofocus is a lens combination that allows the use of wide-angle lenses, i.e. with short focal length, on reflex cameras.

In 1950, there were few 24x36 interchangeable lens cameras, and they were very expensive. They had standard or long focal lengths, but no wide-angle lenses. Creating this type of lens poses two problems: increasing the brightness of the lenses and making the rear lenses fit with the mobile mirror in the reflex camera. Until then, these devices generally had to make do with 40mm lenses, with an f/4.5 aperture. The exception was Angenieux's 35mm for Alpa, with an f/3.5 aperture, a true technical feat, which reached the limits of practicability of a 24x36 reflex camera. There's no doubt that Pierre Angénieux, always abreast of the latest achievements of the optical industry, was inspired by earlier work.
In 1929, Bell & Howell placed a diverging element in front of the lens on a projection image magnifier, reversing the direction of the light. This could suggest the possibility of the opposite result, increasing the field recorded on the film surface and obtaining a wide angle effect.
In 1931, Taylor Hobson released a 35mm for Technicolor motion pictures. With an f/2 aperture, this lens has a long focal length, leaving sufficient free space between the rear lens and the image plane and making room for the glass cube that provided the range of colors.
In 1935, SOM-Berthiot launched the Hypercinor for motion pictures, 8mm, and then 16mm.

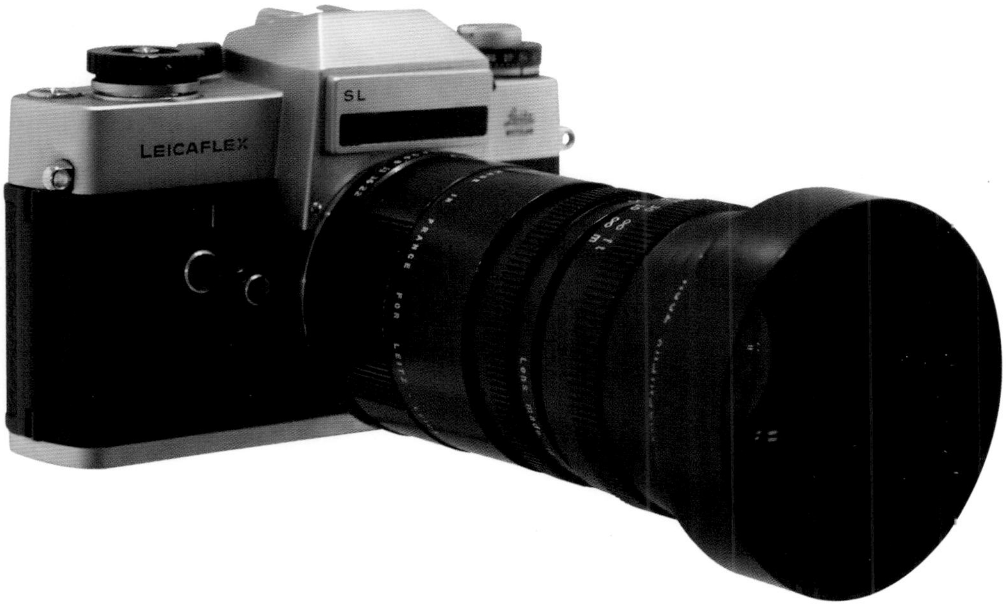

Pierre's 45-90mm zoom, f/2.8 aperture, fitted to a Leicaflex SL camera (Granet family collection, photo by Éric Perrin).

The Hypercinor, a four-lens device, was placed in front of the camera's lens, making it possible to divide the focal length by two without changing the aperture.
Pierre Angénieux would perfect the inverted telephoto method. It consisted of a diverging front element followed by a converging rear element, all forming a homogeneous lens moving as a single block while focusing. This was the Retrofocus, a term Pierre Angénieux invented, and which he would always regret not patenting.
Space was thus freed for the movement of the reflex mirror and, with the increased focal length, the angle of the light offered a sharper image and less vignetting[1]. This progress was another step towards mastering large apertures.

A Full Range of Fixed Focal Length Photographic Lenses
By the end of the 1950s, Angénieux had delivered a vast range of 24x36 lenses to the biggest camera makers of the time: telephoto lenses (F 90mm f/1.8 and f/2.5, F 135mm f/2.5 and f/3.5), retrofocuses (F 35mm f/2.5, F 28mm f/2.5, F 24mm f/3.5) and standards (F 50mm f/1.8, F 50mm f/1.5). These lenses are still enjoyed by many photographers, vintage camera enthusiasts and fans of the Angénieux brand.
From the end of the 1950s, in Japan, the photography industry was one of the most thriving post-war trades. Japanese cameras began to dominate the global market. For these cameras, French lenses were too expensive. Only Alpa and Exakta were left for Angénieux photographic lenses.
Pierre Angénieux would leave still photography for a time to devote himself to motion pictures and to developing a zoom lens for the cinema. The zoom lens was quickly adopted by filmmakers, and this success encouraged Pierre Angénieux to offer this new solution to still photographers.

1964, First Angénieux Photo Zoom
In 1964, Pierre Angénieux offered still photographers his first zoom lens. It was an F 45mm to F 90mm lens with an f/2.8 aperture, for Leicaflex. Leica had refrained from putting a zoom lens in its catalog, aware of the reluctance in the world of photography, which was very attached to the German doctrine that then believed a zoom lens could never be a match for a prime lens. The modest 2x zoom did not necessitate excessive costs to develop. It covered focal lengths between 50mm and 90mm, those most used in photography, and advantageously complemented the Angénieux catalog.

[1] Vignetting is a reduction in brightness or saturation at the edges of an image.

Its quality, and the fact that it bore both the names Angénieux and Leica, finally convinced even the most cautious photographers to use a zoom. Angénieux's 45-90 would contribute to widespread acceptance of zoom lenses in still photography. It remained in production for ten years. At the same time, deliveries of fixed focal lengths continued but were gradually reduced as Japanese cameras invaded the market and Angénieux's range of cinema zoom lenses expanded.

Short-lived Return to Still Photography

At the beginning of the 1980s, the 24x36 photograph experienced a real boom. Jean Moret, Pierre Angénieux's son-in-law, became director of Établissements Pierre Angénieux after the founder retired in 1974, and under his impetus Angénieux would offer the world of photography a complete range of zooms.

First, an F 35-70mm f/2.5-3.3, with an aperture unmatched in its class, then an F 70-210mm f/3.5, produced from 1983 to 1987. In 1986, an F 28-70mm f/2.6 was released, evidently the only French-designed autofocus photographic lens. Offered with fittings for Canon, Nikon, Pentax and Minolta, it was the latest photographic zoom sold by the brand.

The performance of these three zoom lenses is unanimously recognized. The look of their silkscreened matte black Makrolon barrels would inspire other manufacturers. An Angénieux club formed for those with a photo zoom lens, as well as a magazine, *VU*, and a foundation offering an annual contest for young photographers. Angénieux then added to its catalog two new exceptional fixed focal length lenses, the DEM F 180mm f/2.3 APO[2] and the DEM F 200mm f/2.8 ED – DEM[3] (Differential Element Movement), two devices with a floating lens offering better image quality at different focuses.

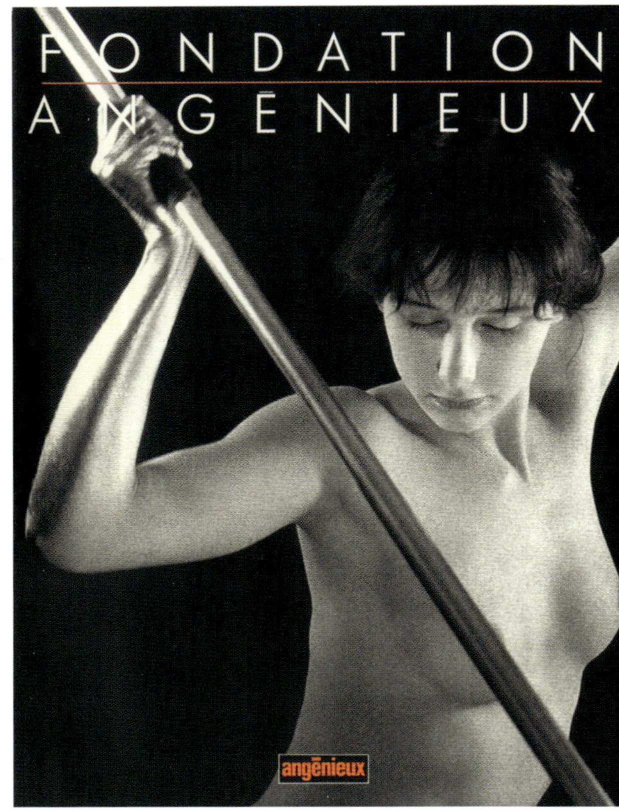

Promotional brochure from the Fondation Angénieux (Arnaud Saudaux collection).

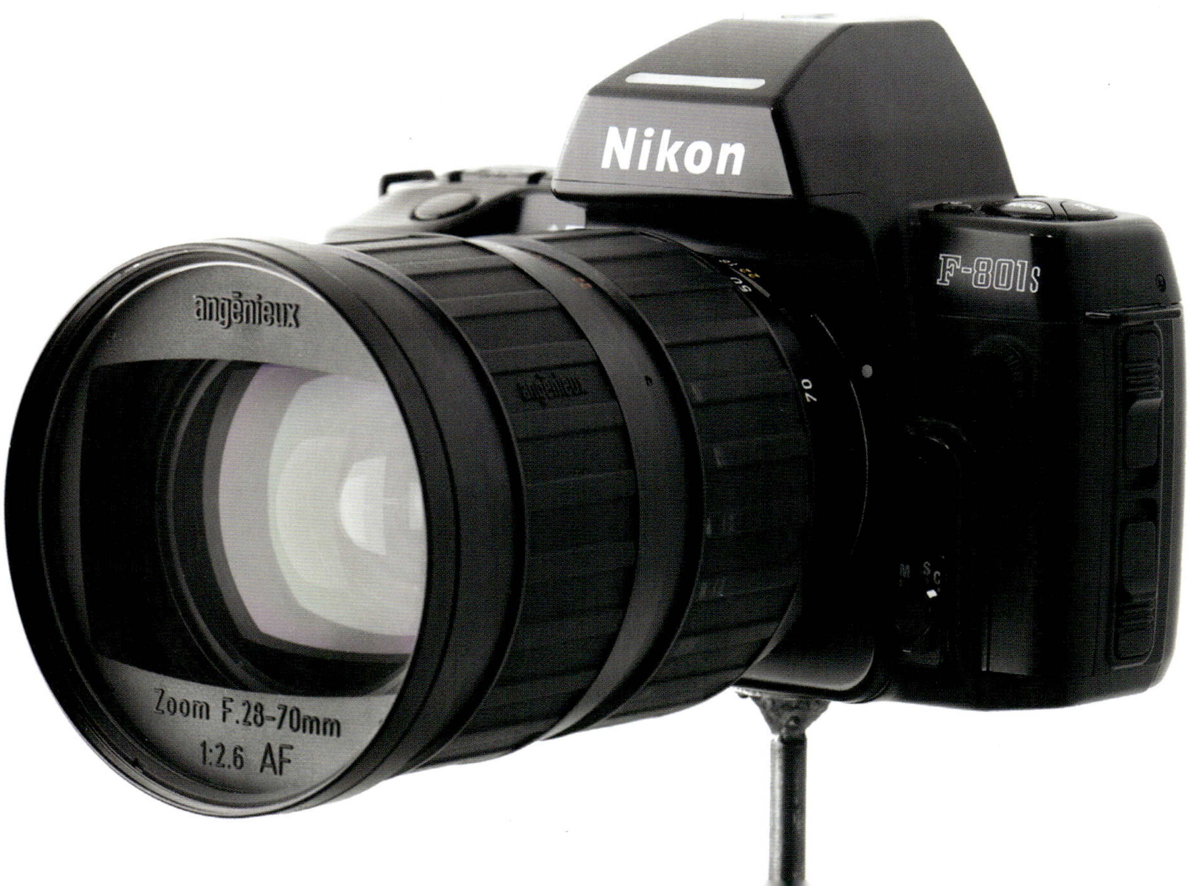

Angénieux 28-70mm zoom, no. 1546081, 1993, mounted on a Nikon F-801 S camera (Granet family collection, photo by Marc Salomon).

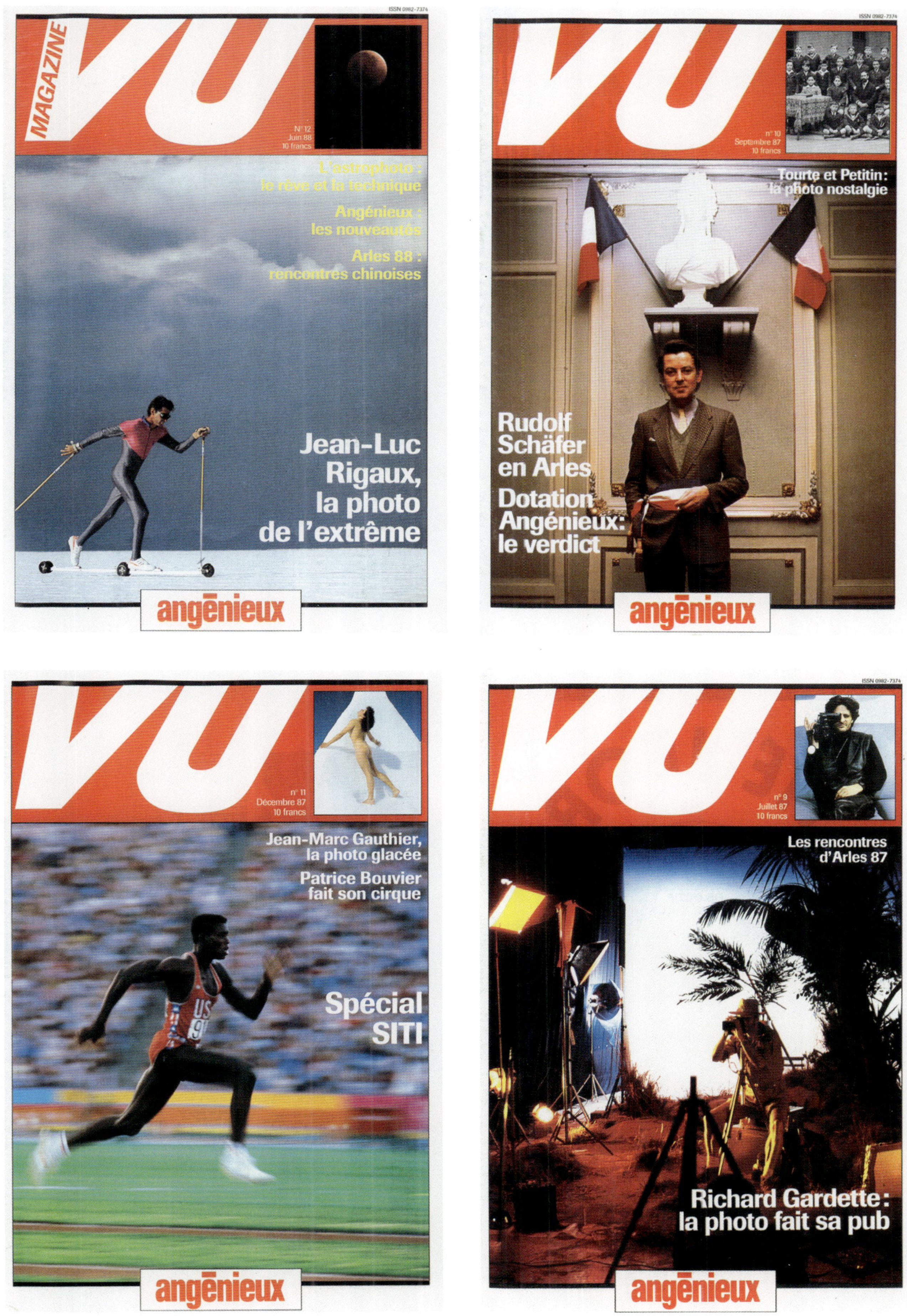

Four covers of the Club Angénieux magazine *VU* (Patrice-Hervé Pont collection).

However, the price of Angénieux lenses put their use by the general public at a disadvantage. They were kept for the discerning amateur or for professional photographers. The appearance of Japanese digital cameras electronically linked to the lens made market access even more difficult. Work in still photography was stopped in the early 1990s.

Angénieux, a Brand still Popular with Photographers

Although it stopped producing still photography zoom lenses for nearly twenty-five years, the brand still enjoys an incredible reputation among the oldest photographers as well as younger generations.

In a world where images are everywhere, vintage is now fashionable. The imperfections of the old optics offer a strong sense of identity. Angénieux's photographic lenses are still solid and still very popular on the second-hand market.

In addition, the new video features of cameras are always being optimized. Videographers, wishing to combine the compactness of DSLR (digital single-lens reflex) cameras with the quality of a motion picture zoom could use Angénieux lenses. Traditionally in PL[4] mount, Angénieux offered Canon EF and Sony E mounting options for its zoom lenses.

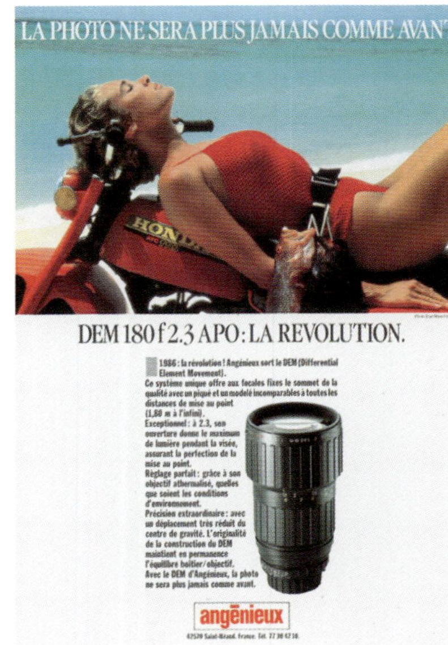

The launch of Angénieux's new still photography lenses was highly publicized.

Other Angénieux Products: Lenses for Enlargement and Projection, Magnifiers, Viewers and Binoculars

As early as 1943, Angénieux produced lenses for enlargers in all the usual focal lengths: 50, 75, 88, 100 and 135mm. Years after ending production, Angénieux returned with a high-end 48mm lens.

Angénieux's earliest products also included small magnifiers, available in two different magnifications, 10x and 15x.

Between 1950 and 1960, the production of projection lenses was an important part of business, involving about twenty different models, for brands such as Alsaphot, Inox, Procolor, and Rob, but especially Kodak Pathé. Thus Angénieux would equip the Kodak Junior, Senior and 300F projectors, made in France, as well as the Kodak Regent and Supermatic. The Kodak Senior model was designed by Lucien Seignol, none other than the architect of the new post-war Angénieux factory. The model was entirely manufactured in Saint-Héand.

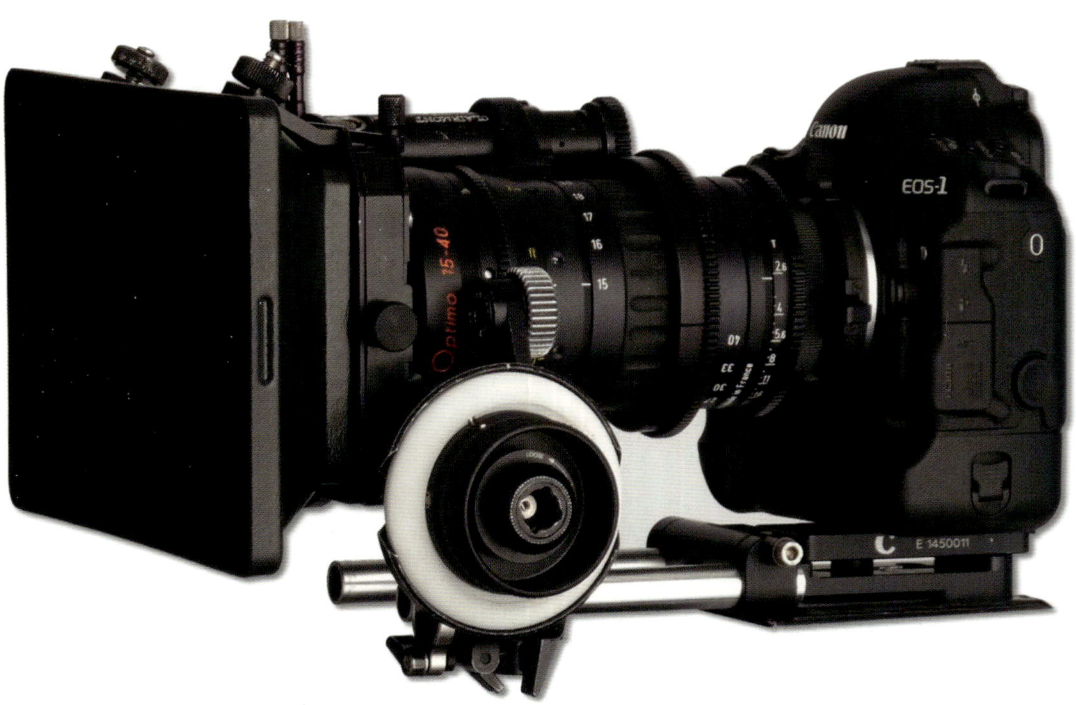

left
Optimo 15-40 on a Canon DSLR camera (Angénieux archives).

[2] APO (apochromat): a lens or optics system corrected for chromatic aberration at three wavelengths and corrected for comatic and spherical aberration at two wavelengths.
[3] DEM (Differential Element Movement): a device with a movable lens that provides excellent image quality at different focuses.
[4] The PL ("positive lock") was a lens mount developed by Arri for its 16mm and 35mm cameras in the 1980s.

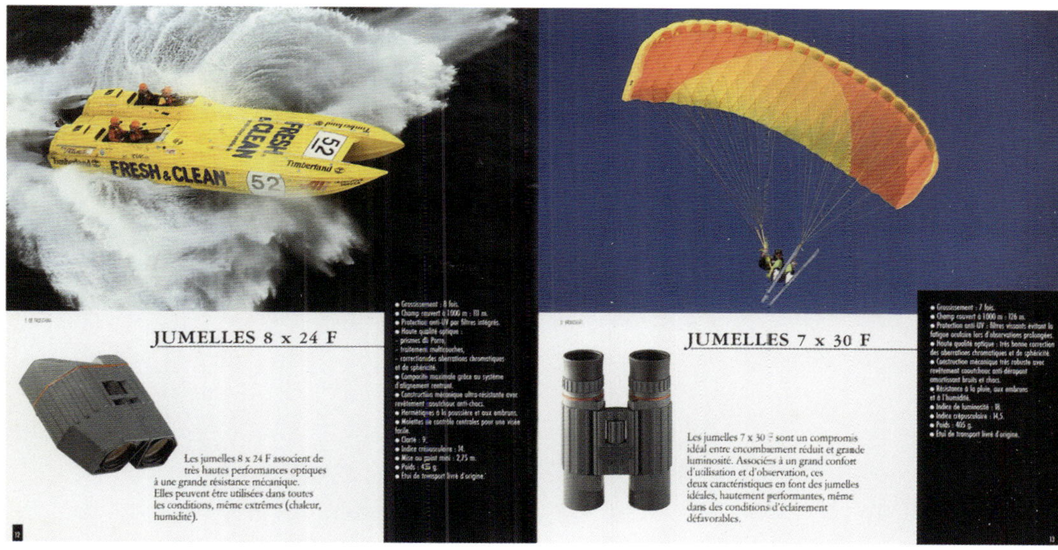

Binoculars Angénieux 7x35, conceived on the principle of Porro prisms, were presented at the Photokina fair in Cologne (Germany) in 1988.

The collaboration with Kodak-Pathé continued until 1964, the date of the appearance of the circular basket carousel built in Germany by Kodak AG, equipped with German lenses. Angénieux would also manufacture two viewer models, the Diffocus (1951) and the Visumax (1967), and equip two Kodak models, the Junior viewer (1953) and the 5x5 folding viewer (1955-1959).

Like all the major lens manufacturers of the time (Leitz, Zeiss, Nikon, Olympus), Angénieux put binoculars on the mainstream market: an 8x24, with a very original design in 1987, followed by a 7x30, with a more classic design, in 1991. The performance of these binoculars, along with that of fifty others, was hailed by the magazine *Chasseur d'images* in August 1992.

For fans of Angénieux photographic lenses, the work of Patrice-Hervé Pont, *Angénieux made in Saint-Héand, Loire, France*, in its second edition in April 2003, gives a complete and detailed list of all Angénieux photographic products.

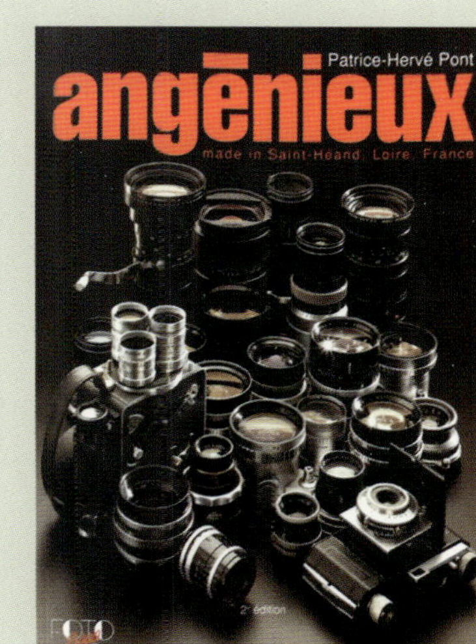

Television

Angénieux Takes Part in the Birth of Color Television

Pierre Angénieux quickly understood the immense potential that the television market could offer. In the late 1950s, he took interest in the subject and adapted his cinema zooms to the new constraints of electronic camera tubes. The company played a major role in the spread of color and high definition television. For decades, it was an important player in the electronics industry and the technical departments of major television networks. Its influence extended worldwide and especially in the United States, as a pioneer in the field.

A New Challenge for Angénieux, from Filmmakers to Electronics Engineers

The first tests of live public television broadcasts date to the 1920s. In the course of the tests dotting the years that would follow, as television was rolled out, the image entered homes, which would now be equipped with a television set.

Pierre Angénieux very quickly understood his company's interest in positioning itself on the new market, but the world of television was quite different from that of the silver screen. Relations with the world of cinematographic production and with the electronics engineers were not the same. However, Angenieux's experience and excellence in calculations and lens processing (thin layers), his mastery of computer and his implementation of the concept of MTF would make the French industrialist the point man for all the major players of the time (Thomson CSF in France; RCA, Ampex, and General Electric in the United States; Marconi and EMI in England; Philips in the Netherlands and Fernseh in Germany) and for the technical departments of major production and distribution organizations (RTF in France; ABC, CBS, and NBC in the United States; the BBC in England; the RAI in Italy; SudWestfunk in Germany and RTB in Belgium). Japan was not yet that all-important part of the conversation, but would be later on, with Sony, Panasonic, Ikegami, JVC, and Hitachi.

From Cinema Zooms to Television Zooms

With television, the image is transmitted remotely. Optics now depended on a concept of waves of light. It was no longer about chemical transformation of a sensitive surface (the film), caused by photons, but about electronic emission to a television set. New cameras emerged. The cameras used analog

The MTF Measurement

The definition of a lens was traditionally evaluated by photographing a test card of parallel black lines grouped increasingly close together and using a microscope to establish how many lines it managed to separate in the space of a millimeter. This approach is subjective and fails to take into account the degree of contrast, which plays a key part in the impression of sharpness.

With the advent of television, the quality of the sensor and the electronic processing of the camera signal were taken into account in the design of lenses for this market. The Modular Transfer Function or MTF of the lens, which represents its contrast at a given frequency, must be adapted to that of the camera. In the 1960s, with a view to developing TV lenses and ensuring their correct calibration at the time of manufacture, Angénieux began to work with the BBC laboratory directed by W.N. Sproson on instruments for the precise measurement of their MTF. These instruments are used today to test lenses for the cinema and surveillance markets.

Good to Know... to Better Understand this Chapter

In a camera, it is necessary to know the differences between the technology of the sensor used (tube, CCD, or CMOS) and the processing of the video signal (analog or digital). The sensor converts photons into electrons. Regardless of type, the electrical signal it generates cannot be used directly because of numerous defects related to the technology in particular. It must be corrected and adapted at the camera's output interface. This is the role of video processing.
It is entirely possible to conceive of a digital tube camera. The order in which broadcast camera technologies appeared was the following:
- tube(s) camera / analog processing → analog camera
- CCD camera / analog processing → analog camera
- CCD(s) camera / digital processing → digital camera
- camera with CMOS sensor(s) / digital processing → digital camera

In motion pictures, film cameras survived until the testing of digital 2/3" tri-CCD HD motion picture cameras such as the Sony F900 CineAlta, the Thomson Viper and the Panasonic Varicam. These cameras, derived from broadcast cameras, did not have the commercial success expected because the use of a prism limited the aperture of the optics, and therefore the creativity of the filmmakers. The real digital turning point in motion pictures came only in the mid-2000s with the arrival of the large-format single-sensor digital cameras. The size, similar to that of 35mm film, made it possible to use cinema optics.

[5] See p. 126.

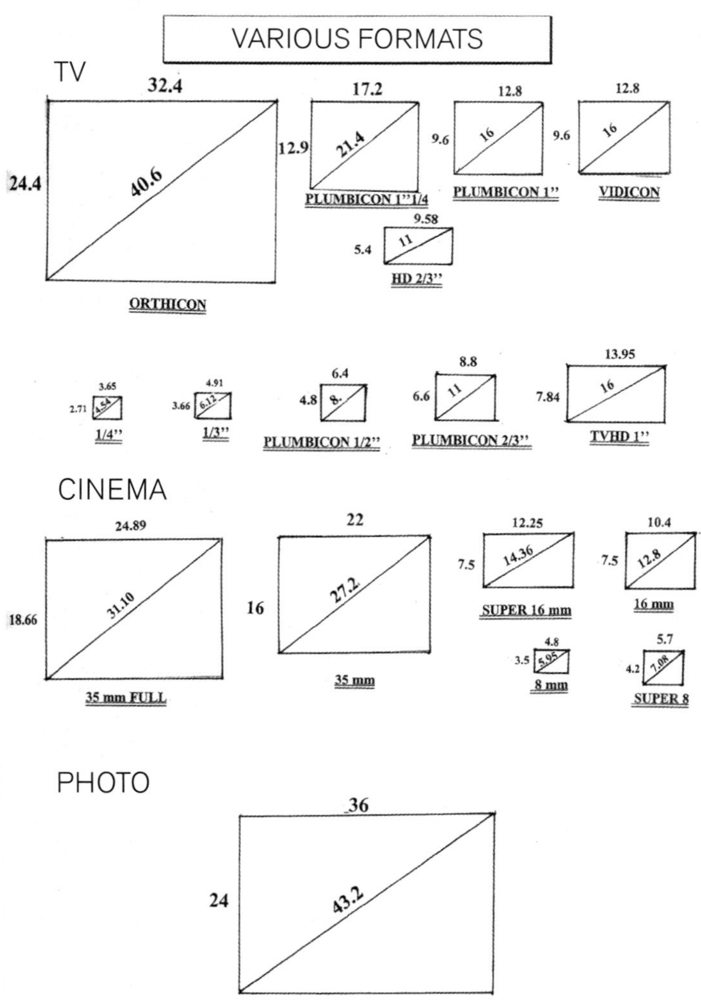

(electrical signal proportional to the brightness of the object filmed) or electronic tubes.
For a lens builder, the impact of the transition to electronic sensors was modest, at first glance. All that was required was to take account of spectral sensitivity, that is, sensitivity as a function of the different wavelengths, the size of the sensor and the thickness of the glass of the tube to reach the sensitive surface.
As for photography and cinema, the first link in the chain of a television image is the lens for shooting. The eye perceives all the points constituting an image, but it is not possible to transmit all these points simultaneously. The transmission must be sequential. Because of retinal persistence, the eye cannot distinguish between two successive illuminations lasting less than 1/25th of a second. In accordance with the frequency of the local alternating current, Europe adopted a standard of 25 images/second, 625 lines; in the United States, 30 images/second, 525 lines.
Working closely with manufacturers, Angénieux optimized its motion picture zoom lenses for the different sensor sizes of these new analog cameras.
As early as 1950, Orthicon tube technology was replacing iconoscopes and provided an image 40.6mm on the diagonal. There was no cinema zoom lens for this format. So Angénieux modified the backs of its 35mm zoom lenses. In 1960, the 35-140mm zoom for 35mm motion picture

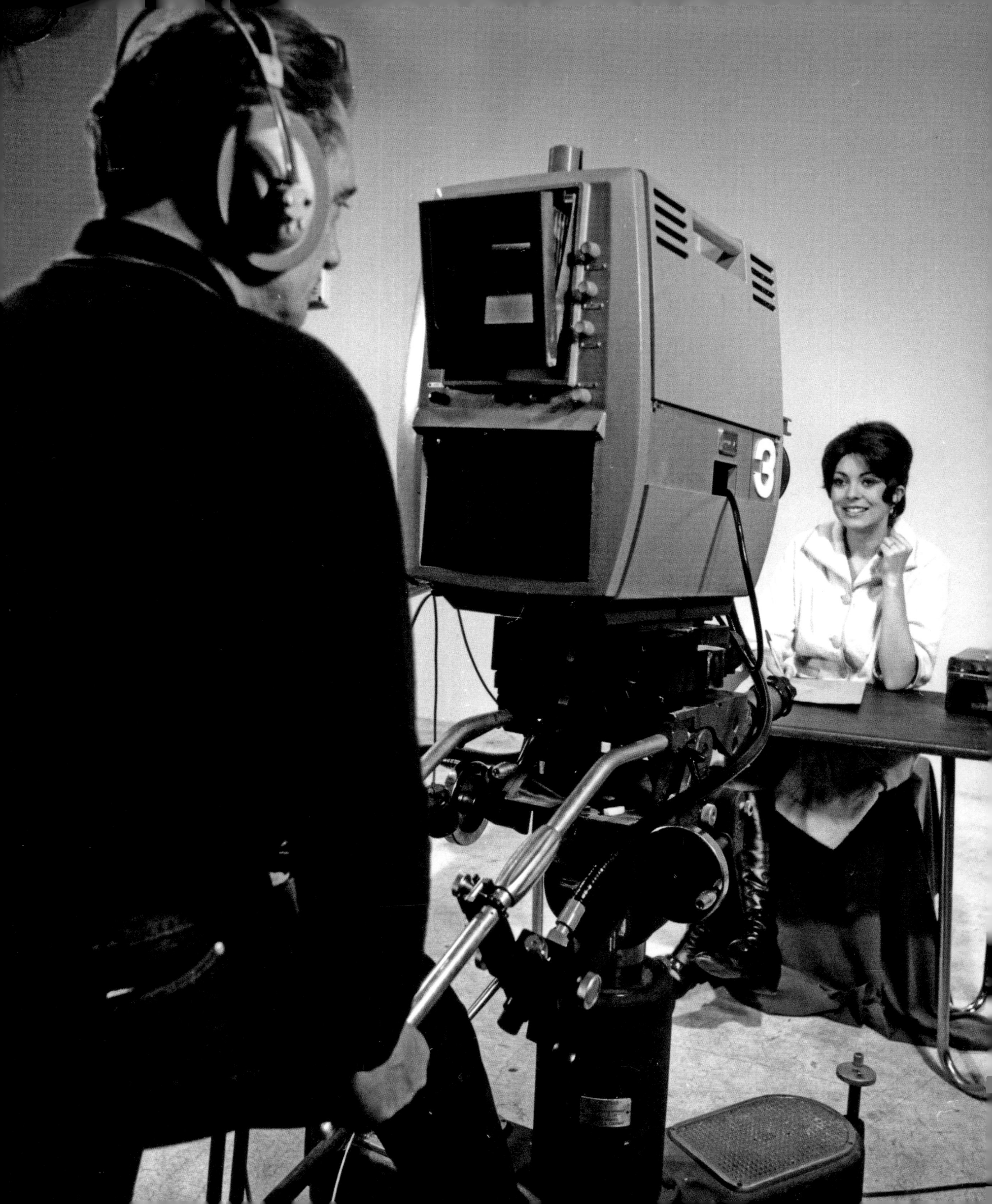
3

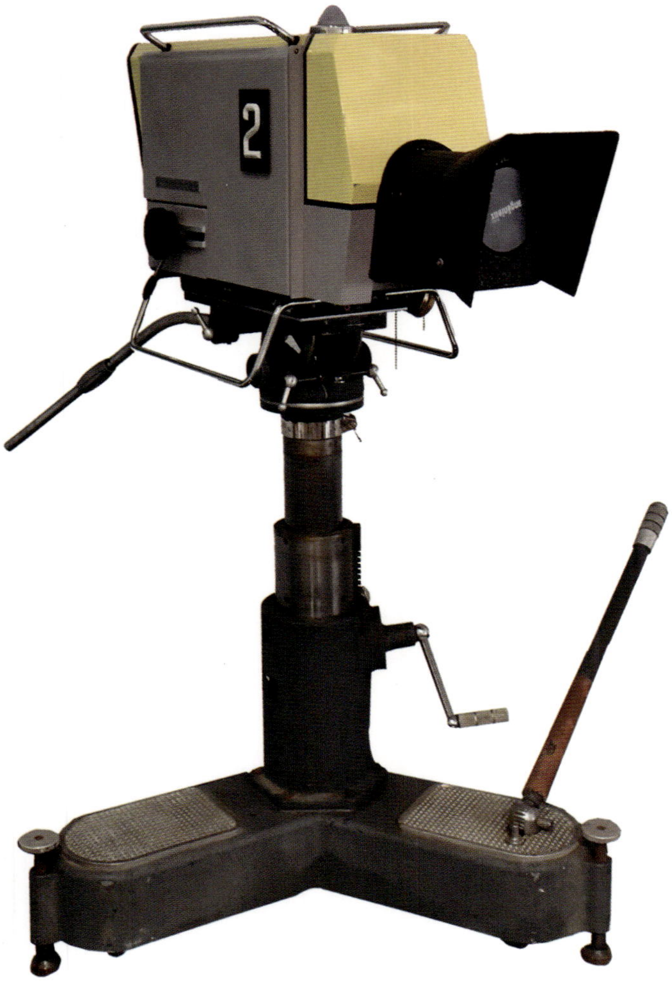

above
Thomson black and white Orthicon THT 628 television camera, France, 1959, on Éclair "Crab" pedestal, with an Angénieux 40-400mm, 1:5, zoom lens type 10x40 T1, no. 1407023, 1974 (Photo by Marc Salomon. Poinsignon collection, Association du Centre Historique de la Diffusion Radiophonique).

camera was transformed into a 40-160mm to cover the Orthicon image format. This was followed in 1961 by a 40-400mm. These two zooms fit cameras including Thomson's such as the 1959 THT 628, which dominated the market at the time.

Many other formats would follow the Orthicon: Vidicon (tube diameter 1", image diagonal 16mm), Plumbicon 1-1/4" (tube diameter 1-1/4", image diagonal 21.4mm), Plumbicon 1" and Plumbicon 2/3" (tube diameter 2/3", image diagonal 11mm). Angénieux was able to adapt perfectly to all these formats and sell lenses to the various camera manufacturers.
Since the first Orthicon tube cameras were produced in the English-speaking world (RCA TK-10 and TK-30, Dumont Marconi MK IV), inches became the standard for giving the diameter of the electron tube.
In the early 1960s, the benefits of zoom lenses for shooting in television were such that neither their weight nor their size was a problem, the cameras also being so large. But for news and documentaries, which required light, portable equipment, 16mm film was still the preferred format. It was then necessary to transfer the film images to electronic receivers, using optical and electronic systems, called telecine.

Angénieux Pioneers Color Separation Technology

While black and white television was conquering the world, he was preparing for the arrival of color television. The Plumbicon tube was invented by Philips in 1960. It offered a better image and quickly replaced the Orthicon. It would make the advent of color television possible.

With the switch to color, the individual electronic tubes had, until then, only had to deal with brightness. It was necessary to modify the cameras and integrate several sensors able to analyze color. In collaboration with the major camera manufacturers, Angénieux would then play a key role in mastering and separating the light at the camera. This separation into three images for red, green, and blue (RGB) required an optical separation system associated with both the lens for shooting and with the color sensors in the camera.

The Battle for Coding Standards

In terms of transmission, the technological challenge consisted in transmitting a color signal in a channel previously used only for black and white. Various analog coding standards for color video emerged in the United States and Europe. In the United States, RCA, under the aegis of David Sarnoff, launched the NTSC (National Television System Committee) process at 525 lines. In France, inventor Henri de France created the SECAM (Sequentiel Couleur A Mémoire) system, with 625 lines.
In Germany, Walter Bruch developed the PAL (Phase Alternating Line) process, also at 625 lines.
Europe and the United States could not agree on a single choice.
For optics designers, these different systems concerned only coding, which occurs downstream of the formation of the images, and therefore did not interfere directly with their work. Their role was at another level.

In 1970, Angénieux designed and manufactured the separator for the TTV 1515 color camera (Plumbicon 1-1/4") for Thomson-CSF. This **separator, called a relay system**, relies on the use of a key component, a dichroic mirror, a complex optical system composed of a stack of thin interference layers – up to 40 layers of oxide and fluoride – which reflect certain colors while allowing others to pass through. These separators were heavy and bulky and could not be used with very open color lenses. The TTV 1515 color camera, which was also very successful because of its triaxial connection between camera and the control unit, would be the only camera to integrate this technology.

All but a few later color cameras on the market (tubes and then, beginning in the 1980s, CCDs) would use the principle of the **three-prism color separator** lens. These new separators consisted of three dichroically treated prisms and were placed between the lens and the focal plane. They were small and made it possible to develop color cameras that could be carried on the shoulder and accepted very open lenses.

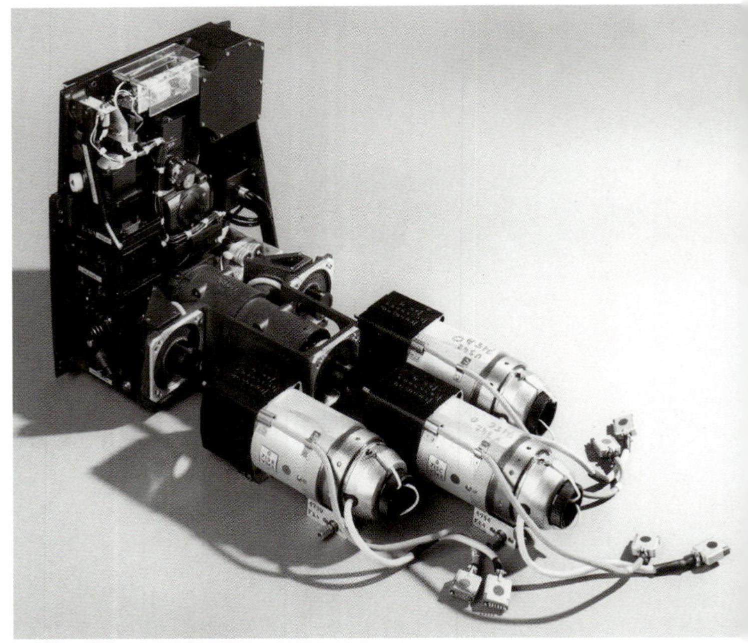

above
Angénieux G05 separator, no. 1403710, for TTV 1515 color television camera (collection of the city of Saint-Héand).

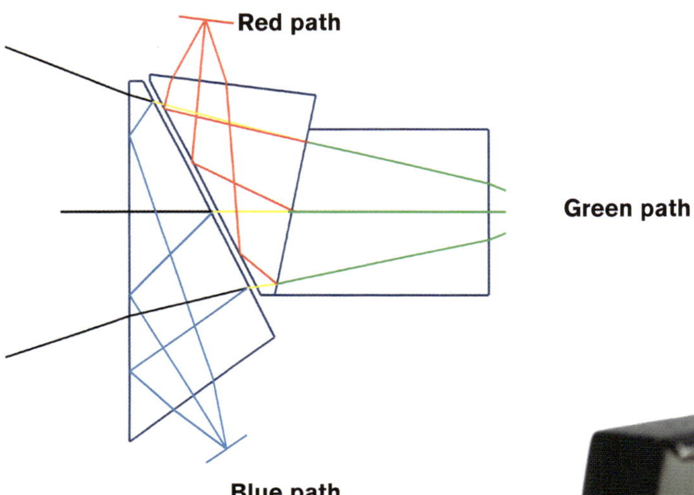

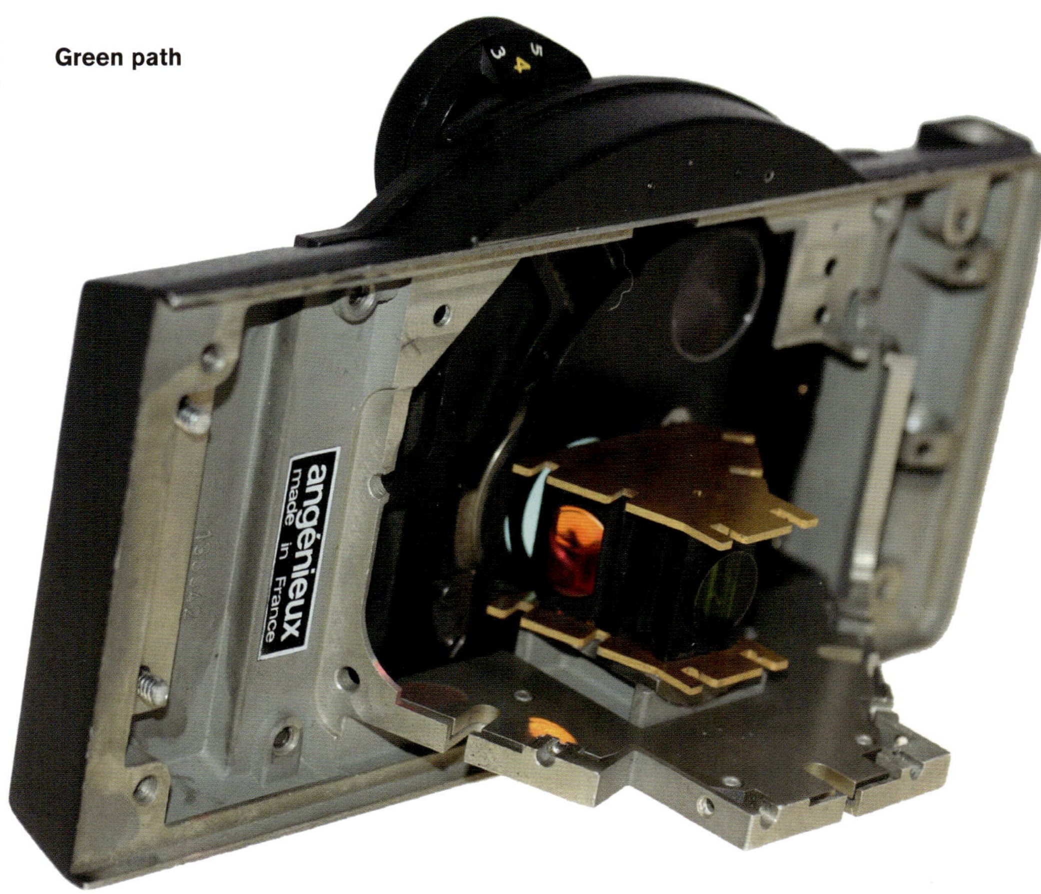

above and right
Diagram of the prism separator with prism and RCA Hawkeye camera equipped with this separator (collection of the city of Saint-Héand).

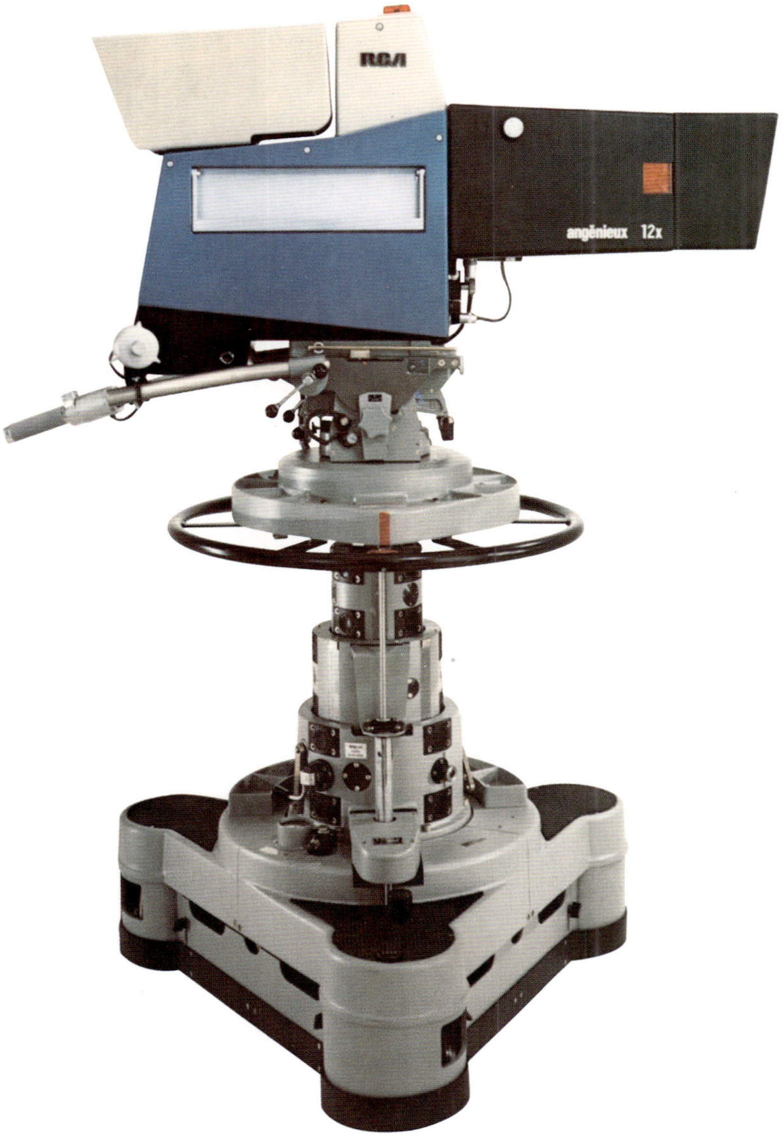

The separators Cameras including the RCA TK47 and TK76 would be equipped with Angénieux prisms. In 1979, Angénieux developed the color separator for the world's first CCD camera, the RCA Hawkeye, and delivered 1,000 prisms to the manufacturer, in the US.

Video Cameras using Tubes and the Preeminence of Angénieux Zoom Lenses

In order to place the color separators, it was necessary to modify black and white television zoom lenses, to change the back end to increase the focal length (free space between the last lens at the back of the zoom and the focal plane) and compensate for the optical aberrations generated by the separator.

above
RCA TK47 camera with Angénieux 12x zoom lens (Bernard Tichit archives).

right
Thomson TTV 1515 color television camera, 1971, on Éclair "Crab" pedestal, with Angénieux 18-180mm, 1:2.5 zoom lens, type 10x18 L, no. 1372468, 1972. Angénieux separator no. 1390434 (Association du Centre Historique de la Diffusion Radiophonique, photo by Marc Salomon).

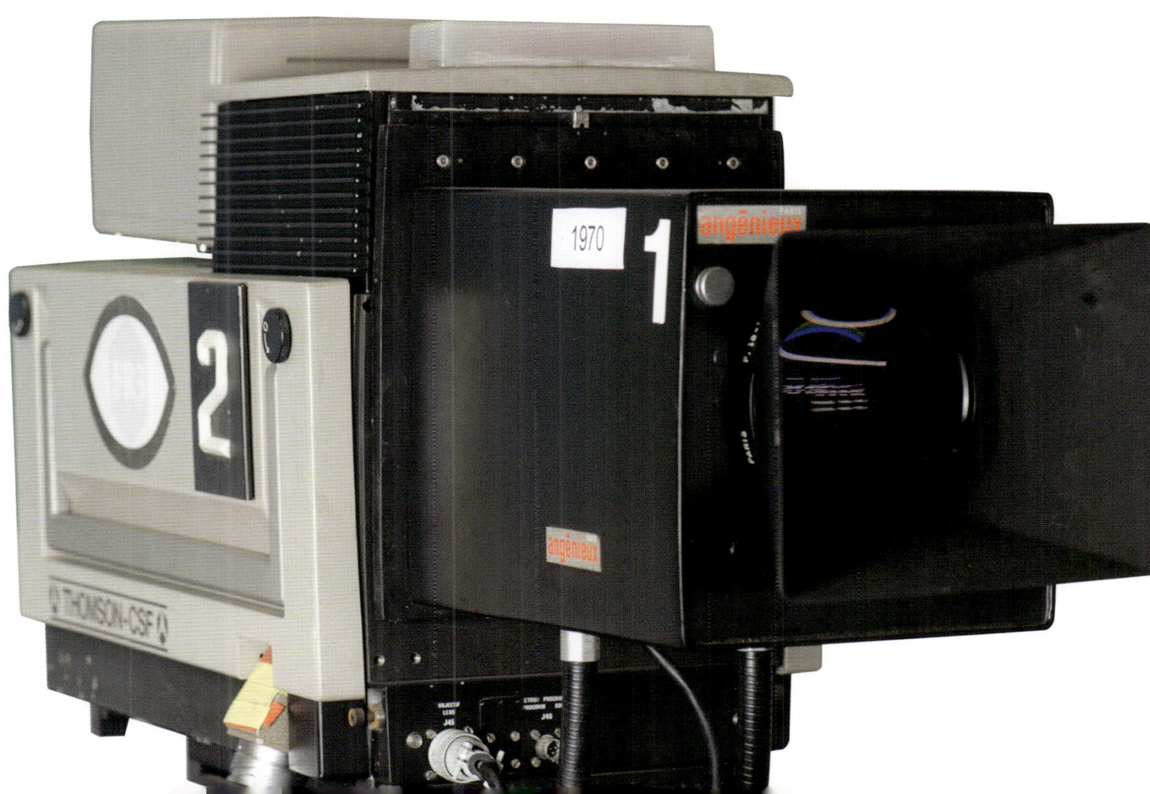

Olympic Winter Games in Grenoble, 1968 (Angénieux archives).

In 1963, Angénieux introduced the first zoom lens for color television, a 10x for the Plumbicon 1-1/4": the F18-180mm, with an f/2.2 aperture and a minimum focus distance of just 0.95m. This zoom lens made it possible to develop the Philips LDK3 and the General Electric PE250 cameras. Two thousand cameras from these brands, equipped with the Angénieux 10x zoom, were sold across the Atlantic.

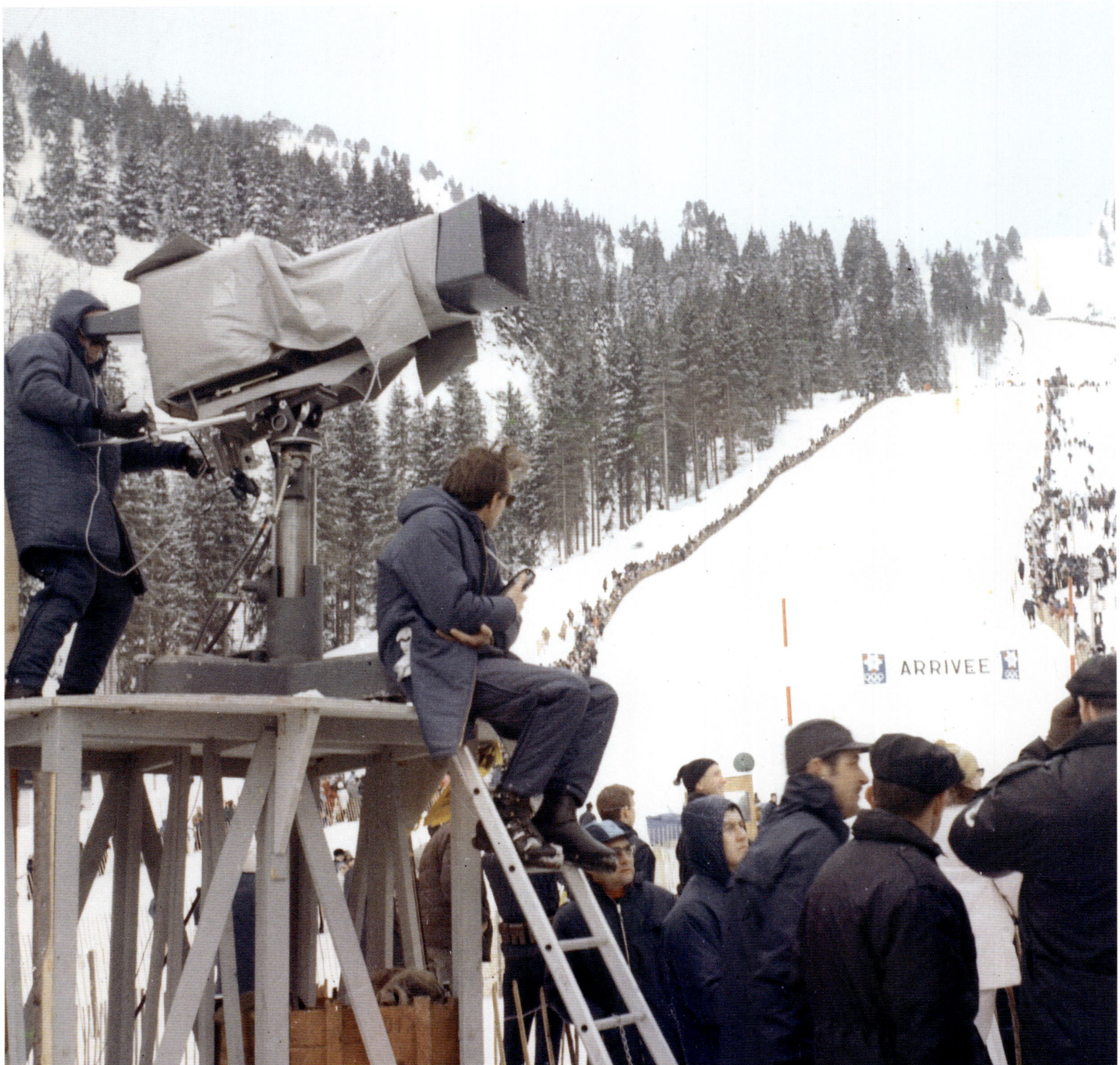

At the Heart of the American Market

« In 1966, Angénieux came to the United States. First to Long Island, New York, where three major television networks were headquartered. Then to Hollywood, because so many programs were made as movies. There was also an office in Ottawa, and one in Miami for Latin America, during the 1968 Olympics in Mexico in particular. In this particular television market, being close at hand was crucial for three reasons. First, it was necessary to insure a technical liaison with the camera manufacturers. With color television, lenses and cameras had to be technically compatible. And the cameras were made in the United States. Or, if they were not, they were Americanized on site because American and European standards were not the same.
There was also a need for a commercial connection, to maintain contacts with those who ordered and used the lenses, with exhibitions, and with advertising. It was also necessary to provide service. For outdoor reporting in particular, the material was often mistreated. And with live broadcasts, repairs needed to be fast.
At that time, the US market accounted for about half of the world market. It was in 1967 that RCA, then the largest manufacturer of television sets, having saturated the market for black-and-white sets, decided it was necessary to launch color TVs. To do so, it needed content. Beginning in 1967, RCA, which also owned NBC, the largest television network, decided to produce two hours of broadcasts daily in color. Suddenly, the two other major networks and some six hundred private stations understood they had to do the same. At that time, a television zoom lens cost about $10,000, which was more than the price of two cars, and a camera cost about $65,000. Many programs also came from Hollywood.
Television series and nationwide commercials were produced there with Angénieux zoom lenses. Our prospects in the United States in the late 1960s and early 1970s were enormous. They accounted for most of the work at the Saint-Héand plant. »

Bernard Angénieux

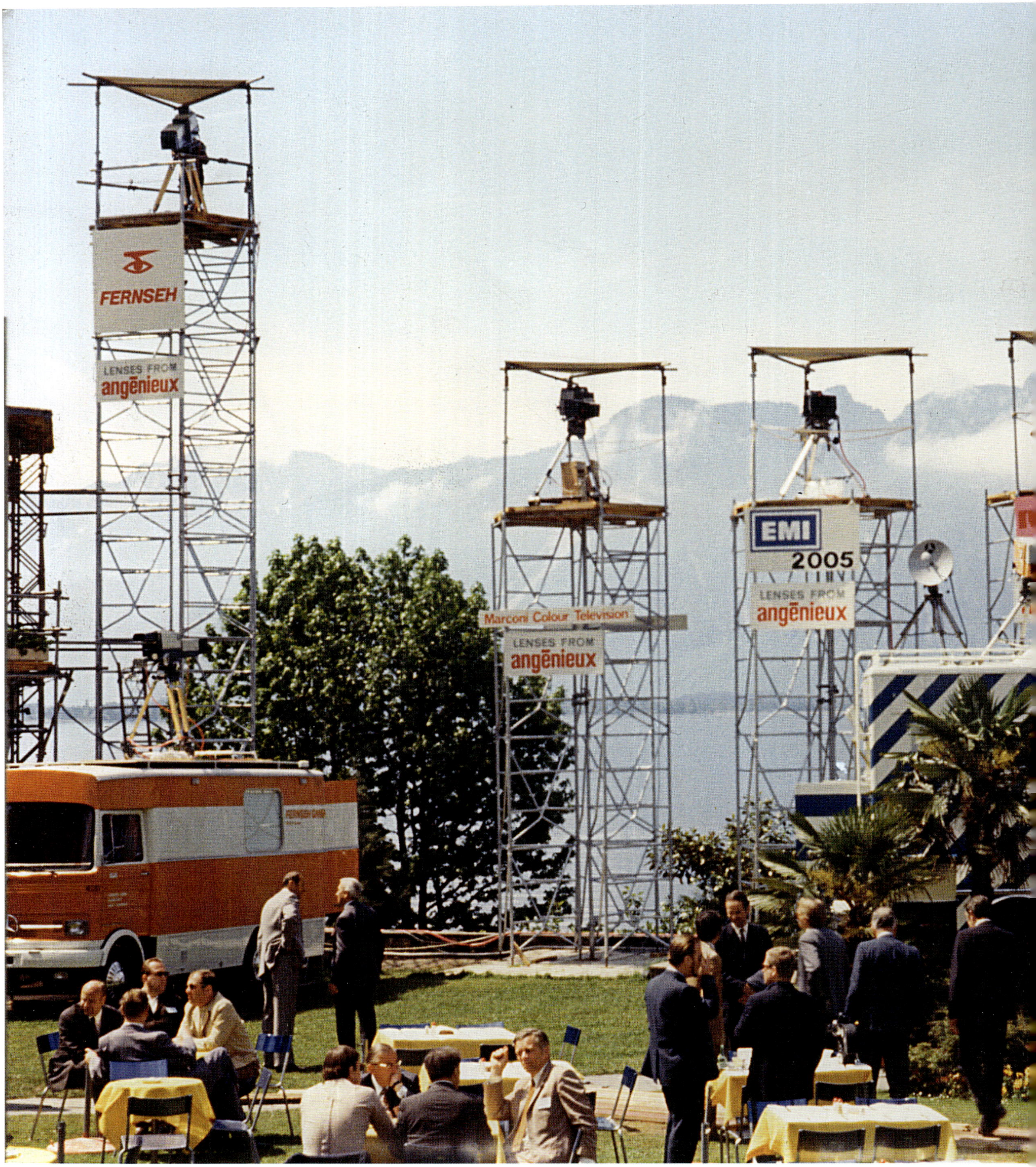
FERNSEH
LENSES FROM angēnieux
Marconi Colour Television
LENSES FROM angēnieux
EMI
2005
LENSES FROM angēnieux

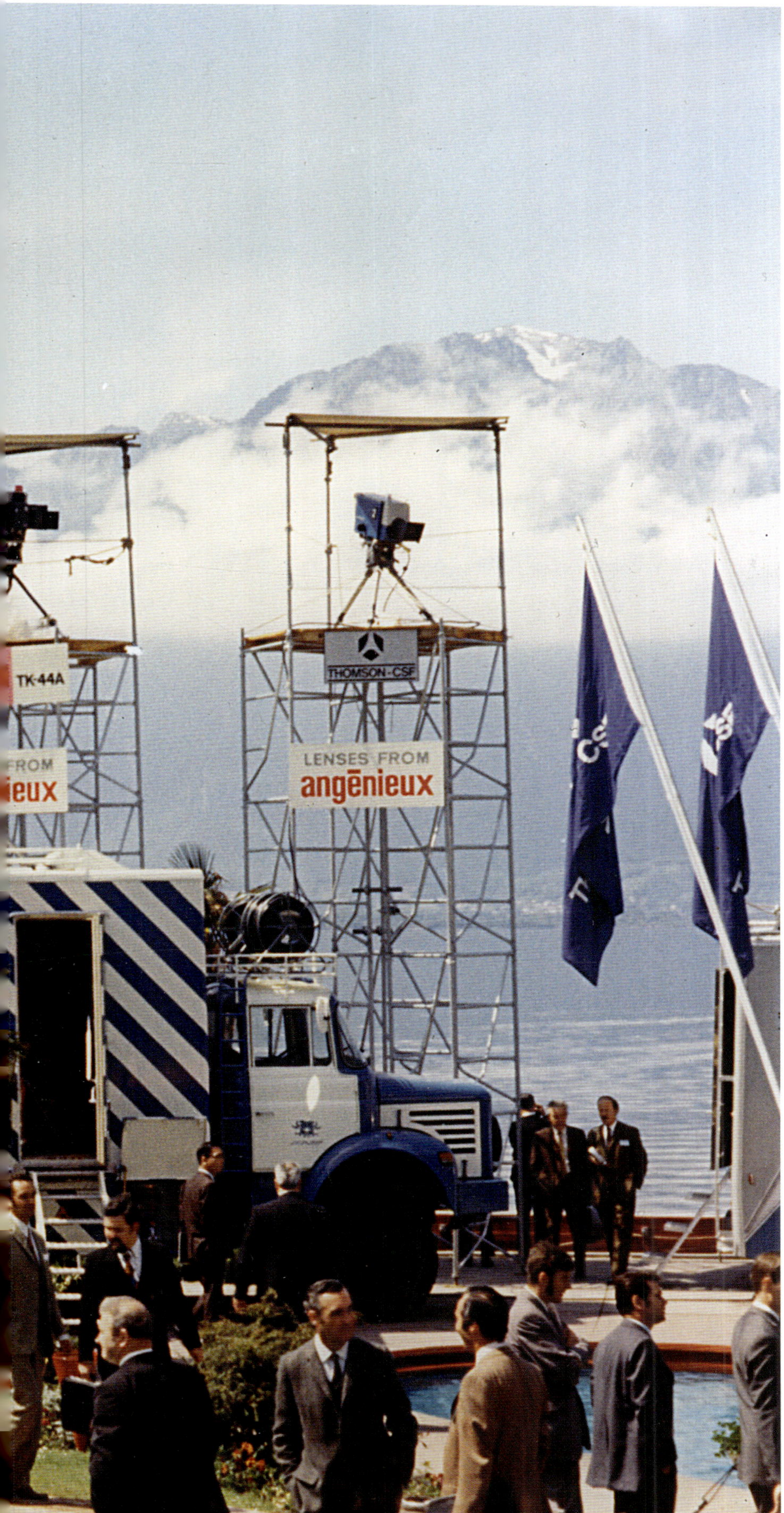

In 1967, Angénieux released the first telephoto lens for Plumbicon 1" 1/4 tubes, an 18x F 27.5-500mm (18x27.5). This zoom lens was followed, in 1969, by a 15x18 F18-270mm for Plumbicon 1" 1/4 tubes. It was the beginning of Angénieux's long-focus lenses for sports and for large studios.

Also in 1969, while Angénieux zooms had a role on every American television screen, the broadcasting of man's first steps on the Moon on worldwide satellite television certified recognition of the brand across the Atlantic. Angenieux zoom lenses were on board the *Apollo 11*, which sent the world the images of mankind's first steps on the moon on July 20, 1969 (see "Angénieux and space exploration," pp. 134-153).

Broadcasting the Olympic Games was a strategic moment for television channels. They upgraded their filming equipment for the occasion, which meant many important contracts for the purchase of Angénieux zoom lenses for both the winter and summer games.

Angénieux's impressive presence at the Montreux International TV Symposium International in Switzerland in 1969. Angénieux had a monopoly on television lens supply (Bernard Tichit archives).

A Designation that became the Universal Naming Convention

Beginning in 1967, with the release of the 18x27.5, Angénieux designated its television zoom lenses by the ratio: magnification by shortest focal length in millimeters, describing the largest angle of the field of the scene covered. Such a designation has become universal.

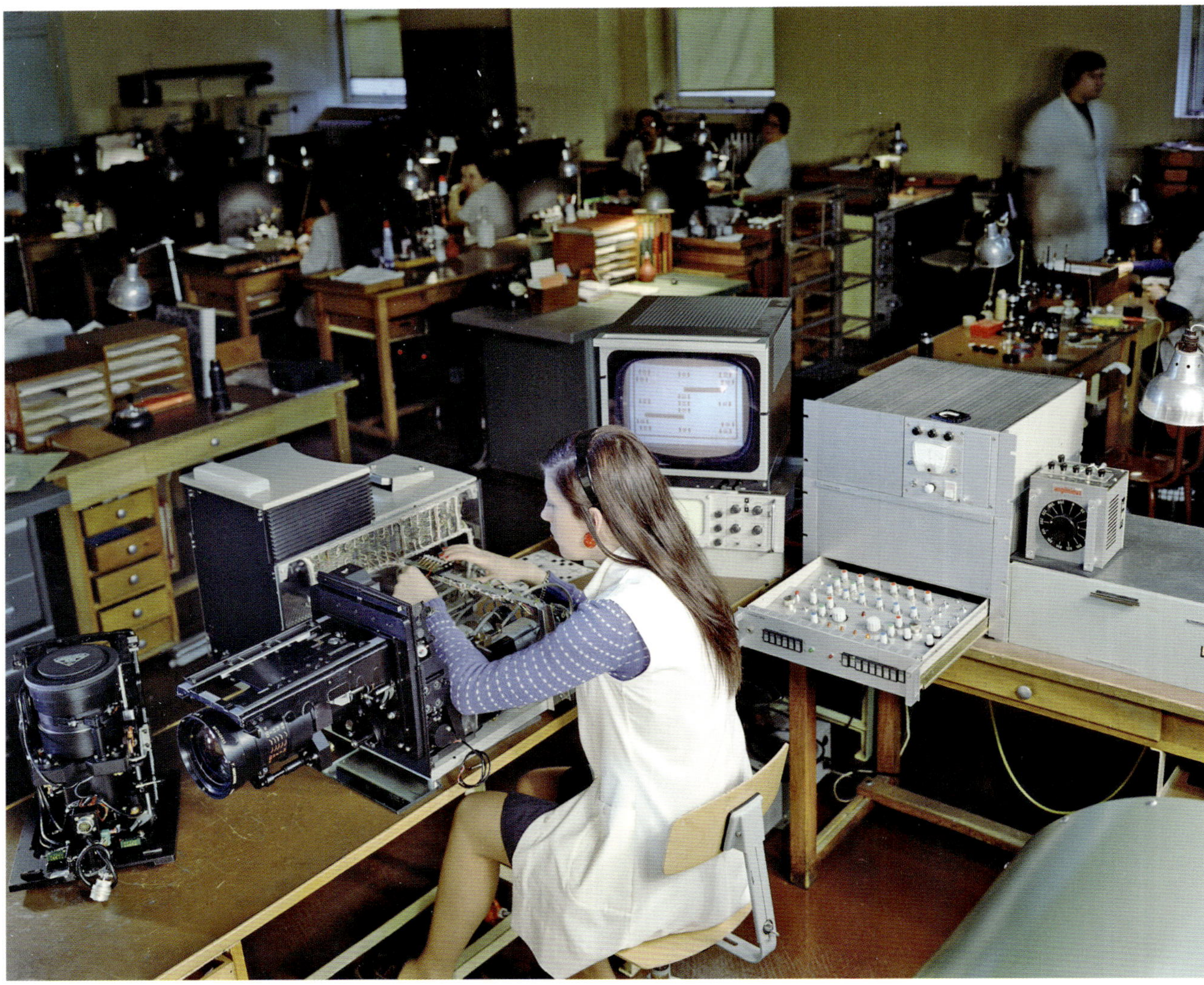

left
Electronic check of a television zoom lens and its optical separator at Angénieux (photo by Ito Josué, 1974).

below
Angénieux diascope, no. 1276807, 1970, in its carrying case (collection of the city of Saint-Héand, photo by Éric Perrin).

The Diascope, an Angénieux Invention.
With the sale of electronic cameras carried on the shoulder, Angénieux initially developed an external diascope. The diascope is a piece of optical equipment that projects a color chart for reference, which makes it possible to regulate the position and scanning of each of the three electronic tubes, to optimize focusing and registration. This system was heavy and bulky.
In 1978, Jacques Debize succeeded in integrating the diascope with the lens. The color control pattern could be projected directly at a constant value, inside the lens. It was no longer necessary to use an external test pattern. The diascope disappeared with the advent of CCD cameras. The CCDs were attached to the separator system and therefore could not be adjusted in position.

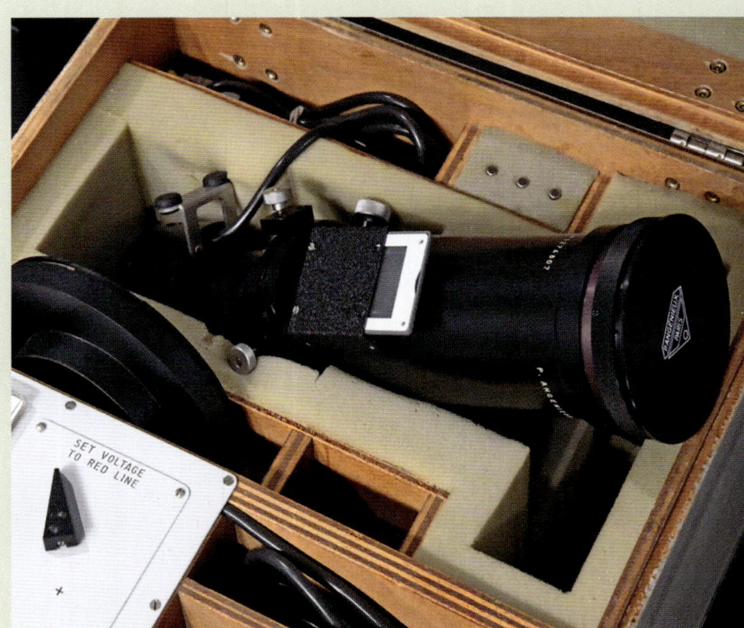

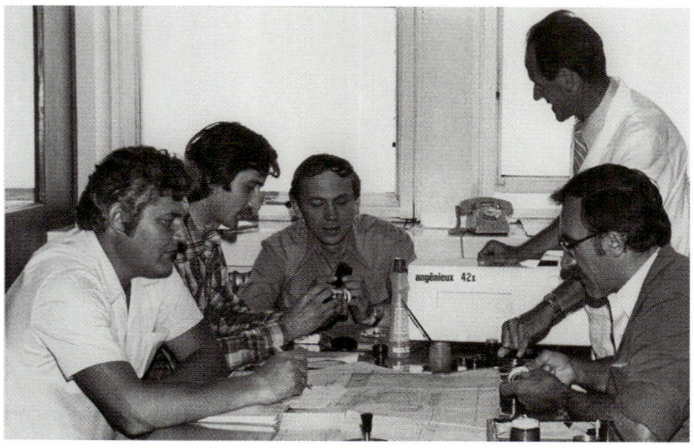

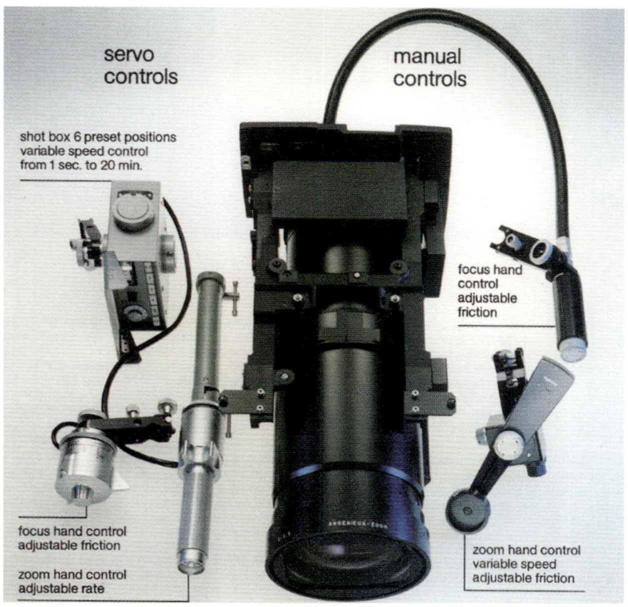

In 1976, Angénieux released a 42x16 for Plumbicon 1-1/4" tubes with an unequalled magnification ratio. This technological feat, which came out of the direct collaboration between Pierre Angénieux and Jacques Debize, offered new possibilities for covering sports events. It was now possible, at 500 meters, to frame the entire stadium as well as half of an athlete's face.

In 1977, Angénieux became involved with the contract THOMSON TSF won to supply television equipment for coverage of the 1980 Moscow Olympics. The job covered 100 42x zoom lenses for THOMSON TTV 1515 cameras. The deliveries were to begin in October 1978. But on January 19, 1978, Mr. Jushevtshus, Deputy Chairman of the Council of Ministers of the USSR and Chairman of the State Committee for television, sent word that the Soviet cameramen's labor organization was refusing to work with the lens on the grounds that it weighed 26.5 kg (58 lbs), while according to the regulations, it could not weigh more than 20kg (44 lbs). The contract would be cancelled if Angénieux did not find a solution by April 15! A special team was formed. Each of the 2000 parts was reviewed to see how a few grams could be saved by changing materials, thinning walls, or drilling holes, etc

On April 11, a new "light' 42x was born: it weighed only 19.3 kg (42-1/2 lbs). Mission accomplished.

In 1986, Angénieux supplied the major Mexican television channel Televisa with all the lenses (15x13 zooms) for the one hundred and sixty Philips LDK 6 1" tube cameras for the World Cup in Mexico City. It was the most significant television contract in Angénieux's history.

above
Work session on the 42x zoom at Angénieux and presentation of the 42x zoom (Angénieux archives).

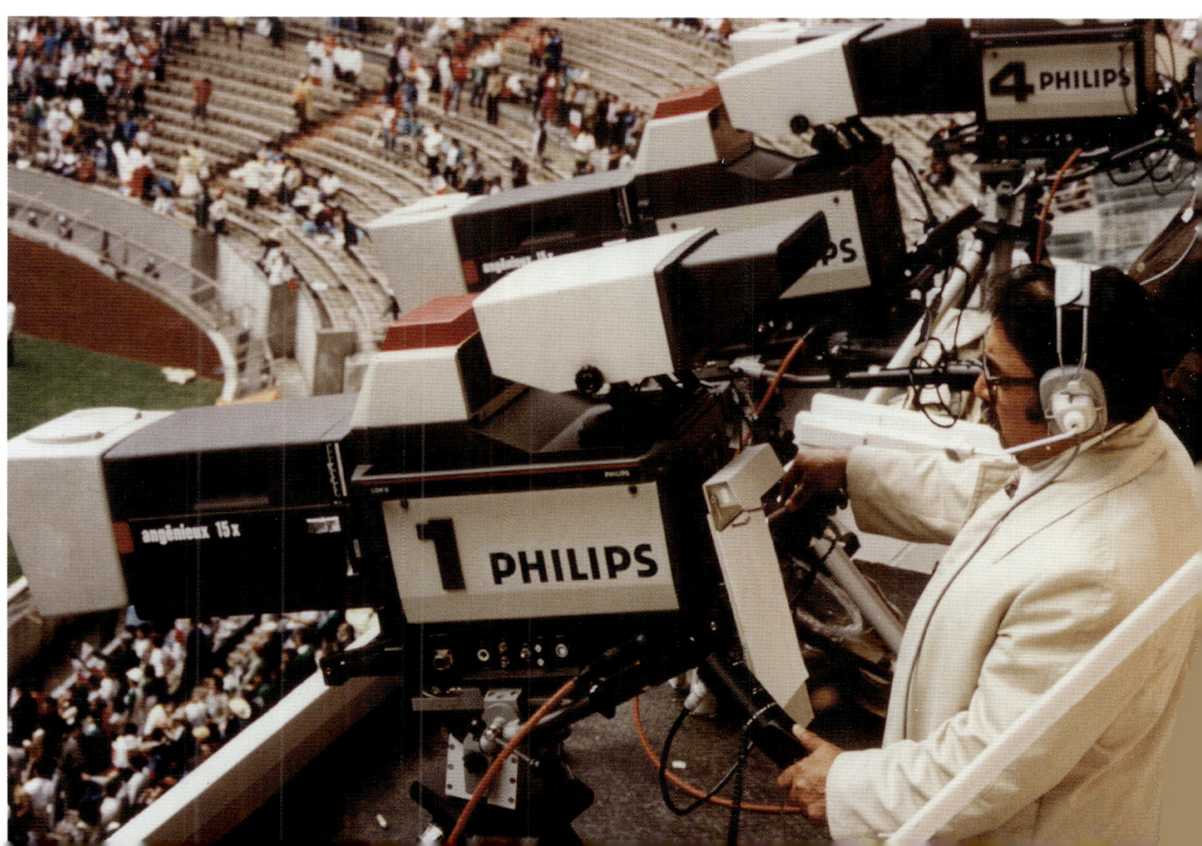

right
Mexico 1986. Angénieux zoom lenses capture images of the World Cup soccer games for the whole world.

CCD (Charge Coupled Device) Video Cameras: New Challenges Overcome by Angénieux Lenses

The mid-1980s saw electronic tube cameras gradually replaced by CCD cameras. The first professional CCD camera, the Sony DXC-3000, appeared in 1986.

The technology of prism separators, together with the CCD, offered new possibilities for reducing the weight of the cameras. With the advent of ENG (Electronic News Gathering) cameras for news broadcasts and reports, 16mm film gradually disappeared from the world of television. It would remain in use until the 1990s in documentaries and independent films. ENG cameras encouraged use of the 2/3" format. Though the first ENG shooting and recording equipment was heavy and bulky, the Betacam recording format developed by Sony beginning in 1982 made way for one-piece cameras that could be used by a single journalist (without needing an audio recorder) for video reporting.

Television lenses for CCD cameras required a new adaptation. It was necessary to go even further in increasing focus distance and in correcting aberrations. Angénieux responded to the demand by placing movable elements in the lens. Now "back focus" made it possible to adjust the zoom and the camera very precisely without loss of image quality.

opposite, top
72x zoom lens at the 1999 Roland-Garros (Angénieux archives).

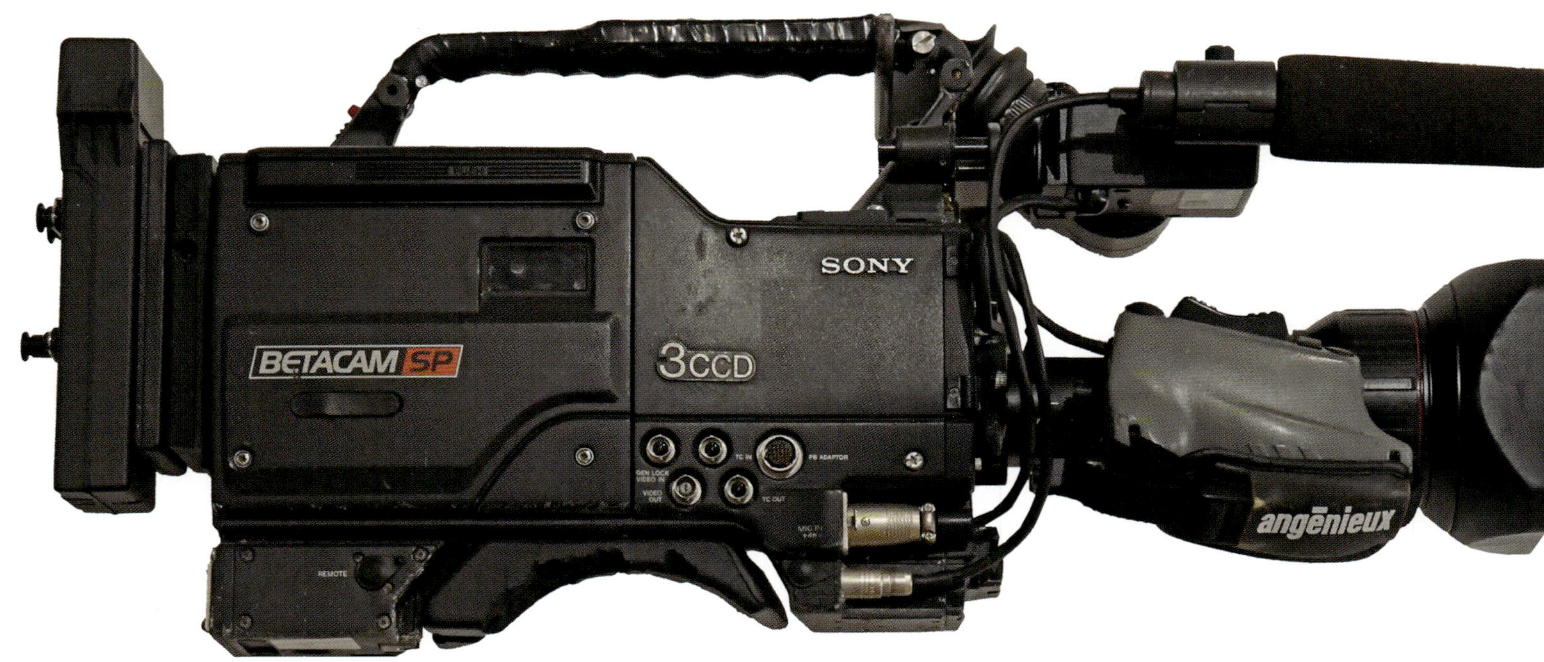

below
Sony Betacam SP color two-in-one camera, 1990, with Angénieux 8.3-125mm, 1:1.7 zoom, type 15x8.3 B, 1 ESM AIF+ (Association du Centre historique de la diffusion radiophonique, photo by Éric Perrin).

Angénieux released a full range of SD (Standard Definition) lenses for 2/3" cameras for applications, as well ENGs 14x9 (1983), 15x6.5 (1988), 20x8.5 (1989), 14x8.5 (1990), 15x8.3 (1992), 22x7.5 (1994), and 12x5.3 (1998), and OB (Outside Broadcast) with higher ratios for sports recordings, including: 40x9.5 (1985), 62x9.5 (1992), and 72x9.5 (1994).

In this range, the 22x, 62x and 72x zooms were particularly innovative. They were able to store position settings (aperture, focal length, focus) and offered improvements in the control system: temperature corrections and back-focus control.

To mitigate the issues of fogging related to significant variations in temperature, Angénieux invented a solution for watertight protection by using an inert gas inside the optics. This solution, tested on the roof of the Saint-Héand factory, was a welcome advantage against the already well-established Japanese competition.

opposite, below
12x5.3 zoom lens at the 1999 Cannes Film Festival, and 62x zoom lens at the Maritime Festival in Brest in 2000 (Angénieux archives, photos by Jacques Durand).

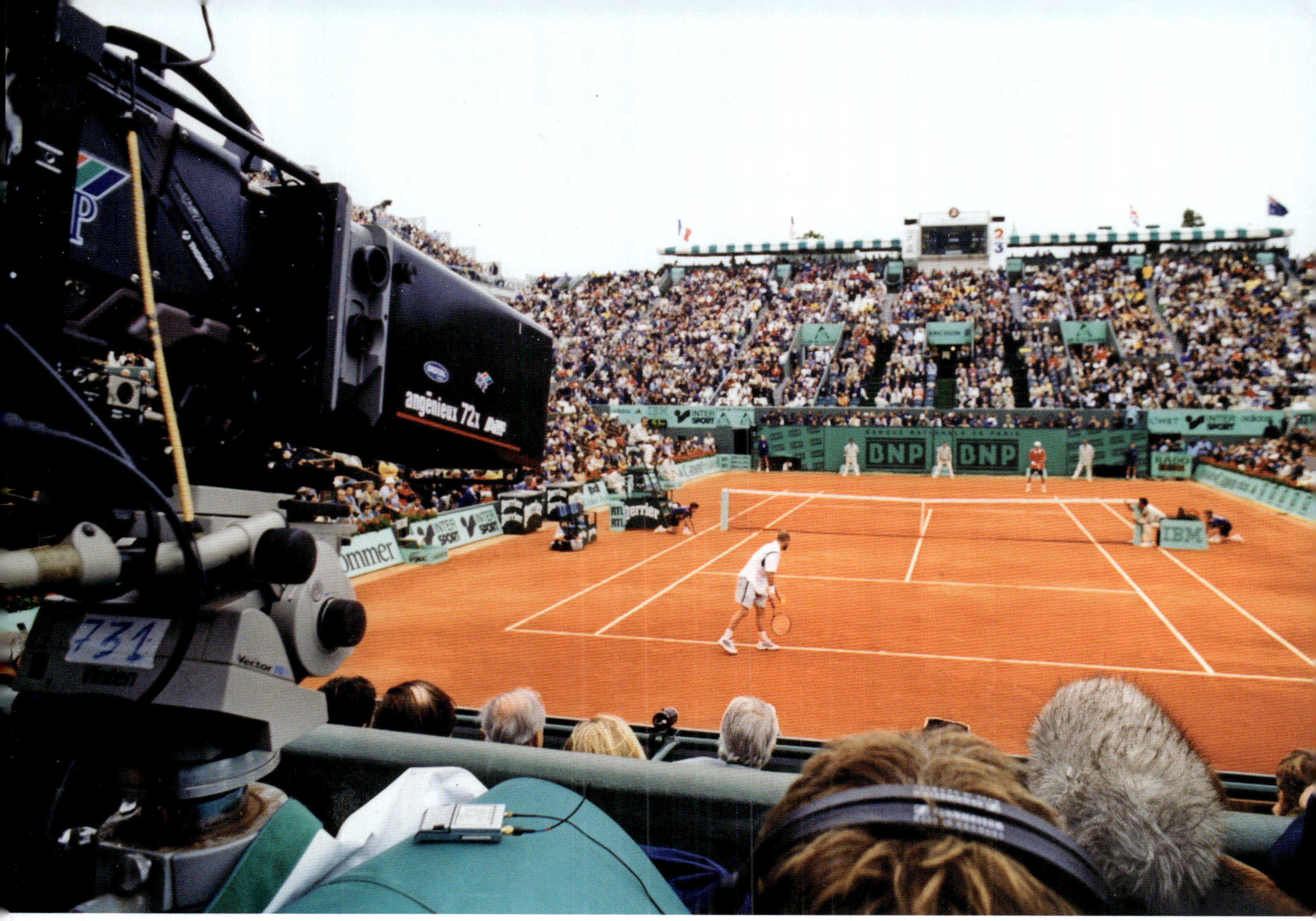
angénieux 72x
BNP
BNP
IBM
731
INTERSPORT

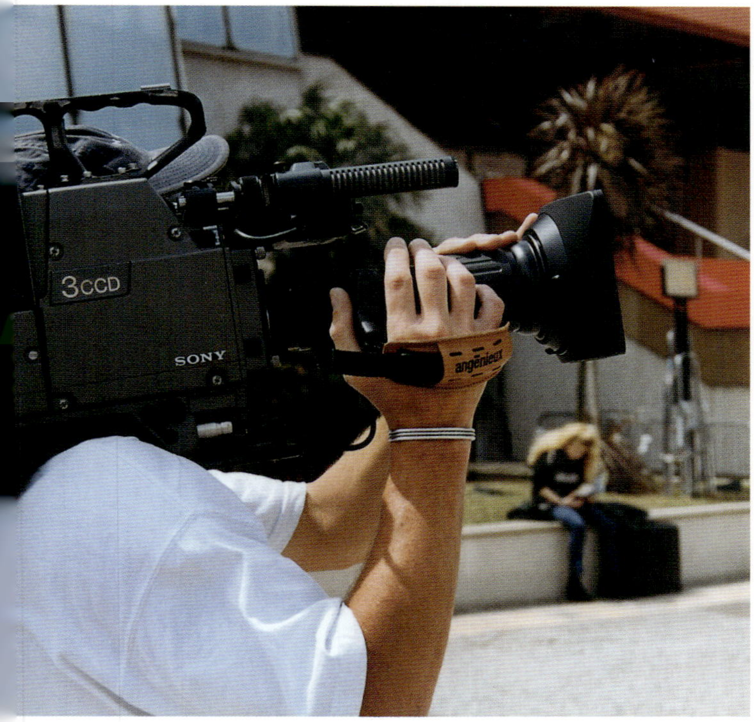
3CCD
SONY
angénieux

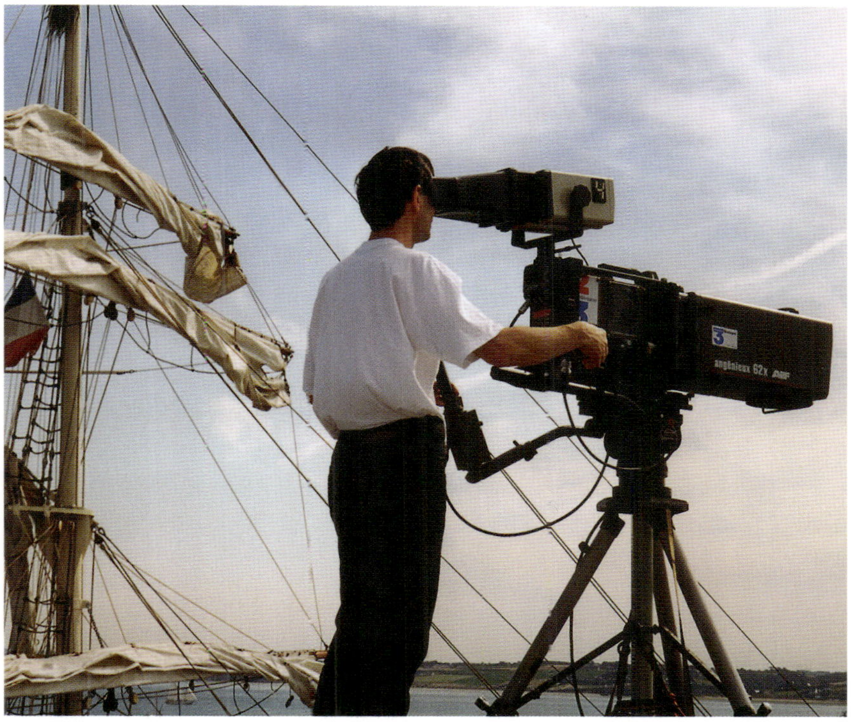
angénieux 62x

CCD and CMOS Sensors
In 1970 Boyle and Smith of Bells Labs completed ten years of research to integrate surfaces that convert light into electricity (optoelectronics) and shift registers for reading in the same chip: the CCD, or *Coupled Charge Device*, was born. It was then possible to imagine a camera that could analyze a video image without the use of a tube and its scanning beam. The first black and white images from a solid-state sensor prototype were introduced in 1973, and the first professional color camera equipped with CCD sensors, from RCA, in 1984. The same year, in Tokyo, Sony introduced a CCD sensor that could finally be described as "usable in broadcasting." From then on progress moved very rapidly, but there were still obstacles to its use on set. Its resolution was still considered too low and, most importantly, it performed poorly under strong illumination. Studio cameras would have to wait until the late 1980s for the performance of solid state sensors to reach that of the tubes used at the time.
Today, production of CCD sensors has disappeared. The semiconductor industry has shifted to CMOS technology, which now equips almost all television and motion picture cameras.

The Japanese mastered the technology of color separators, CCDs, and, before long, motorization (rocker switches).
In 1989, 48 zoom lenses (twenty-four 40x and twenty-four 20x), were delivered to the RAI, the Italian national channel, for the athletics championships in Rome. The lenses were equipped with a control system using a microprocessor, a world first for television lenses. They made it possible, among other things, to have a focus distance of 0, against the front lens.

New Context in the Era of HDTV (High Definition Television)
Since 1972, NHK (the public company that manages the Japanese public radio and television stations) embarked on research on what would become HDTV, financed by public funds.
In 1985, RCA left television as the Japanese competition took the lead. The HD market would now play out between the Europeans and Japanese.
In May of 1986, the Japanese addressed the assembly of the CCIR (International Radio Consultative Committee) in Dubrovnik to propose their HiVision production system, at 1125 lines and 60 Hz, and their Muse distribution standard as the global standard for HDTV. To forestall this proposal, Europe undertook to adopt an HD standard at 1250 lines (2x625) and 50 Hz.
In 1988, at the IBC conference in Brighton, with the help of Thomson, Philips, and the European Eureka program, a high definition system (HD1250 16/9) was proposed, to challenge the Muse broadcast standard. Angénieux was also involved in the Eureka program. The HD experimentation was done on analog cameras with 1" tubes. In fact, CCD sensors were not yet able to accept HD, and would not be until 1992. The first 1" HD CCD cameras were the Sony HDC-500 and the Philips LDK 9000. The first 2/3" HD CCD camera was the Sony HDC-700/750, in 1996.
In 1986, Angénieux tasked Gérard Corbasson, then technical director of the company,

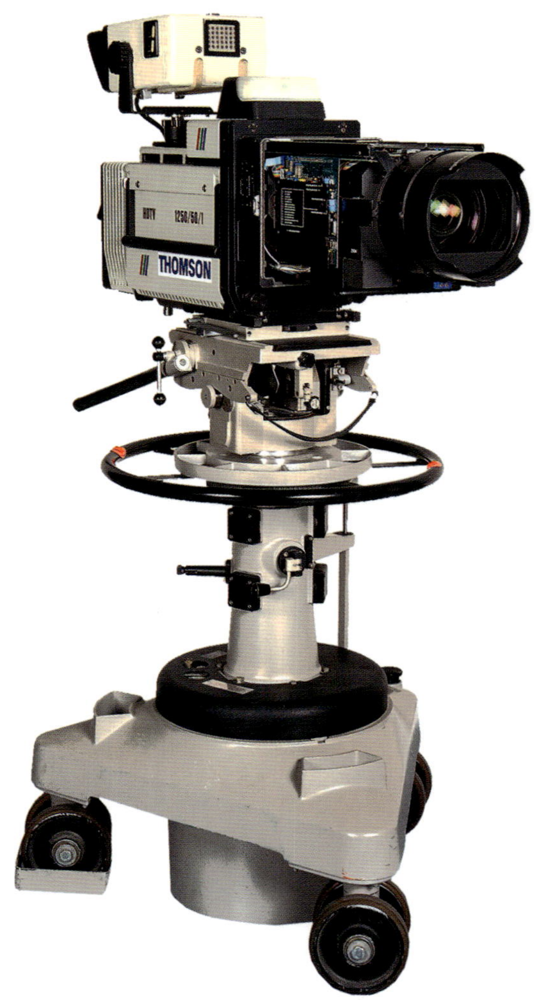

Angénieux E6 HD 15x12 zoom lens, no. 1 539 618, F 12-180mm f. 1.5 on a Thomson camera (collection of the Cinémathèque Française, photo by Stéphane Dabrowski).

The Transition to 16:9

Beginning in the 1990s, TV display format changed from 4:3 to 16:9.
16:9 was adopted to offer a compromise that would benefit in broadcasting films on television. This format was better suited to the human eye's "panoramic" vision.
Angénieux filed a patent on November 5, 1990, for an "anamorphic lens device with variable focal length."
This zoom lens made it possible to widen the field of view to a 16:9 ratio by means of three cylindrical lenses. The image on the CCD was thus compressed horizontally and then re-expanded electronically when reproduced.

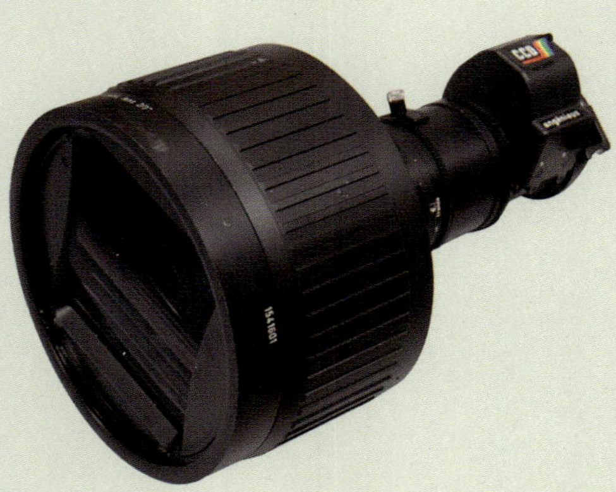

with creating a subsidiary in Japan and thus promoting industrial and commercial relations with the Japanese camera manufacturers. For these cameras, Angénieux developed a 15x12 HD. This prototype, as well as many 10x40 SDs, also fitted for the cameras of the SFP (Société Française de Production) would be used to broadcast the Olympic Games in Albertville in 1992.
A few months later, the Summer Olympic Games in Barcelona led to new experimentation in HD.
For Angénieux as a lens manufacturer, the move to HD imposed new quality requirements for zoom lenses: better selection of glass lens elements, lower levels of aberrations, finer tolerances, and

The First HD Video Zoom Lens for Movies

In the early 1990s, the 16mm format was disappearing from television. Along with its 2/3" range, Angénieux also developed an 11.5-138 HR zoom lens for the Super 16mm format, which was appreciated for its better image quality. Then, in 1994, came a wide-angle version, the 7-81 HR.
This 11.5x zoom lens could be fitted with a rear group compatible with 2/3" HD CCD television cameras, thus creating the first HD video zoom lens for the movies, the 11.5x5.3, with film-style ergonomics, designed for new HD video cameras such as the Sony F900 CineAlta and the Thomson Viper which were expected to replace 35mm film cameras.

The Angénieux CinéVidéo Camera

In this changing world, Angénieux even designed its own camera, in collaboration with Thomson. For those accustomed to 16mm, the brand sought to offer a standard television solution closer to their filming style. A digital camera prototype was introduced in 1998 at the IBC exhibition in Amsterdam and in 1999 at the MITIC (Marché International des Techniques et de l'Innovation du Cinéma) in Cannes. The CinéVidéo was developed under the ADTT (Advanced Digital Television Technology) program. The camera's ergonomics were closer to those of film cameras than camcorders. It combined a 2/3" format with a prism, an optical converter for film lenses and a PL mount. The use of the prism limited the aperture for the optics, and therefore the creative options for filmmakers. 2/3" would not become an established format for movies. The camera would never be marketed.

First presentation
of the Ciné Digital
camera at IBC 1998.

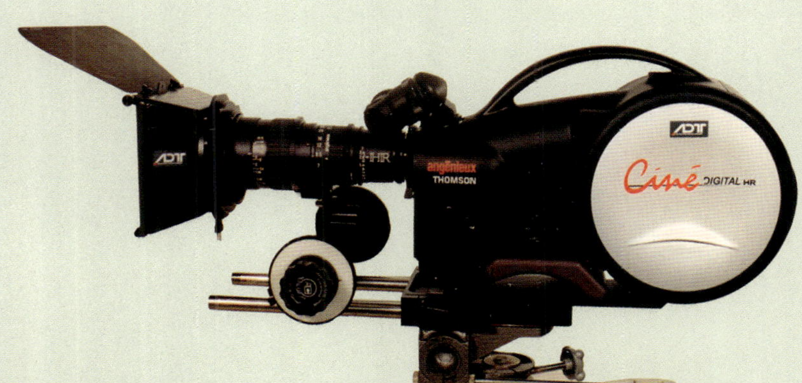

below and right
1" HD 18x12 zoom lens on Philips camera and 40x HD zoom on Thomson camera at the Olympic Games in Barcelona in 1992 (Angénieux archives, photo by Jean-Marc Bouchut).

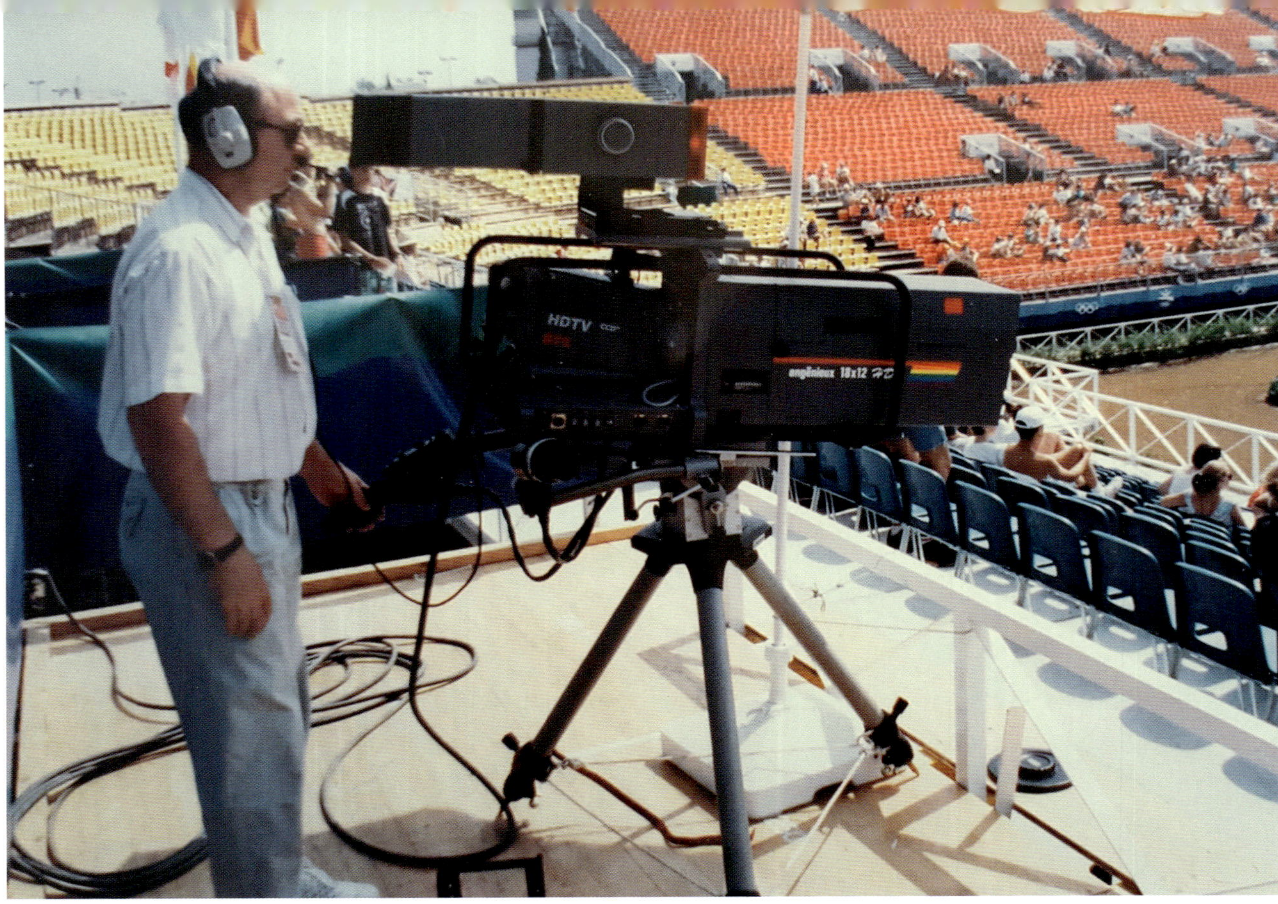

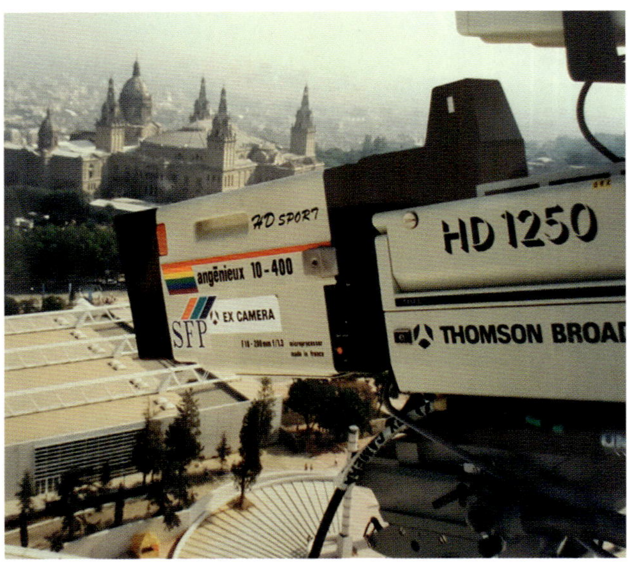

below
22x and 62x zoom lenses at the World Championships in Athletics in Athens in 1997 (photos by Marc Thelisson).

optimization of the manufacturing process. Jacques Debize felt that the Japanese, who had control of the market, would set the HD color standards that would give them the advantage. "Japanese manufacturers have set HD color standards compatible with the quality of their fixed-focus lenses, which had residual longitudinal chromatism. Without those standards, HD quality could have been better."

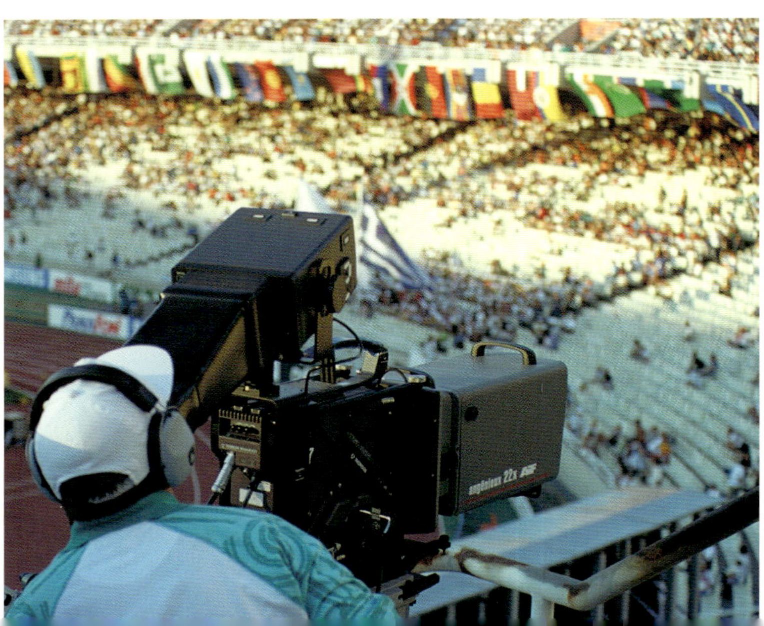

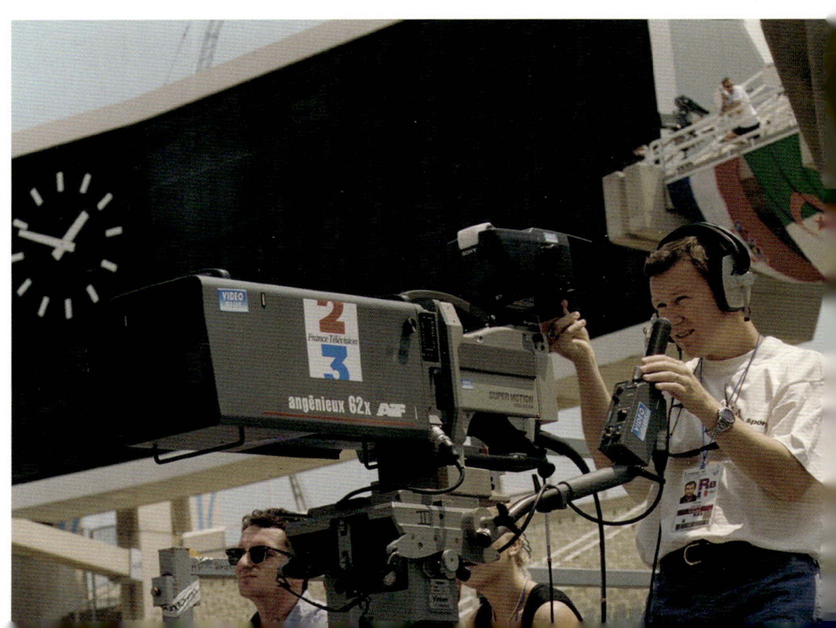

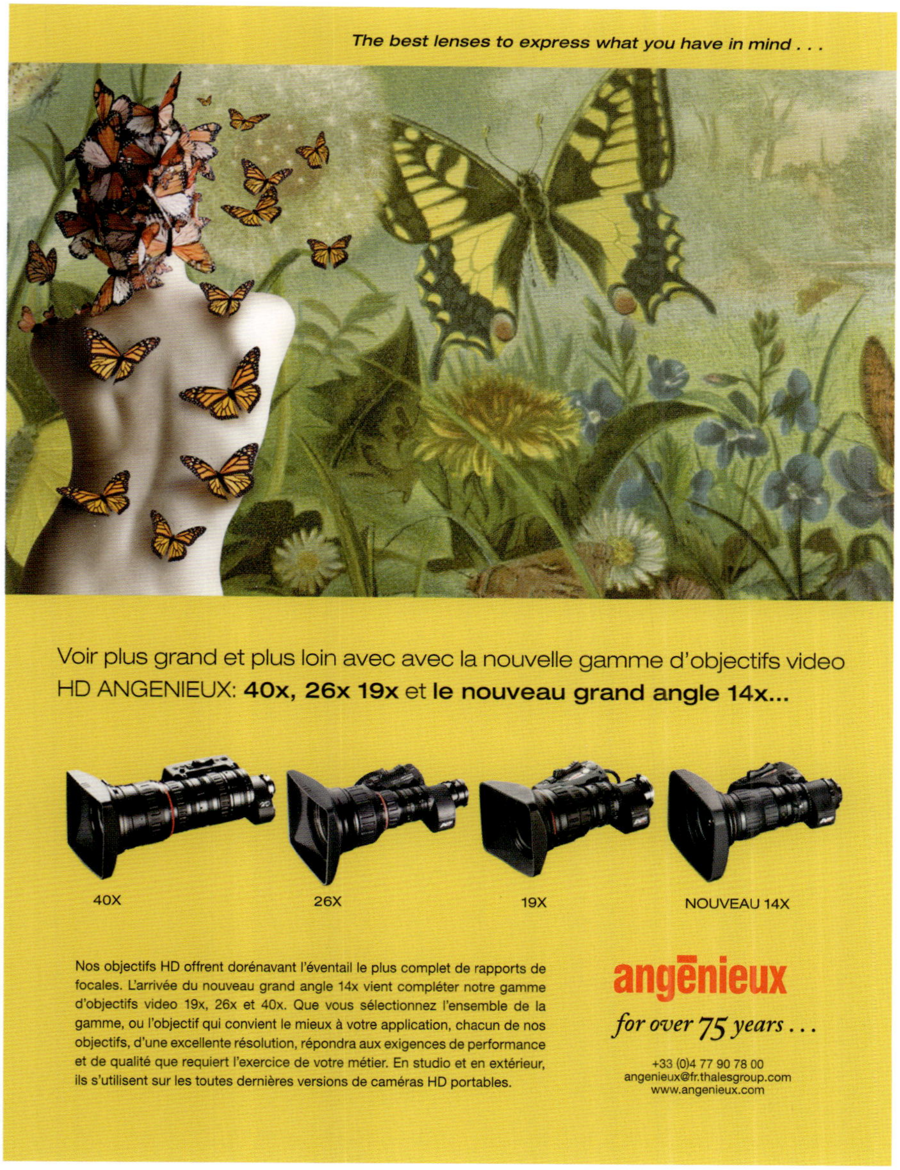

Angénieux advertisement, 2009.

A Line for HD Television

Between 1990 and 1994, Angénieux created a complete range of HD lenses for 1" CCD cameras: 18x12, 10x12, and 32x15.5. Once 2/3" CCD HD technology was brought to completion, in the early 2000s, Angénieux would have a full range of SD and HD lenses for 2/3" portable cameras: 10x5.3 HD wide-angle (adapted from the 12x5.3 SD), 13x4.1 HD (2009) and 19x7.3 HD (2005) standards, 26x7.8 HD telephoto (2000), and 40x15 HD (adapted from the 40x11 SD).

Angénieux also optimized its range of SD OB (Outside Broadcast) 22x, 62x, and 72x lenses and offered the 20x, 60x, and 70x for 2/3" HDTV.

The Challenging Switch to Digital HDTV

The move to HD took place amid a technological transition. The simultaneous arrival of digital technology (digitization and compression of the electrical signal from the light transmitted by the lens) disrupted the spread of high definition analog. The reasons for the switch to digital television in standard definition – while analog HDTV was ready – were primarily economic. The cluttered landscape of analogue terrestrial television did not permit rolling out a large bundle of channels. Americans, unhappy with their NTSC analog standard – which some gleefully referred to as "Never Twice the Same Color" – were also pushing for the switch to digital television. Strategies were redirected.

The analog HD TV of the 1990s was much too expensive and would remain at the experimental level. In 1994, the Japanese abandoned their Muse HDTV distribution standard and Angénieux closed its Japanese subsidiary.

After the 1992 Albertville Olympics, the manufacturers Thomson and Philips gave up their project. But they remained with Europe's Eureka program and now focused on image compression, which was necessary for the expansion of digital television.

While the transition to digital television did not impact Angénieux in technological terms, it delayed the establishment of high definition standards and created a wait-and-see policy on the part of television channels, which no longer knew which equipment to invest in. It was only in 1998 that a single global standard for 1080 line digital HDTV was adopted. By the mid-2000s, HDTV was up and running practically everywhere in the world.

This situation, coupled with Japan's preeminence in the world of broadcast cameras, which were ever more frequently being fitted with Japanese lenses, brought Angénieux to a strategic decision. Production of broadcast lenses was stopped once and for all at the end of 2010.

Angénieux maintained an interest in the 3D HDTV market.

In collaboration with the French start-up Binocle, Angénieux worked on selling an ABLive

 system to facilitate and optimize the recording and processing of live 3D content. This completely innovative solution now made it possible to expertly handle all issues related to stereoscopy, especially in night vision.
Development at Angénieux's centered once again on motion pictures.

The Worlds of Television and Film Reunited in the All-Digital Era

In the age of digital television, 2/3" remains the preferred format in broadcasting (news, reality TV, game shows, sports broadcasts).
The high-end television sectors (films and commercials), the prerogative of filmmakers, remained faithful to 35mm film.
It was not until the mid-2000s, with the arrival of digital cameras with large sensors equivalent in size to the film format, that we would see movies make the jump from film to digital (see chapter 6).
All 35mm film optics were now compatible with these new generations of digital cameras. In this new universe, the boundaries between television and cinema were blurring, and evolving extremely quickly.
The look of film had become accessible to the digital world; it was no longer reserved only for movies and advertisements.

In 2005, Angénieux received an **Emmy Award** from the National Academy of Television Arts and Sciences (NATAS) for its work on the standardization of SD and then HD zoom lens/camera interfaces, in particular within the SMPTE (Society of Motion Picture and Television Engineers) committees.

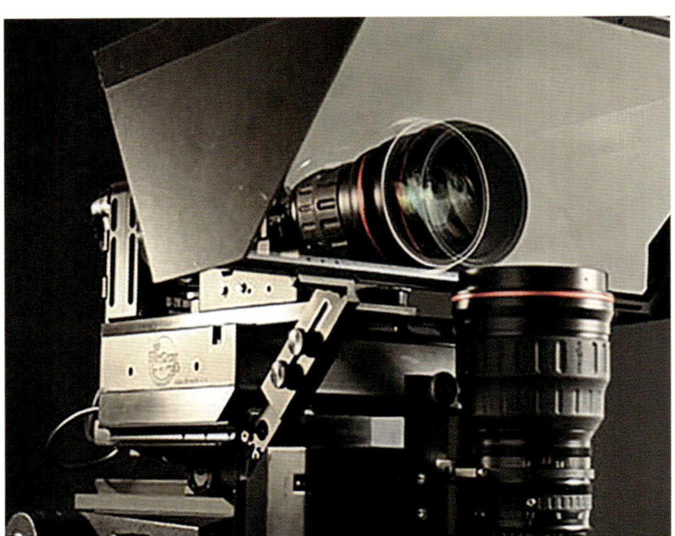

left
3D shooting system with Angénieux zoom lenses (Binocle and Angénieux archives).

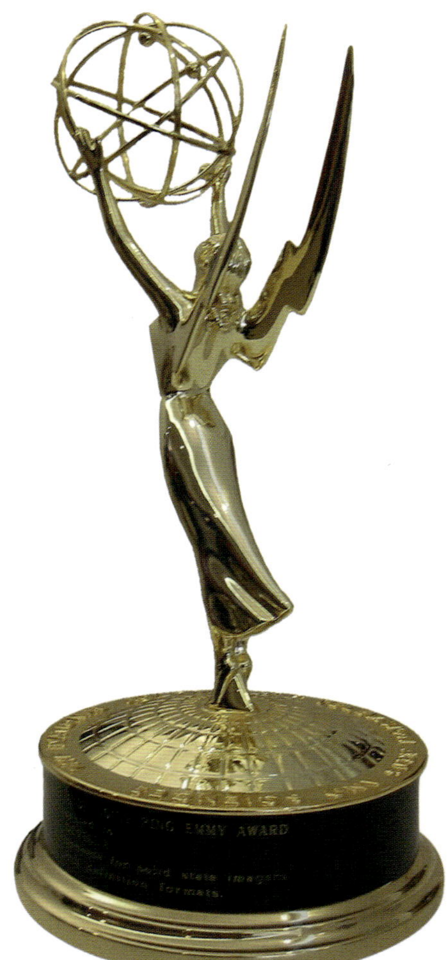

right
Emmy Award
© NATAS

Optimo 24-290 on the set of *Forever* Season 1 - "The Pugilist Break", a series broadcasted on ABC television channel. (photo by John Clifford/ABC via Getty Images).

Optimo Angénieux zooms are part of the shooting of television series: *Game Of Thrones, Westworld, Doctor Who, Better Call Saul, American Horror Story, Criminal Minds, Preacher, Homeland, Veep, The Magicians, Star Trek Discovery.* They were also on *Breaking Bad, Dr. House, Bones, 11.22.63, Community, The Strain, Downtown Abbey, Luther, Rise* and many others.

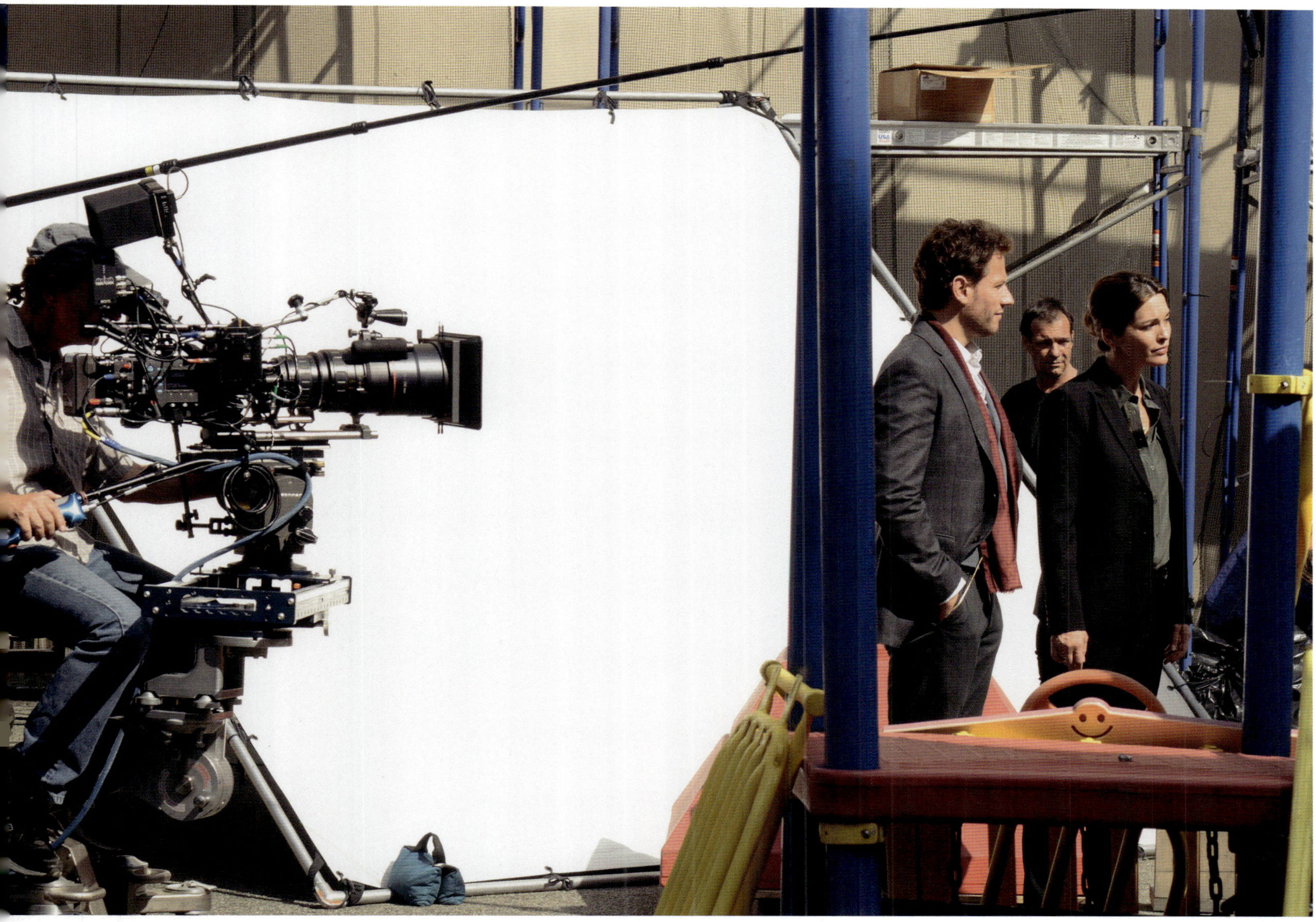

« We cover all kinds of high-end TV production from documentary to music video. A fare portion of our activity is focused on live concert production, be it live on-air, or pre-recorded. We work with a lot of high-profile artists who are always looking for a better visual expression, to differentiate their work from others. Ever since digital cinema cameras became available in the market, I have been exploring how to blend cinematic expressions with live production. My first attempt was to shoot live events in 1080/24P rather than the normal 1080/60i, and then came larger sensor cameras based on S35 format. Obviously S35 cameras allow a significantly better control over depth of field, but at the same time it reveals the different characteristics of each lens we use. Focusing becomes more critical, even though we are primarily shooting in 1080 HD.
We do receive demands for 4K & sometimes 8K production, which make focusing even more difficult. In 2018 we are seeing full frame cameras coming into the market. Perhaps my next challenge is to produce an entire concert in anamorphic. It would be very interesting to see how the anamorphic look can augment live visual experience.
I am a big fan of Angénieux zooms for various reasons. Since we are covering live events, use of prime lenses is not an option. We have tried S35 zooms from all manufacturers, and Angénieux lenses perfectly meet my criteria which are; 1) A high zoom ratio with constant T-stop, 2) beautiful flare characteristics, and 3) a gentle roll off between the area in focus and out of focus.
The flare characteristics are particularly important for live stage production, as we have very little control over lighting positions. The rainbow-like flare on the Optimo 17-80 and 19.5-94 are simply beautiful, a distinctive feature unique to Angénieux lenses. The whole production crew and even the artists can tell the shot was made on quality optics. We own two 12x Optimos, both the 24-290 and the 28-430.
We also have a veteran 25-250HR, which still works fine for HDTV production. Recently we shot live concerts with the band Oasis, B`z, Mister Children, and the MTV Japan Awards. »

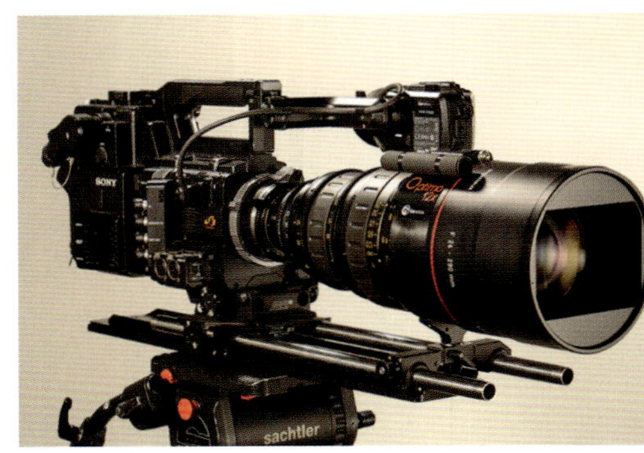

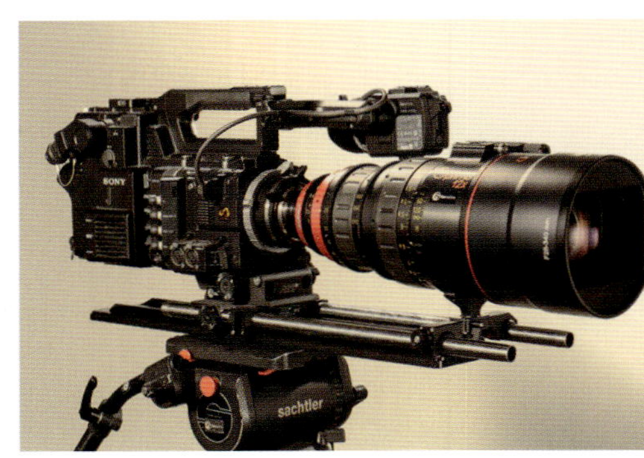

Yuji Morihara

Executive Vice President / Director of Photography, Gonshiro co. Ltd.

Cinema optics and Angénieux zoom lenses would be used to shoot a broader range of television content: documentaries, filmings of cultural events (concerts, operas, etc.), and video clips, but also and above all, television series, whose quality now appealed to an ever expanding audience.

At the 2006 NAB Show, the newcomer American camera manufacturer Red Digital Cinema presented the Red One camera with a 35mm CMOS sensor. The success of this camera was the real digital turning point for the cinema. All 35mm film optics are now compatible with this new generation of digital cameras.

4K, 8K, the Race to Higher Resolution

The progress of digital camera sensors from CCD to CMOS, combined with increased speed and capacity of smaller and affordable memory cards, is helping to significantly increase the performance and reduce the price of digital equipment. With the increase in sizes of the sensors and the decrease in pixel pitch, the resolution of the camera system has increased, making HD (equivalent to 2K), 4K, and even 8K possible. These new digital cameras were more affordable and became extremely popular replacing traditional film cameras, improving the production quality across the industry from high end feature production, to budget conscious independent production.

In parallel to camera development, the methods to deliver content are also changing. New producers and content providers, known as OTT (Over The Top) media services, such as Amazon, Netflix, Apple and many others, has accelerated the demand for high quality cinematic content. As many distributors are mandating 4K resolution, the need for high-performance cinema optics has increased significantly.

Optimo 24-290 and 28-340 became an industry favorite for high quality 4K live stage productions. Here they are attached to the Sony F55 camera at a leading production company in Tokyo: Gonshiro. (photo by Gonshiro).

However, the proliferation of digital cameras caused a problem. Given the lack of an industry standard for camera sensor size, every manufacturer, and sometimes different camera models manufactured by the same company, came with sensors of slightly different sizes. This was a major problem for equipment owners and rental companies, since a lens that used to work on a certain camera may not work on the next camera that has a slightly larger sensor.

To respond to this problem, in 2016 Angénieux began marketing a completely new range of lenses designed around an innovative solution called IRO Technology. IRO (Interchangeable Rear Optics) Technology made it possible, simply by changing the rear group, to adapt the lens to different image formats, such as S35 or Full Frame Format. The market immediately embraced this solution, and this key technology has contributed to the brand's recent success. The IRO technology was first used in Type EZ zoom lenses, two specially designed affordable compact zooms destined to small budget productions. It is also found in the brand new 12x zoom in the Optimo range, the Optimo Ultra 12x, which offers three different coverage options: S35, U35 and Full Frame formats[5].

Today in broadcasting, the race to 8K has started. 8K satellite broadcasting has started in Japan since December 2018, and the 2020 summer Olympic games will be covered in 8K. Neighboring countries Korea and China are also experimenting 8K technology. Angénieux is well prepared to continue supporting the industry, since the resolution of Optimo zoom lenses exceeds the resolving power of today's 8K cameras.

[5] The Optimo Ultra 12x can be configured for S35, U35, or Full Format, to provide coverage of images 31.1mm, 34.6mm, or 46.3mm on the diagonal.

Space

Angénieux and Space Exploration

The fame that the Angénieux brand had won across the Atlantic, its mastery of the f/0.95 aperture, and its television work since 1958 with American camera manufacturer RCA explain the reasons that made NASA choose lenses from the small French company.
On July 31, 1964, Angénieux fixed-focal-length optics travelling aboard the *Ranger* 7 took the first close-range images of the Moon.
On July 20, 1969, Angénieux optics were aboard the *Apollo 11* mission and took part in the most remarkable television event of all time: mankind's first step on the Moon. Angénieux's contribution to American space exploration definitively made history for the brand.

The adventure of American space exploration began in 1957 with the Soviet launch of the first satellite, known as *Sputnik*. The Americans realized then that they had fallen behind their rivals. On July 29, 1958, President Eisenhower signed the law creating the National Aeronautics and Space Administration (NASA) in order to win the "space race" against the USSR. On May 21, 1961, President John F. Kennedy announced the launch of the *Apollo* program, whose mission was to put a man on the moon before the end of the decade. To accomplish this feat, NASA launched several preparatory programs for future lunar expeditions. The *Gemini* program was to develop techniques for space flight. Other reconnaissance programs, including *Surveyor* and *Ranger* were assigned with mapping landing zones and determining the consistency of the lunar soil. To reach the Moon, NASA officials decided to use the daring method of a rendezvous in lunar orbit, which requires two spacecraft: a Command Module (CM) and a Lunar Module (LM); only the second of these would land on the Moon.

For Angénieux, it all began with the Ranger Program

The US's 1961-1965 *Ranger* program provided for a series of nine unmanned missions in preparation for the *Apollo* program. Its objective was to obtain the first close-range images of the surface of the Moon and to provide extremely important topographical information for the *Surveyor* and *Apollo* projects.
The *Ranger* probes were created to take images of the surface of the Moon and transmit them to the Earth before they were destroyed on impact with the lunar surface. A series of incidents led to the failure of the first six flights.
The *Ranger* 7 probe was launched on July 28, 1964. It was equipped with six cameras from RCA Astro Electronics (a division of RCA).
NASA needed ultra-wide-angle lenses for these cameras in order to get quality images while the probe approached the Moon at great speed. Angénieux's 25mm lens with f/0.95 was selected. These lenses would equip three of the six RCA cameras to make the trip.
Before the *Ranger* 7 crashed onto the Moon on July 31, 1964, its cameras and optics succeeded in sending over 4,300 shots to the station on Earth. Since the signal received was seriously disrupted by the electromagnetic emissions of the sun, NASA digitized these

below
25mm f/0.95 lens on board *Ranger* 7 (photo by Matthew Leeg).

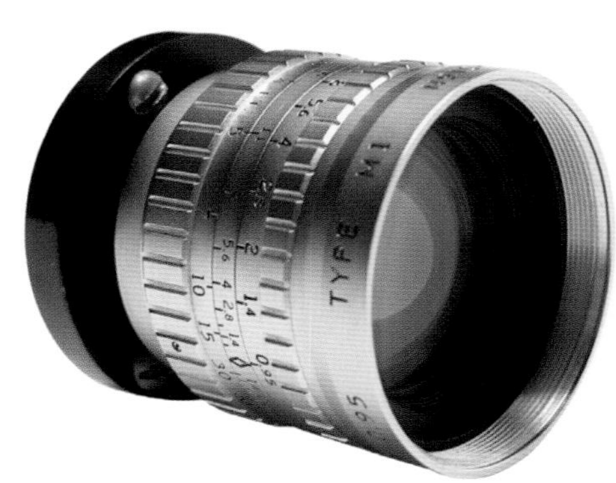

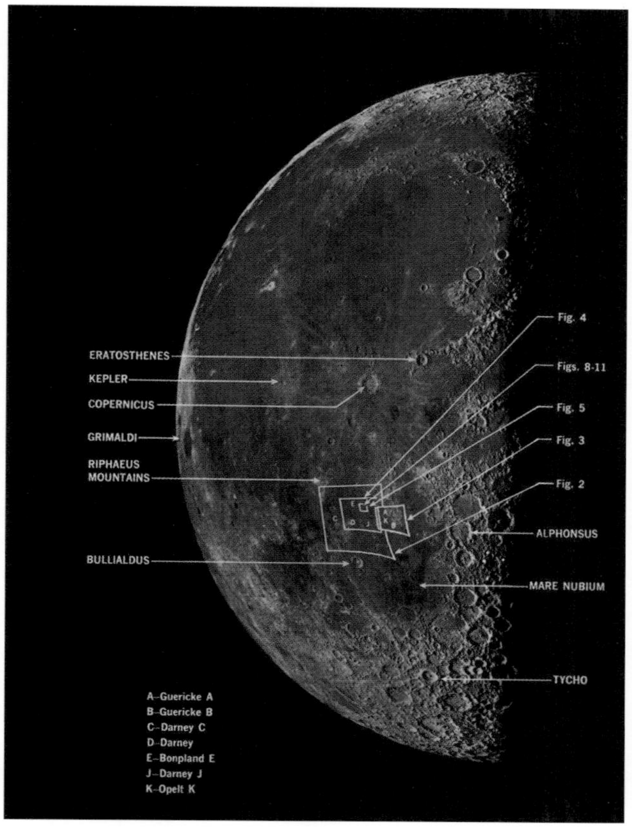

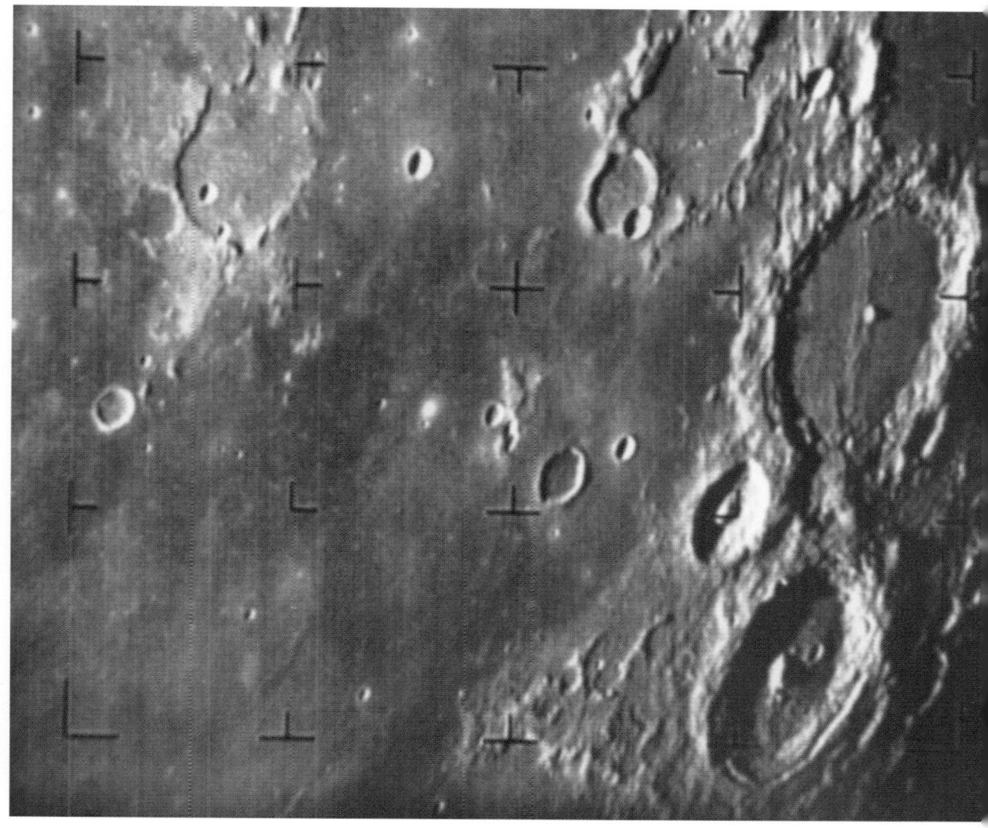

above
On July 31, 1964, Angénieux's 25mm optics captured their first images of the Moon (© NASA).

right
Three of the six cameras on board *Ranger 7* (the bottom ones) have Angénieux 25mm f/0.95 optics. The three cameras above had a B&L 76mm f/2 Super-Baltar lens previously used on *Ranger 6* (photo by Elmer Fred, RCA, published in *RCA Astro-Electronics*, 1965).

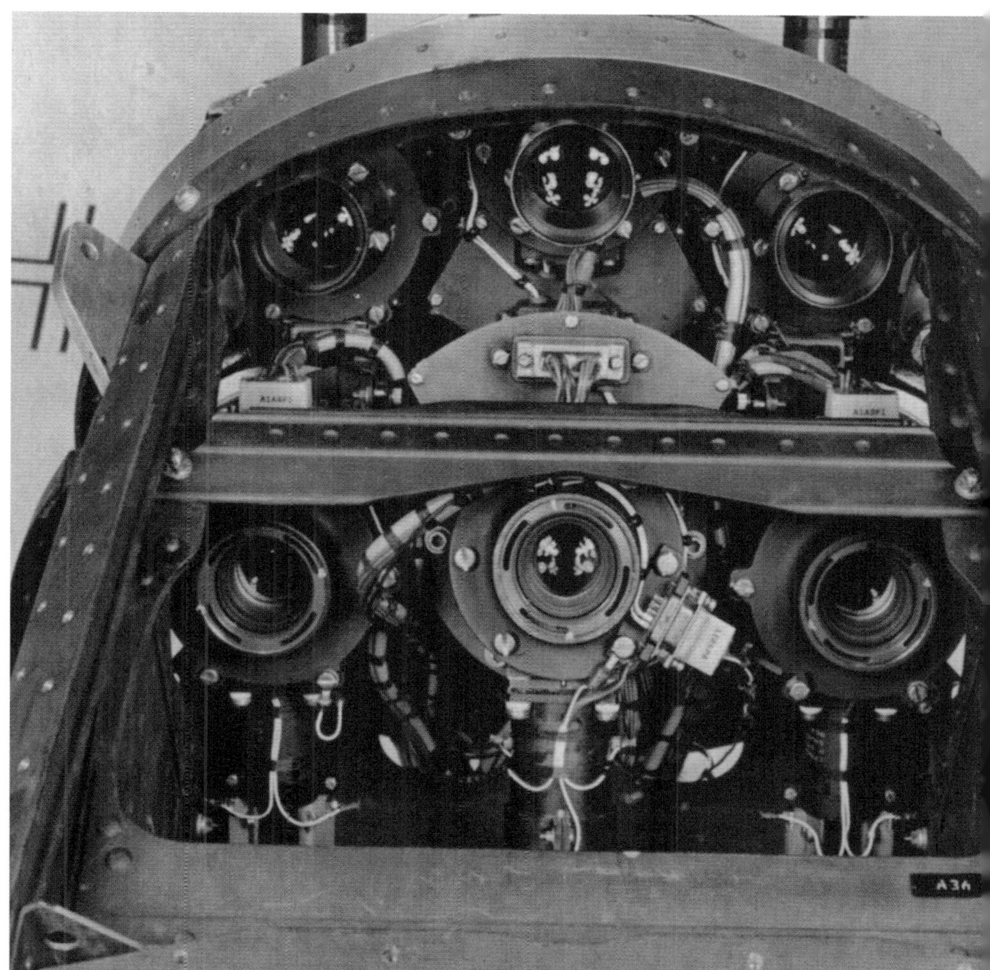

images and processed them with the help of a powerful computer to improve their quality. This operation was a first in digital image processing. The resolution of those images was 1000 times higher than that of images from telescopes on Earth, and revealed information crucial to the missions that would follow. Angénieux optics were again selected for the *Ranger 8* and *9* missions, which also achieved their objectives. The lenses were part of all the successes of the *Ranger* program.

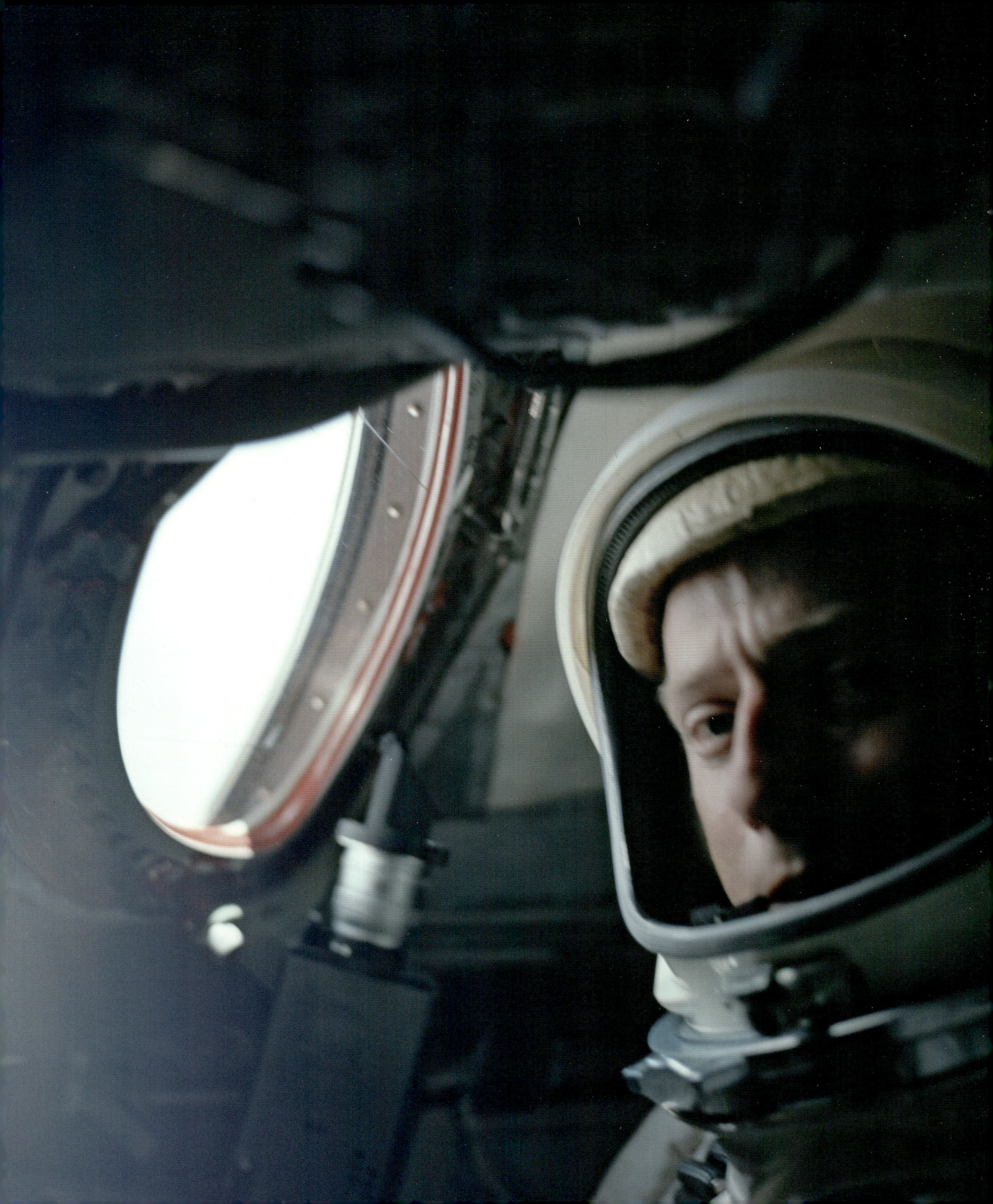

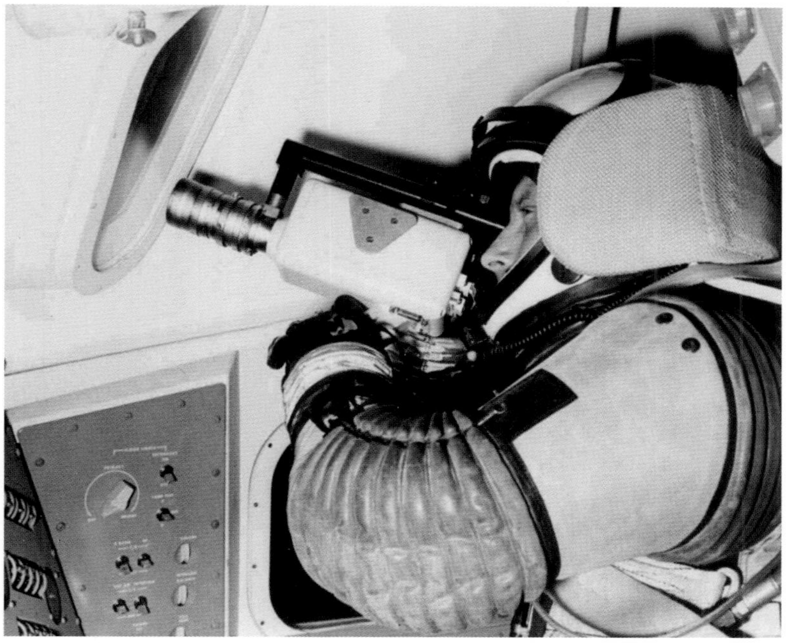

opposite
Charles Conrad (*Gemini 5* mission) with the Maurer 16mm camera with an Angénieux 18mm f/2 lens (© NASA).

above
Apollo 1, photo taken during testing on the ground of the 6x12.5 zoom on a Maurer 16mm camera. *Apollo 1* caught fire before its take-off on January 27, 1967 (© NASA).

Angénieux's Role in the Manned Gemini and Apollo Programs

The first manned flight project NASA developed was the *Mercury* program, which began in 1958 even before the agency was created. It would make it possible to send the first American into space. On May 5, 1961, Alan Shepard completed the first fifteen-minute flight in the capsule *Freedom 7*. This was only a simple suborbital flight because at that time NASA did not have a rocket powerful enough to put a probe into orbit. It was not until the mission *Mercury-Atlas 6* on February 20, 1962, that John Glenn became the first American astronaut to orbit the Earth. Three other manned flights took place in 1962 and 1963.

When the *Mercury* program came to an end in 1963, important aspects of space flight, necessary to complete flights to the Moon, had still not been mastered.
NASA's directors launched the *Gemini* program well ahead of the development of the highly sophisticated vessel for the lunar mission. A dozen capsules were launched between April, 1964, and November, 1966. This intermediate program was to fulfill three objectives: mastering the techniques of positioning, maneuvering and space rendezvous; perfecting working in space during extra-vehicular activity; and studying the consequences of weightlessness on human physiology during long flights.
Onboard these capsules were 16mm Maurer cameras to film the crew's response to weightlessness and in piloting situations. Angeniéux's 25mm f/0.95, 18mm f/2, and 75mm f/2.5 lenses were among the equipment selected for this camera. Notably, the 25mm lens would film the solar eclipse observed from the *Gemini 12* capsule.
At the completion of the *Gemini* program, the United States had caught up with the USSR.

The *Apollo* program began in 1961 and ended in 1975. Its mission: to send a crew of three men to the moon, two of them reaching the surface of the Moon, and to bring them all back safe and sound, using the largest rocket ever built, the *Saturn V* with the ability to launch two space modules (the Command Module and the Lunar Module). The program began badly with the fire in the *Apollo 1*, still on the ground, on January 27, 1967, and the tragic deaths of its crew members, astronauts Virgil Grisson, Edward White, and Roger Chaffee. *Apollo 7*, in October, 1968, was the first new manned mission after *Apollo 1*. Six of the *Apollo* missions (11, 12, 14, 15, 16, and 17) reached the Moon; the disastrous explosion of an oxygen reservoir in the *Apollo 13*'s Service Module forced the astronauts to return to Earth.

July 20, 1969, marked the incredible success of the *Apollo 11* mission. The rocket lifted off from Kennedy Space Center in Florida on July 16, 1969. It carried a crew consisting of Neil Armstrong, commander of the mission and pilot of the Lunar Module, Edwin "Buzz" Aldrin, who would go with Armstrong to the lunar surface, and Michael Collins, pilot of the Control Module that would remain in lunar orbit. Armstrong and Aldrin, after a bit of a hard landing, spent 21 hours and 30 minutes on the Moon and made one two-and-a-half-hour excursion outside the module. After re-launch, the Lunar Module rejoined the Command Module still in lunar orbit. The *Apollo* spacecraft then returned to the route toward the Earth and splashed down, without incident, in the Pacific Ocean, after a flight that had lasted 195 hours.

Angénieux Lenses Have Taken Part in Space Exploration since 1964

The first Angénieux equipment selected by NASA, Angénieux's 25mm f/0.95 wide-angle, ultra-bright prime lens was introduced in 1964 on the *Ranger 7, Ranger 8,* and *Ranger 9* missions, the three successful *Ranger* missions. These lenses recorded their first photos of the Moon on July 31, 1964.
Along with the *Ranger* program, NASA launched the *Gemini* program in 1964. The *Gemini* missions used Maurer 16mm cameras with 18mm f/2.25mm f/0.95 and 75mm f/2.5 prime optics. These optics would again equip the Maurer cameras on the *Apollo 7, 8, 9,* and *10* missions and on the *Apollo 11*, which gave the world the images of Neil Armstrong taking mankind's first steps on the Moon.
Angénieux's 6x film zoom lens was adapted for the NASA missions.
The 6x12.5 was the equipment chosen for the RCA black and white tube camera on *Apollo 1* in 1967. This lens was also fitted to the RCA color camera on the lunar vehicles on the *Apollo 15, 16,* and *17* missions.
The 6x25 zoom was the Angénieux lens most often selected by NASA. It participated in an incredible number of *Apollo* missions. It was on board the Command Modules of the *Apollo 10, 11, 12, 13, 14, 15, 16,* and *17* missions, and also on the Lunar Module for the *Apollo 12, 13,* and *14* missions. On the missions that followed, the transmissions from the lunar surface were made from the lunar rover. The lens was also on the launchpad of the Saturn V rocket for *Apollo 13*.
The 6x25 was again used on the Westinghouse cameras on the *Skylab*, the first US space station, launched in May, 1973. It also played a role in the *Apollo-Soyuz* project in 1975.
The first 6x25s were actually 6x12.5s with a 2x teleconverter. These lenses had focusing rings marked from 12.5 to 75. In 1969 and 1970, to answer NASA's requirements, Jacques Debize redesigned the back of the 6x to make it into a shorter 6x25 to limit the risk of bumping into the astronaut's helmet. This new version was released in July, 1970. The zoom rings were then engraved from 25 to 150.

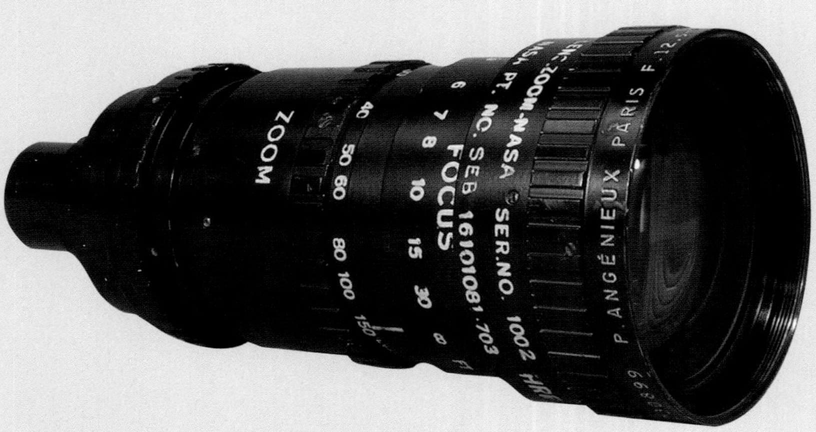

left
Angénieux 6x12.5 zoom lens for NASA, no. 1002 (© NASA).

opposite
Angénieux 6x25 B lens for NASA, no. 1266082, 1970 (Angénieux archives).

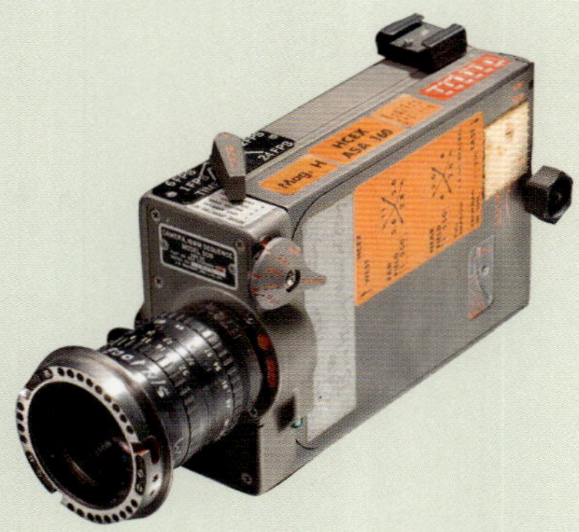

Maurer 16mm film camera with Angénieux 75mm f/2.5 prime optics (© NASA).

To address the space-specific condition of the vacuum, Angénieux developed a concept for lubricating the mechanics, as the usual grease would evaporate in a vacuum and fog the lens elements. New optics treatments were developed to combat solar radiation.

In the early 1980s, Angénieux designed a 3x8.2 and, ten years later, a 15x8.5 for the RCA cameras on board US space shuttles from 1981 to 2011. The 3x8.2 was placed inside the cockpit of the shuttles. It was also used to broadcast Spacelab experiments live. The 15x8.3 was used on the cargo bay's intensified black and white cameras. After the *Columbia* explosion in 2003, one of these cameras was positioned on the Canadian arm to inspect the thermal tiles during the mission. This arm is now used on the ISS, the International Space Station.

In 2005, Angénieux was asked to produce four 150mm prime lenses for NASA's *Dawn* mission to explore Vesta and Ceres, located between Mars and Jupiter.

The Dawn mission ended on November 1, 2018.

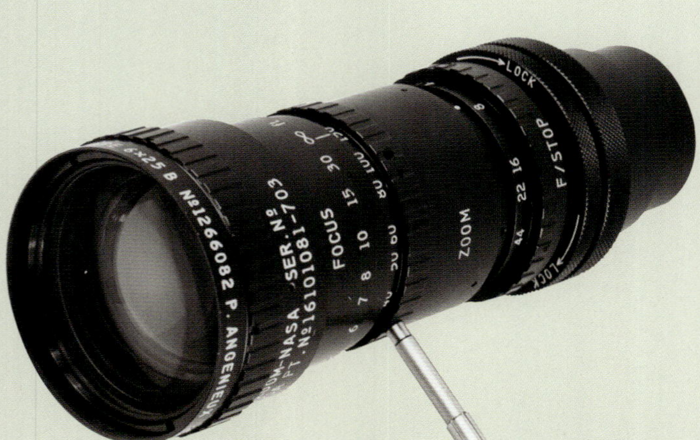

The Choice of Imaging Equipment, a Strategic Communication Issue

In 1961, NASA announced a call for tenders for television camera manufacturers for its *Apollo* missions, so that viewers around the world would be able to follow the American astronauts' achievement live. The demands on such a camera would be many. NASA wanted a camera that was small, reliable in the lunar environment, low in energy consumption and easy for astronauts to operate. RCA Astro Electronics designed a small black and white portable camera called Block 1 for the Control Module. It fitted this camera for **Apollo1** with a dual optics kit: wide-angle fixed-focus optics supplied by Taylor Hobson for images inside the module (take-off and flight phases) and an Angénieux 6x12.5 zoom lens equipped with a magnifier to show either the crew's work inside the module, or the exterior, through the porthole window.

After the *Apollo 1* launchpad fire, NASA concentrated on the project's dependability. The transmission of images for television became secondary or even problematic because of the extra weight to carry.

Manned missions resumed with **Apollo 7**. Black and white cameras from RCA and Maurer were selected for the **Apollo 7**, **8** and **9** missions. On these missions, the RCA camera was simplified (Block 2). It was provided only with American-made fixed-focal-length Fairchild and Argus optics. The Maurer cameras were placed on board the Control Modules. They had French fixed-focal-length 18mm, 25mm, or 75mm optics from Angénieux. The brand's space adventure was continuing!

Color images came with the **Apollo 10**. The manufacturer Westinghouse would join NASA's lunar project, replacing RCA. Indeed, Westinghouse had developed a new color camera using a single SEC (Secondary Electron Conduction) tube originally developed for the US military. This tube was more sensitive than those of earlier cameras and made it possible to extend transmissions of televised images from the lunar morning to the evening. It offered a better view in the shadowy zones around the Lunar Module.

This camera had a filter wheel located behind the tube, allowing RGB color selection. This system, called "Field Sequential," had been developed by

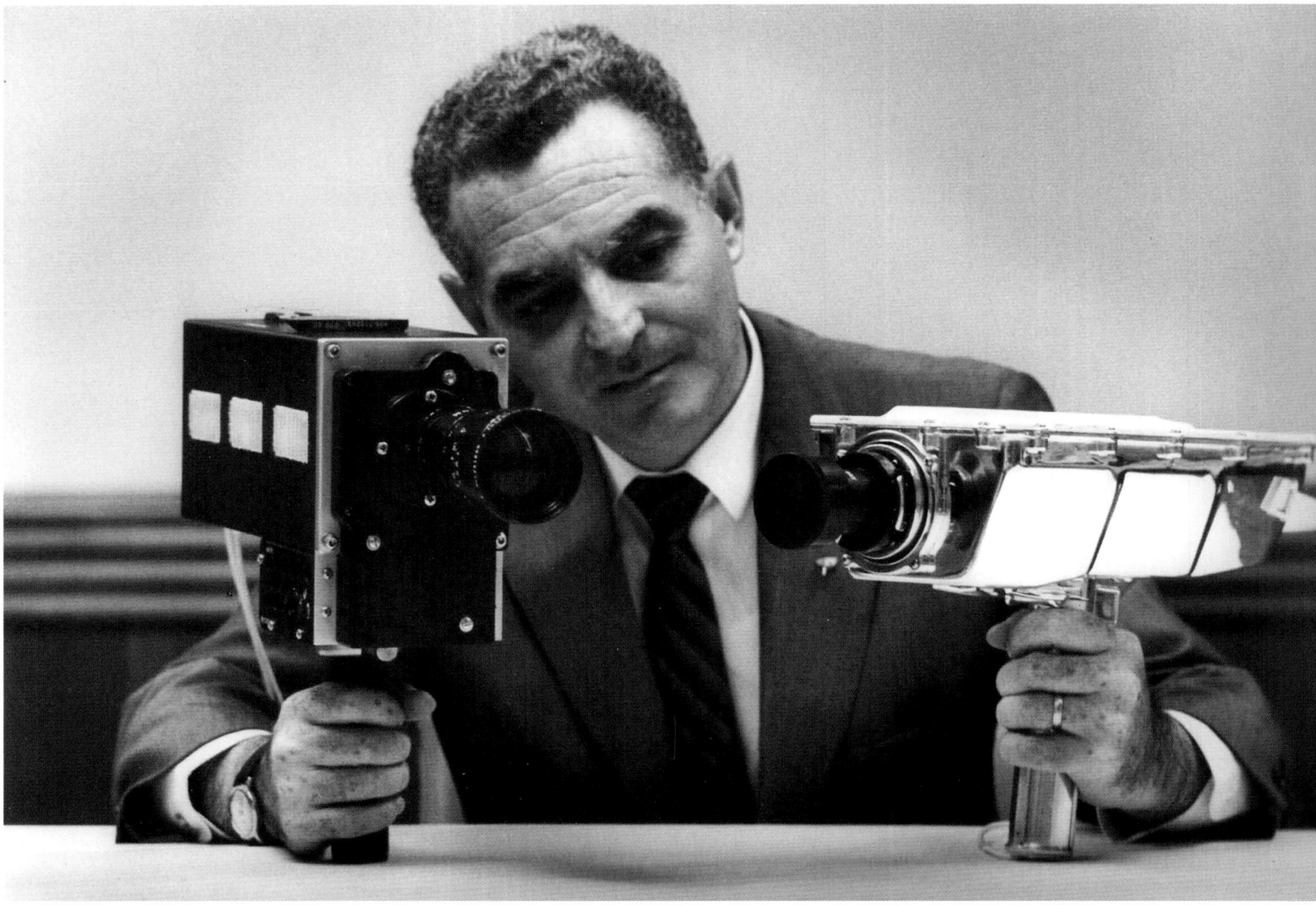

CBS in the 1940s. The camera was more compact and consumed less energy than traditional three-tube color cameras. There was still one problem: the TV signal had to be reconstructed to be compatible with the NTSC signal used for American television broadcasts. NASA therefore used a system with a dual magnetic reader/recorder coupled together, one writing on the magnetic tape from the signal coming from the camera, the other reading it and converting it to the NTSC format of American color television. To facilitate the use of this camera, Westinghouse equipped it with a portable monitor and a 6x (6x25, f/4.4) zoom lens adapted from a 16mm film zoom lens. This adaptation was necessary to cover the sensitive area of the 25mm (on the diagonal) tube.
The first tests of the Westinghouse color camera with an Angénieux 6x25 zoom lens on **Apollo 10** in May, 1969, were successful. The equipment provided the first color images filmed of the crew and the Moon, up close.

For **Apollo 11**, however, NASA decided to shoot man's first steps on the moon live from the lunar module with a black and white Westinghouse camera (equipped with Fairchild optics) to limit the risk of transmission problems. The color camera with the 6x25 stayed in the Control Module and took live color images of Neil Armstrong and Buzz Aldrin on their trip to and from the moon. A 16mm Mauer camera with Angénieux 75mm optics was also on board the Command Module taking images of the Lunar Module's separation and docking maneuvers from the porthole.

above
Westinghouse engineer Stan Lebar holding, in his right hand, a color camera with an Angénieux 6x25mm zoom lens and, in his left hand, a black and white camera (photo provided by Dwight Steven, author of *Live TV from the Moon*).

opposite
On July 21, 1969, Angénieux's entire staff watched the success of the *Apollo 11* mission live on a color television (Angénieux archives).

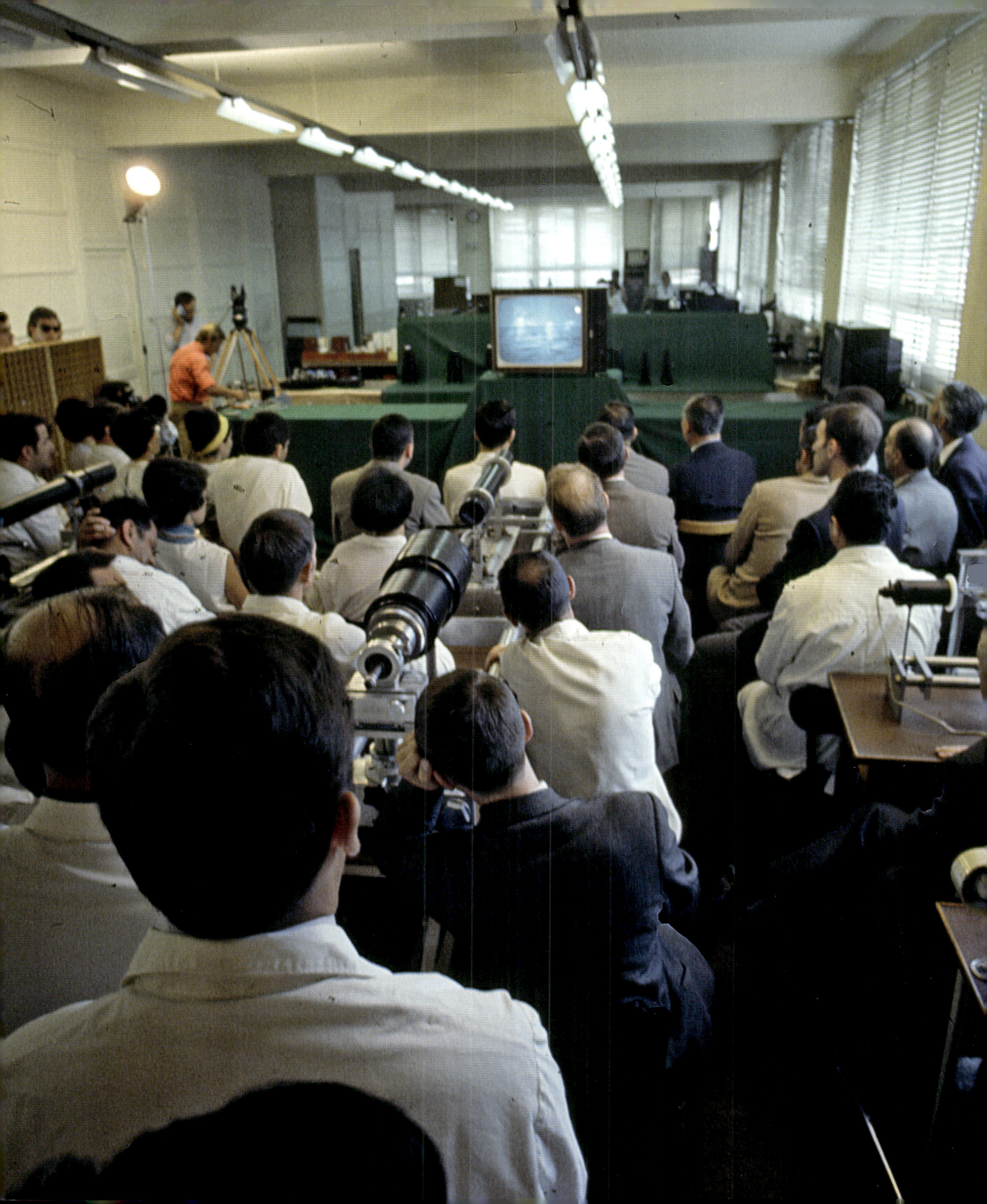

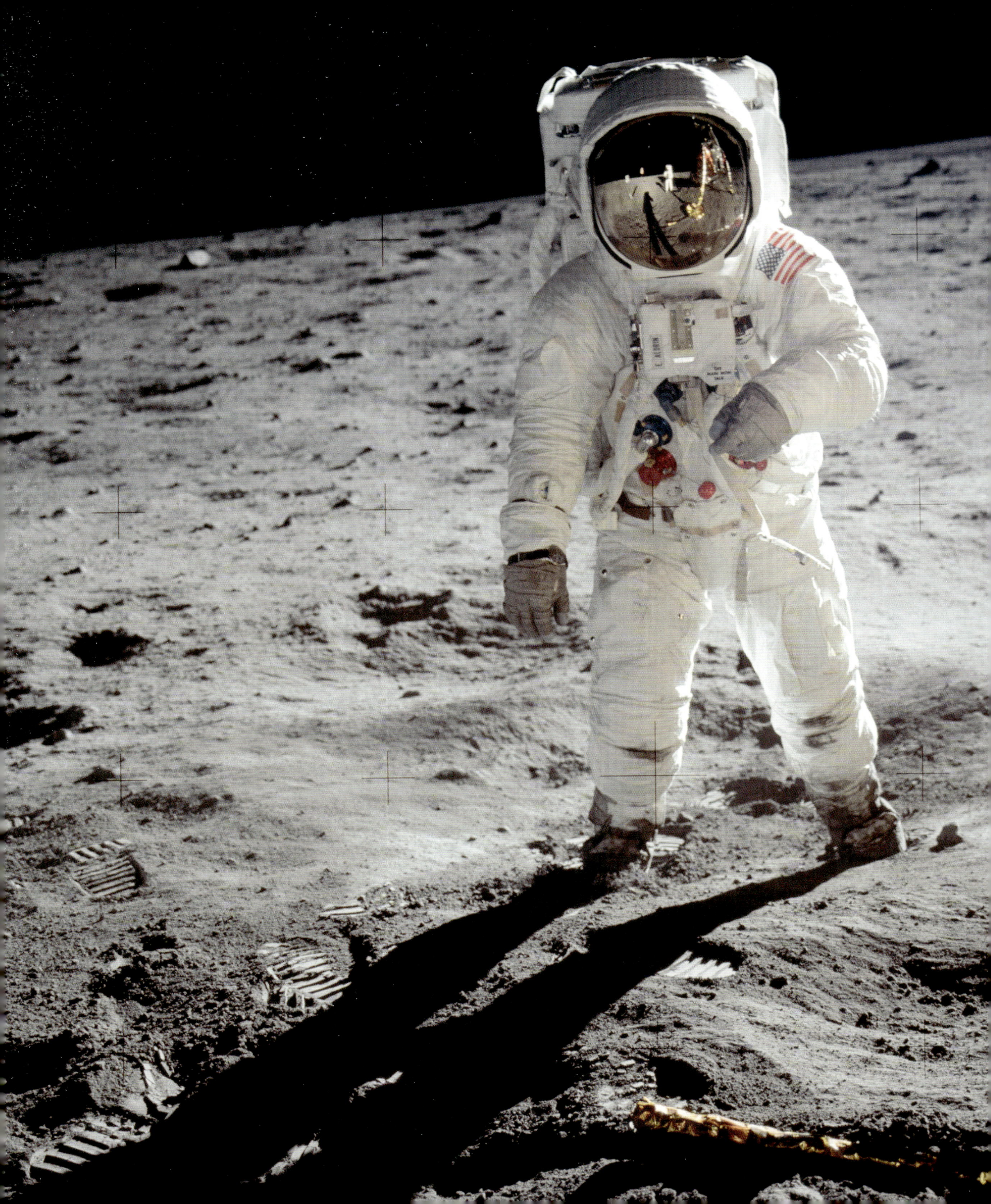

opposite and right
Apollo 11 mission, July 20, 1969. The Angénieux optics remained on board the Command Module. In the image on the right, the 6x25 zoom on a Westinghouse color camera (© NASA).

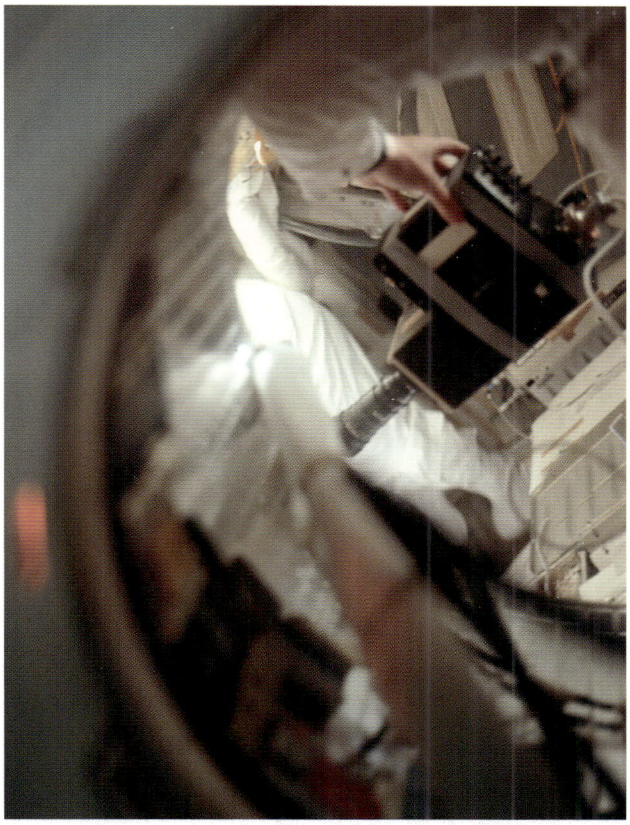

On **Apollo 12**, the Westinghouse color camera was now on board the Lunar Module. In order to withstand the environment, the 6x25 zoom lens and the camera were painted white and the plastic parts were replaced with aluminum parts. The camera was first placed in the equipment storage compartment on the side of the Lunar Module to make it possible to film the astronauts going down the ladder, and was later placed on the surface of the Moon. When astronaut Charles Conrad was setting the camera on the ground, he pointed it at the sun, damaging the tube, and making it unusable for the rest of the mission.

right
Apollo 12 mission, November 14-24, 1969. The first color television images taken from the Moon. The Westinghouse camera with an Angénieux 6x25 zoom, protected by an aluminum box, was positioned on the side of the Lunar Module to be able to film Charles Conrad getting out of the module. The camera was then moved to the surface of the Moon (© NASA).

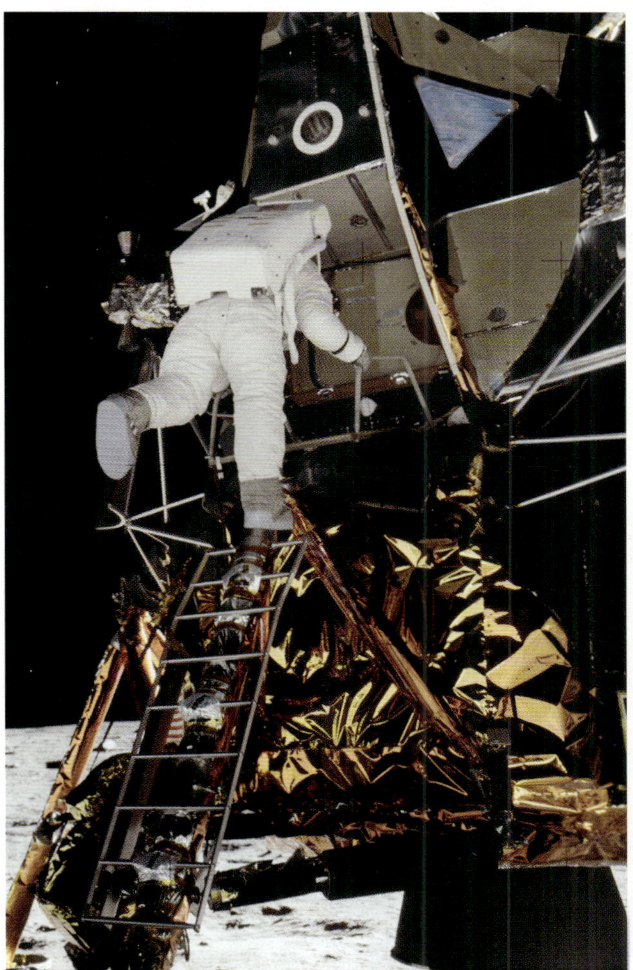

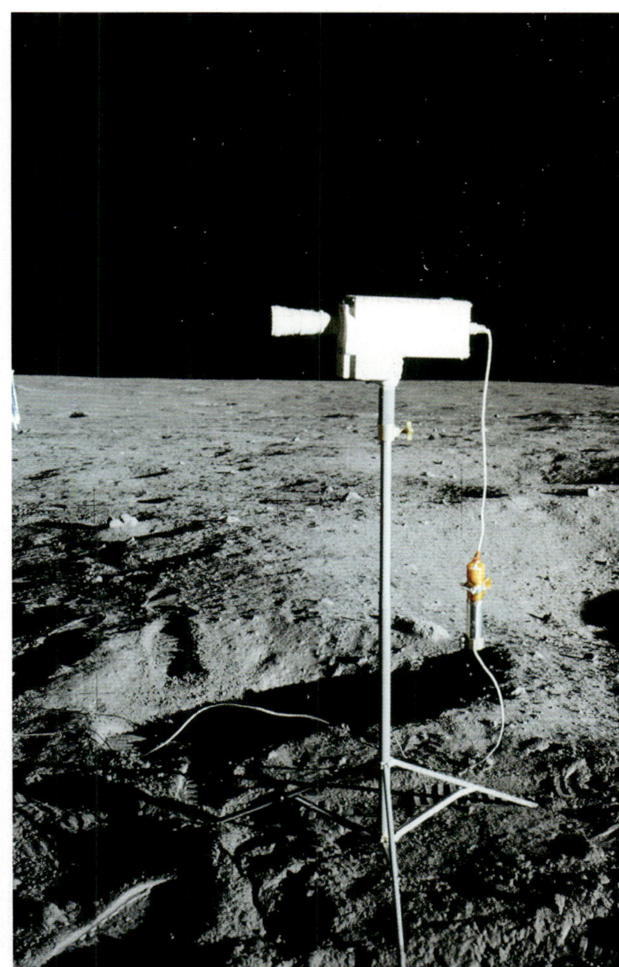

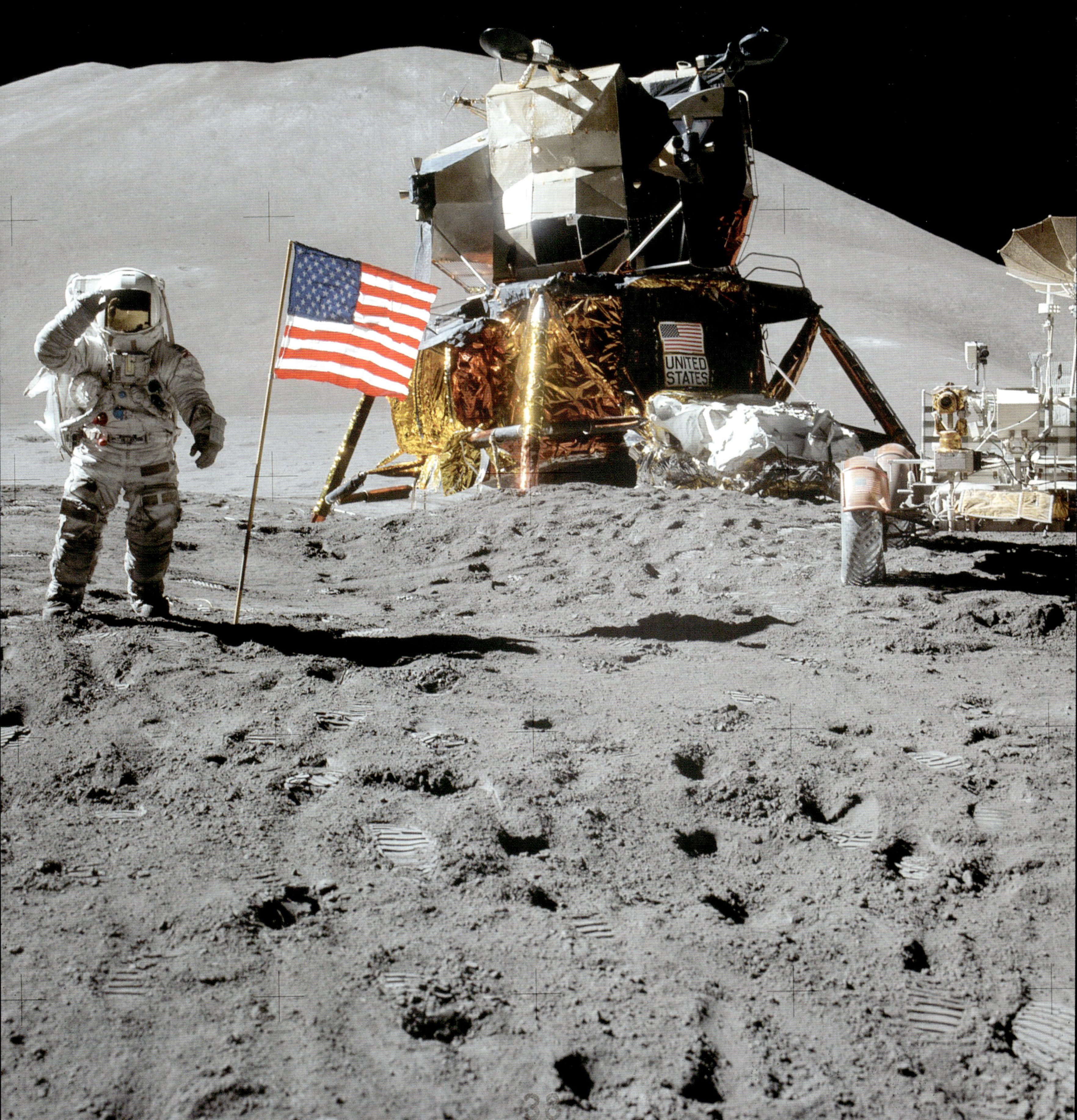
UNITED
STATES

opposite
Apollo 15 program. First mission with a lunar Rover equipped with an RCA camera, with Angénieux's 6x12.5 zoom lens (© NASA).

right
Lunar Rover recording system for the *Apollo 15, 16,* and *17* missions (© NASA).

below
The 6x12.5 zoom integrated into the Lunar Rover recording system (© NASA).

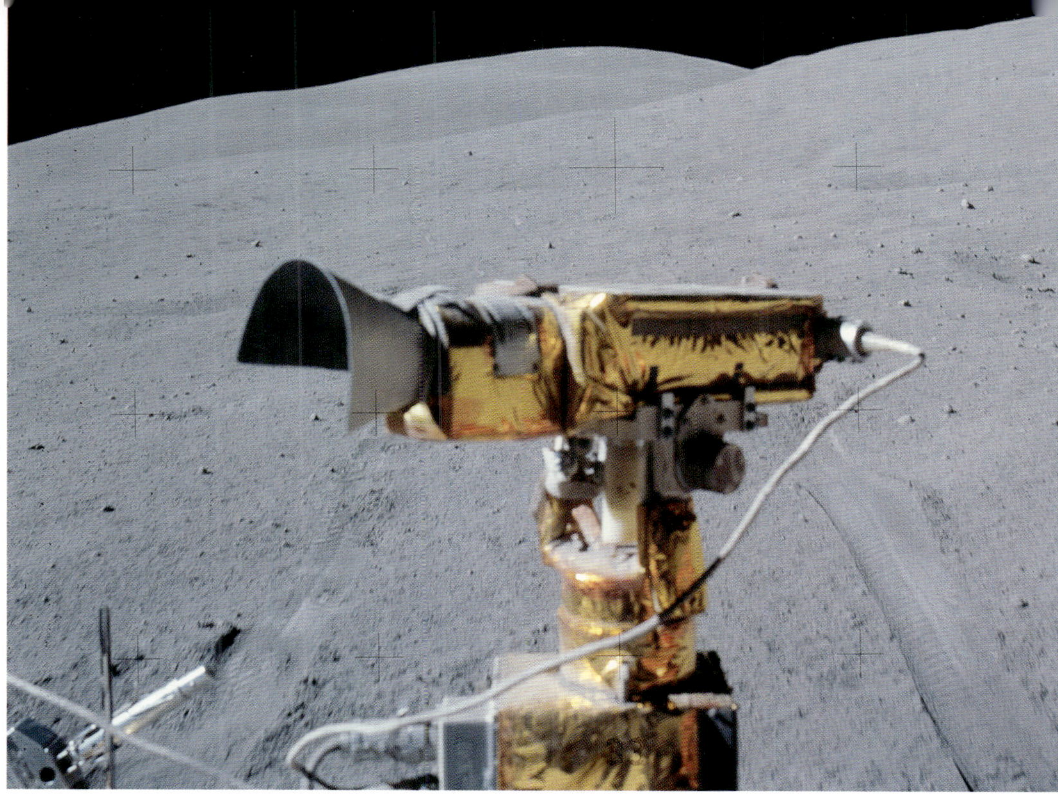

After this incident, Westinghouse upgraded the lunar camera for the **Apollo 13** by incorporating switches to disconnect the tube while the camera was running and adding a lens cap. This camera would make it possible to shoot beautiful images live from the trip to the Moon. Unfortunately, a failure forced the module to turn back before reaching its goal. Also for *Apollo 13*, NASA placed a Westinghouse color camera, once again equipped with the 6x25 Angénieux zoom lens, over the *Saturn V* launchpad to film its takeoff live. The camera was installed in a protective box to prevent overheating during the launch.

After the failure of *Apollo 13*, the **Apollo 14** mission was a success and reached the Moon. The Angénieux 6x25 zoom on the Westinghouse camera installed on the Lunar Module made it possible to send back the first color images of astronauts working on the lunar surface near the module. This recording equipment would remain unchanged for future *Apollo* missions.
To explore the Moon in greater detail, NASA designed a small all-terrain vehicle that would allow astronauts to travel away from the capsule. This vehicle, named *Rover*, was used on the **Apollo 15, 16, and 17** missions. RCA returned to the *Apollo* program, supplying the *Rover*'s camera (Ground Commanded Television Assembly). Once again the camera was fitted with an Angénieux 6x12.5 with a 2x teleconverter. It was motorized and operated by remote control from the Earth. Viewers around the world could once again follow the astronauts' various expeditions on the lunar surface live. At the end of each mission, the *Rover* was left on the Moon, a few hundred yards from the Lunar Module, to broadcast its takeoff live.

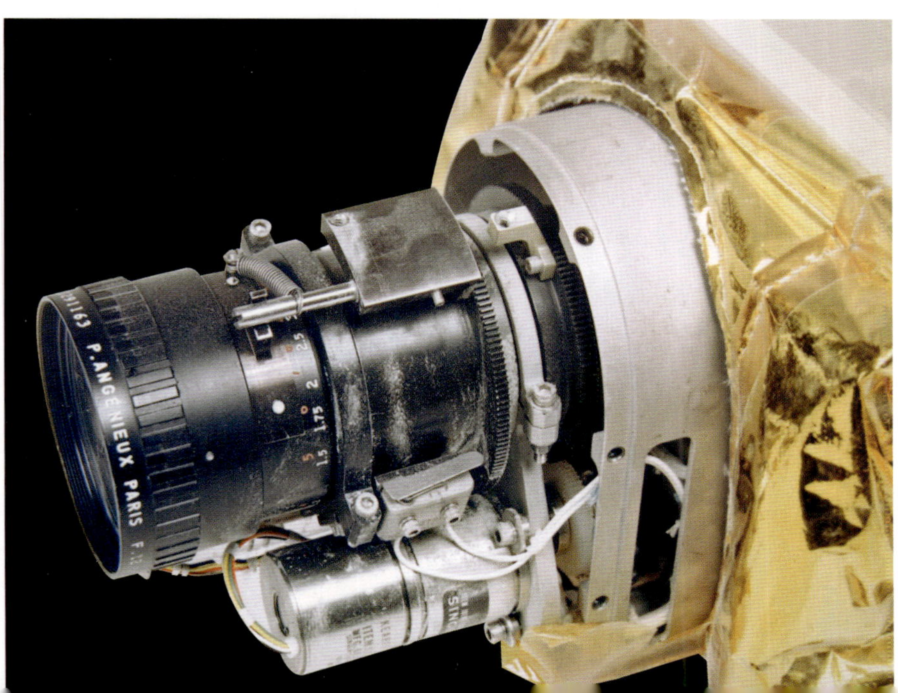

« In 1970, NASA, under director Mr. Von Braun, decided to undertake the *Skylab* program. It would consist of sending a 'space laboratory' into orbit around the Earth, to host three teams of three astronauts each, for a total of nine months. Given the importance of this mission, NASA decided to sell broadcasting rights to television networks. As it had before, NASA turned to Angénieux to develop an Angénieux zoom lens for *Skylab*. The specifications defined by NASA were extreme: vibrations, mechanical shocks, thermal shocks, temperatures from -80°C (-112°F) to 100°C (212°F), wear during use, etc. Despite these requirements, Angénieux accepted the challenge. Given the financial impact for NASA of selling the images for broadcasting to television channels, NASA decided to send their representative Mr. Bauknecht to Saint-Héand, to report directly to Von Braun, to check that Angénieux's capabilities complied with the specifications. It was our great honor to receive him at Saint-Héand, in October, 1972. »

René Michalet

René Michalet, ESO engineer (class of 1955), joined Angénieux in 1958. He would spend thirty years there. He started in the fabrication of optical parts, then went on to the inspection department where, in 1964, he witnessed the first evaluation of Angénieux by American military standards for NASA. In 1970, the inspection department became the quality department. Michalet was the director of this department until his retirement in 1988. He obtained the RAQ 2 qualification in 1981 and the "Prix Qualité" (Quality Award) in 1985.

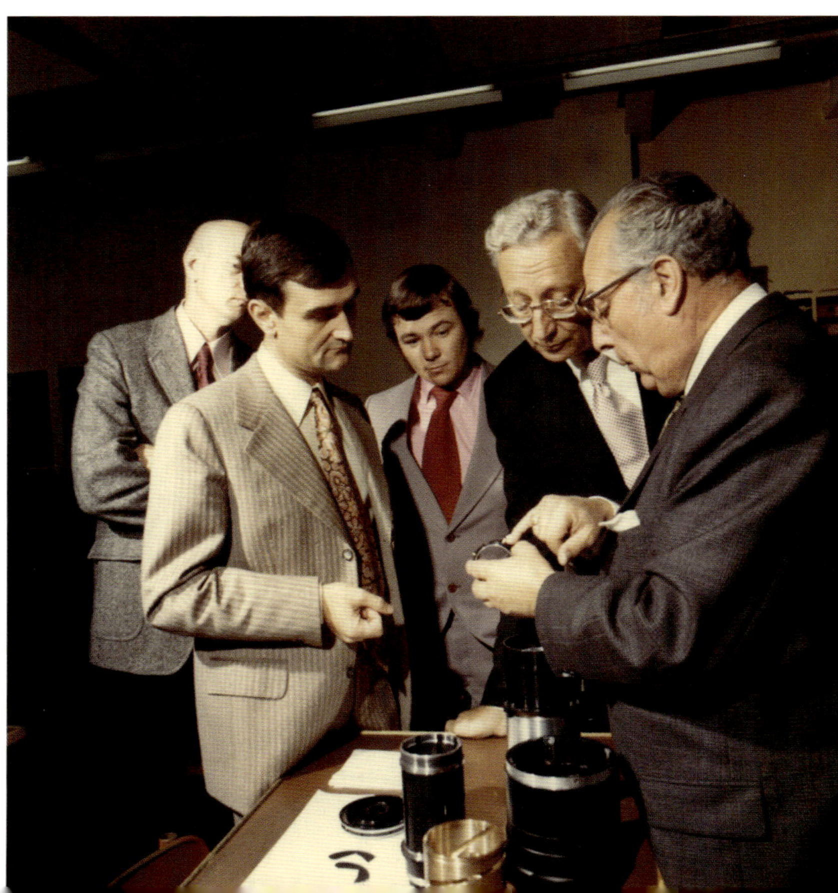

In the foreground, from left to right, Jean Moret, Mr. Bauknecht and Lucien Granet (Angénieux archives).

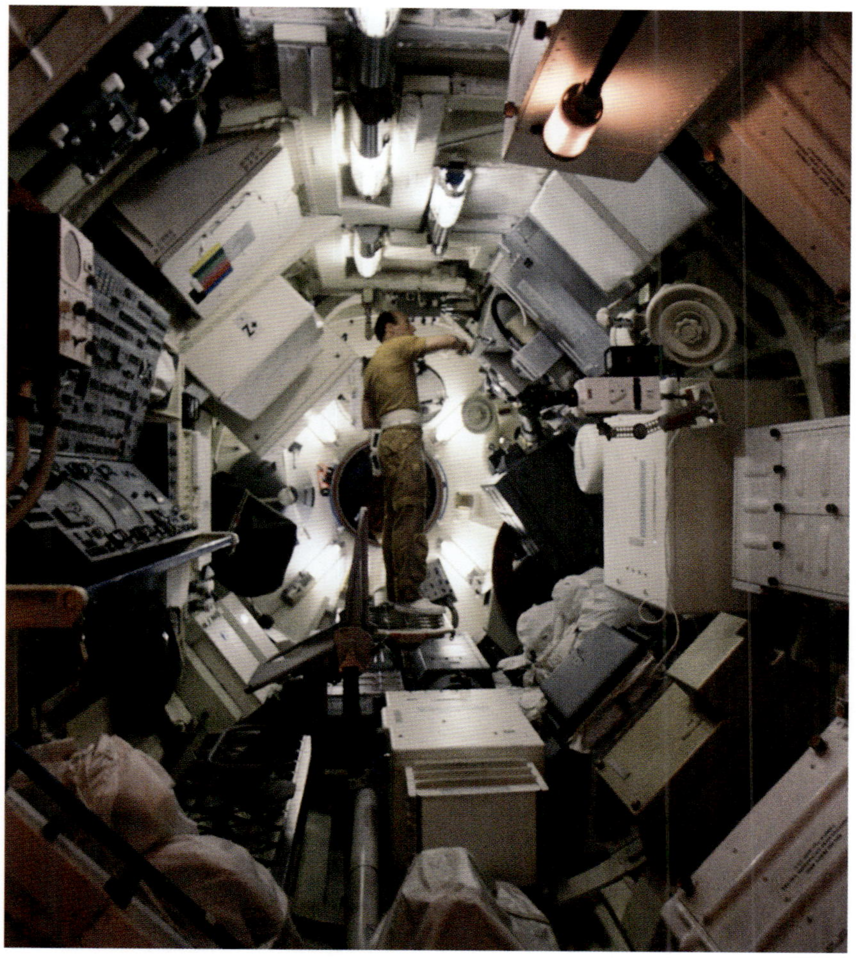

Angénieux, at the Heart of the *Skylab* and *Apollo-Soyuz* Projects

Because of budget problems, the *Apollo* 18, 19, and 20 missions were canceled. NASA's lunar adventure had come to an end.

The six successful *Apollo* missions yielded more than 800 lbs of rock and dust samples from the Moon. American astronauts spent a total of 300 hours on the Moon.

Once the *Apollo* missions were completed, the Westinghouse color cameras with their Angénieux 6x25 zooms were put on board the *Skylab* project. *Skylab* (Sky Laboratory) was the first space station NASA launched. It was sent into orbit on May 14, 1973. There was a camera inside the station to send images of the life and work of the astronauts on board. Three crews stayed there during 1973 and 1974. Another camera with an Angénieux 6x25 motorized zoom lens, was mounted on the articulated arm located outside the station. The zoom lens, which could be operated from inside the station, provided images of the astronauts exiting the station into space. The *Skylab* station, damaged during launch, disintegrated on reentering the atmosphere on July 11, 1979.

A motorized version of the 6x25 zoom was used on the *Skylab* orbiting station (right). Operated from the inside (above), the zoom lens was mounted on the exterior articulated arm to film astronauts exiting the station into space (© NASA).

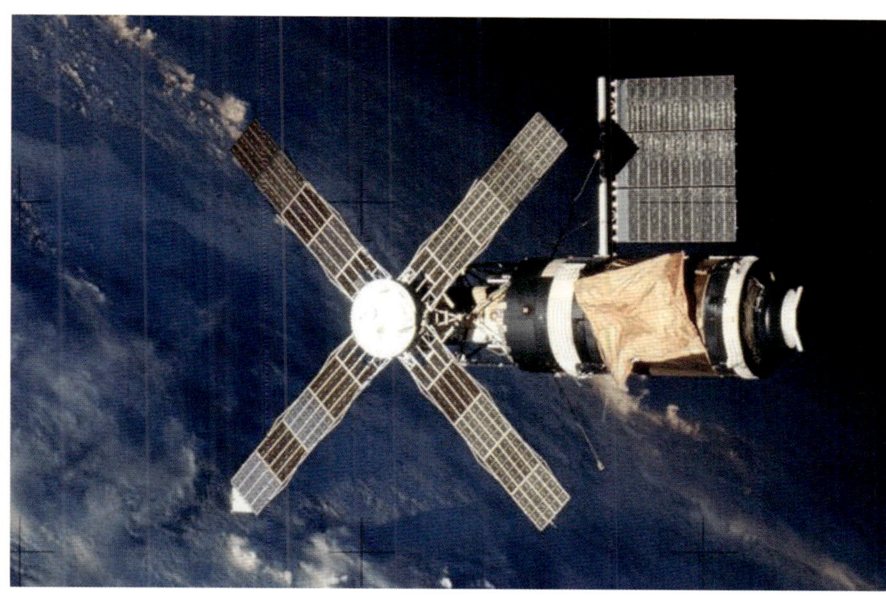

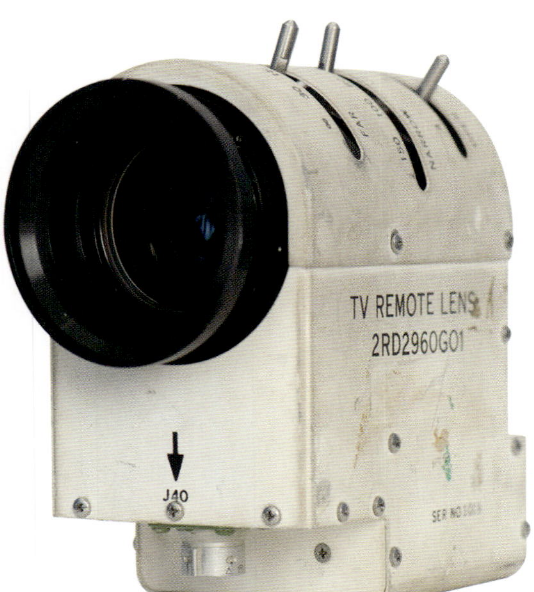

left
Angénieux 25-150mm 1:4.4 zoom lens, no. 1372458, 1972 (Granet family collection, photo by Marc Salomon).

The Angénieux 6x25 zoom lenses on the Westinghouse cameras would be used again on the **Apollo-Soyuz** project in 1975. Apollo-Soyuz was the first joint space mission between the Soviet Union and the United States after the race to the Moon during the Cold War.

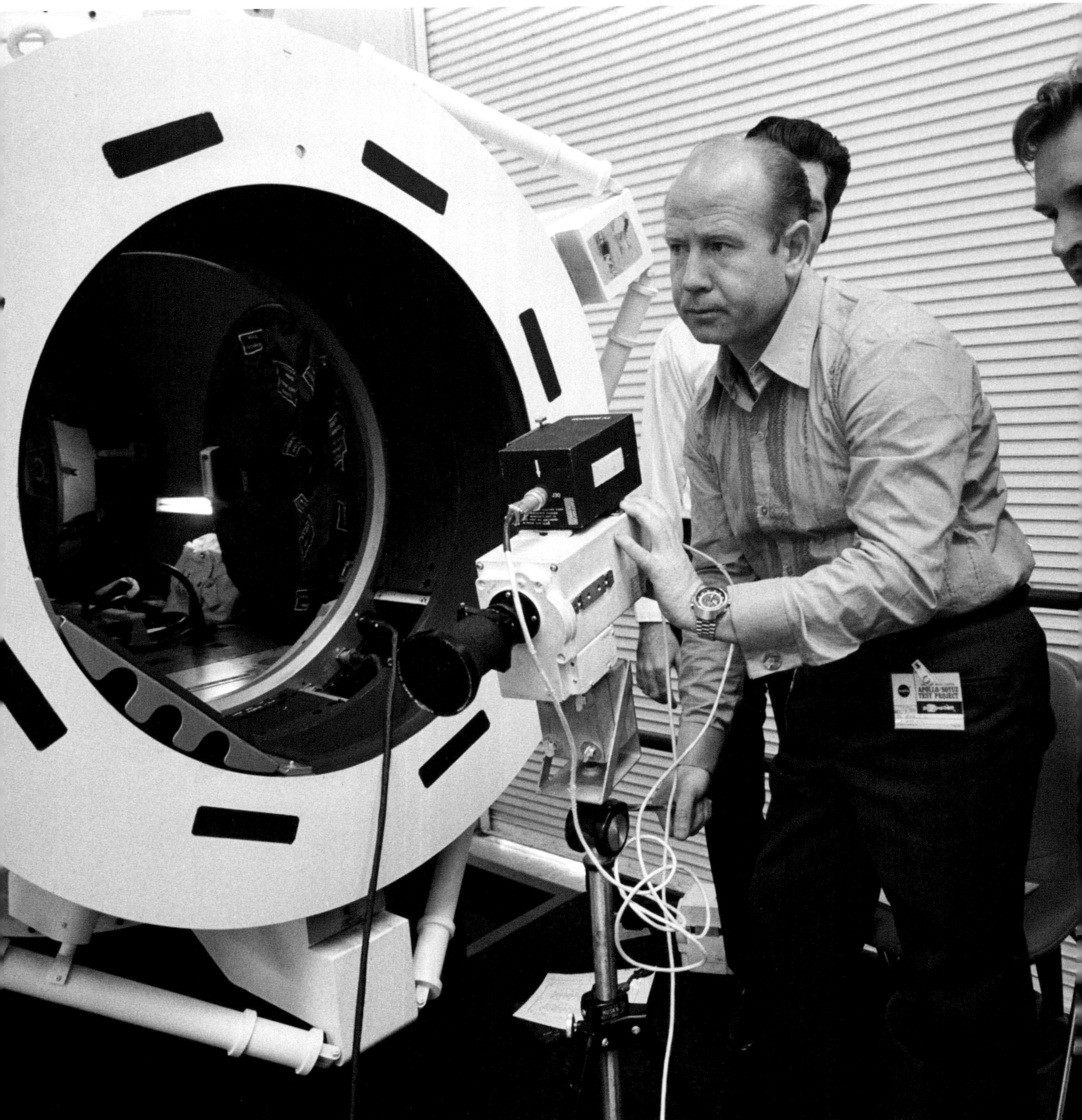
APOLLO-SOYUZ
TEST PROJECT

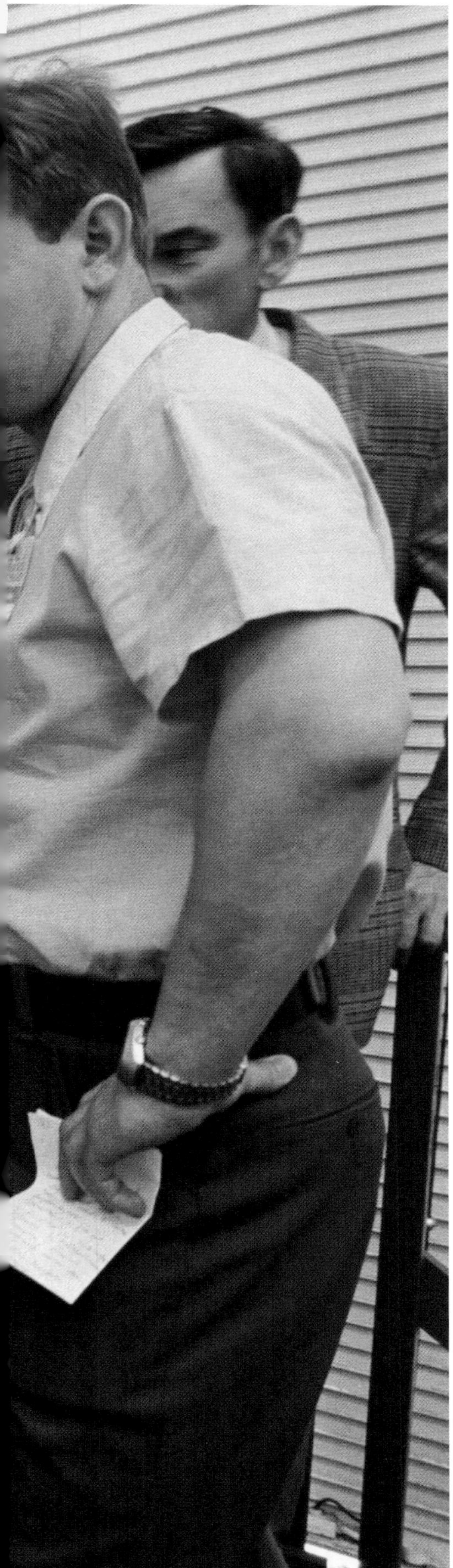

« For the *Soyuz* mission, I was at an exhibition in Moscow and I was asked if Angénieux could rather quickly provide a lens with certain characteristics I was given. I had to call the factory right away. The deal was made, and when it was concluded, I asked them why it was so urgent. They said, well, it's for the docking of *Apollo* and *Soyuz*. This lens would be on board the *Soyuz*. The American *Apollo*, for its part, also had an Angénieux lens. Our two lenses met in space. »

Bernard Angénieux

The 6x25 zoom from the Russian-American *Apollo-Soyuz* program in 1975 (© NASA).

3x8.2 and 15x8.5 Zoom Lenses for the US Space Shuttle

In the early 1980s, tensions between the US and the United States cooled. President Richard Nixon and the US Congress refused to extend financing for the *Apollo* program. NASA was pushing for an ambitious human space program but had to settle for plans for a space shuttle, a reusable craft that aimed to significantly lower the cost of launching into orbit. Its development took longer than expected, into the beginning of the next decade.
Columbia, the first of six space shuttles, made its first flight on April 12, 1981. Following the success of television broadcasts from *Apollo* and *Skylab*, television naturally appeared on scene with the space shuttle project. This time, NASA chose to work with RCA.
Angénieux designed a 3x8.2 wide-angle zoom lens covering an image field 16mm on the diagonal. This lens was equipped with a “wide dynamic iris” system to avoid overexposing the camera tube with the movement of the sun across the field of the optics. Angénieux supplied a total of twenty of these lenses. They were used in the color cameras (CTVC) in the compartment where the crew stayed. They were also mounted on the video cameras in the European Space Agency (ESA)'s Spacelab space shuttle laboratory. On Earth, everyone could watch the astronauts experiencing zero gravity. Towards the end of the 1980s, tube cameras were replaced by cameras with CCD sensors, which would then be fitted with the lenses of competitors.
In 1990, NASA decided to improve the quality and sensitivity of the black and white cameras in the hold. These cameras, called ITVCs (Intensified Television Cameras) were developed by GE Astro, the new name of RCA Astro Electronics. They were able to provide an image in low light (as low as 0.03 lux) because they had an optical intensifier system (system of optical fibers in front of the CCD). Angénieux was again asked to develop a zoom lens adapted to the sensors of these new cameras, and offered a 15x8.5 zoom created from a 15x6 for 8mm cinema. The lens elements had been specially treated to allow infrared light to pass through them. The first 15x8.5s were used on the *Columbia* shuttle launched on June 20, 1996. The number of ITVC cameras varied with each mission. There was usually at least one ITVC at each of the four corners of the cargo bay.
After the explosion of the space shuttle *Columbia* in February, 2003, NASA designed an arm equipped with an analysis unit (ITVC, Angénieux zoom lens, and laser analysis system) to inspect the shuttle's thermal protection tiles once in space. During the last American space shuttle mission, this arm was transferred to the International Space Station (ISS) where it is still used for maintenance operations. Between April, 1981, and July, 2011, American space shuttles had made 135 flights.

top left
The articulated arm with a 15x8.5 zoom lens could inspect the space shuttle's thermal tiles (© NASA).

top right
After the fire on the shuttle Columbia in 2003, an articulated arm, also called "Canadian arm" with a 15x8,5 zoom lens was fitted to US space shuttles (© NASA).

opposite
On the space shuttle's last trip, the articulated arm was transferred to the International Space Station (© NASA).

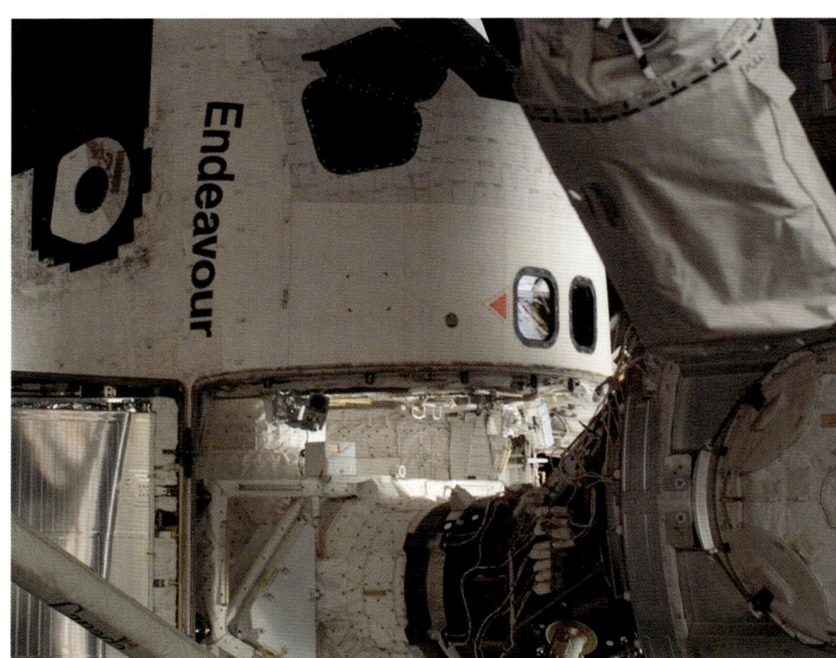

below
ITVC camera and Angénieux zoom aboard the shuttle *Endeavor* during the *sts-123* mission (March 2008).
In this image the shuttle at the International Space Station (© NASA).

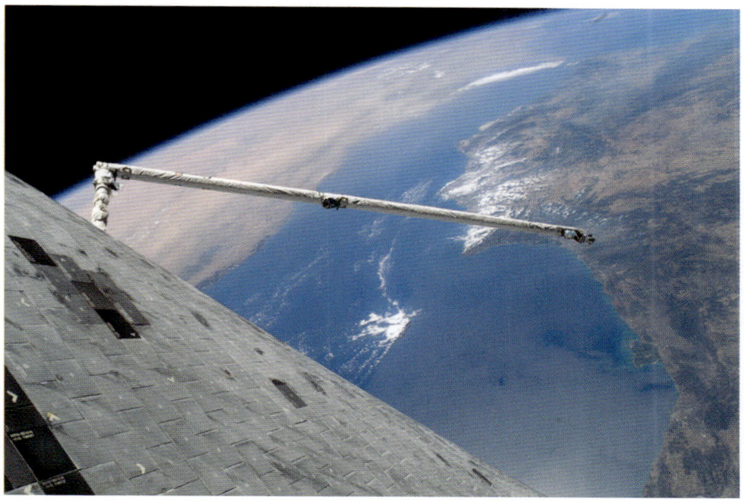

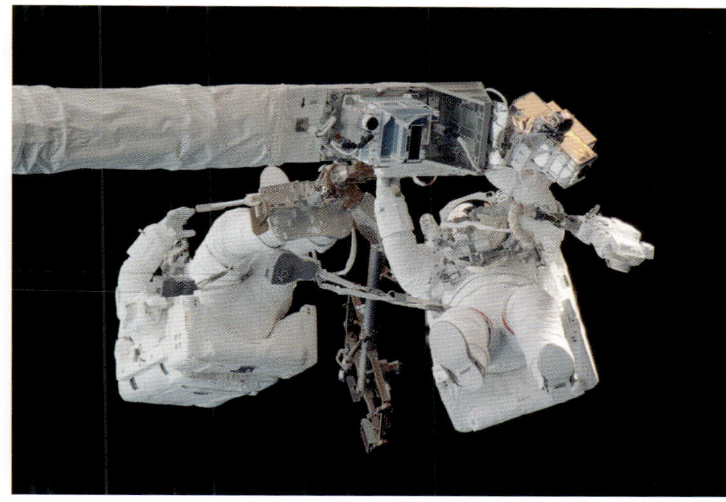

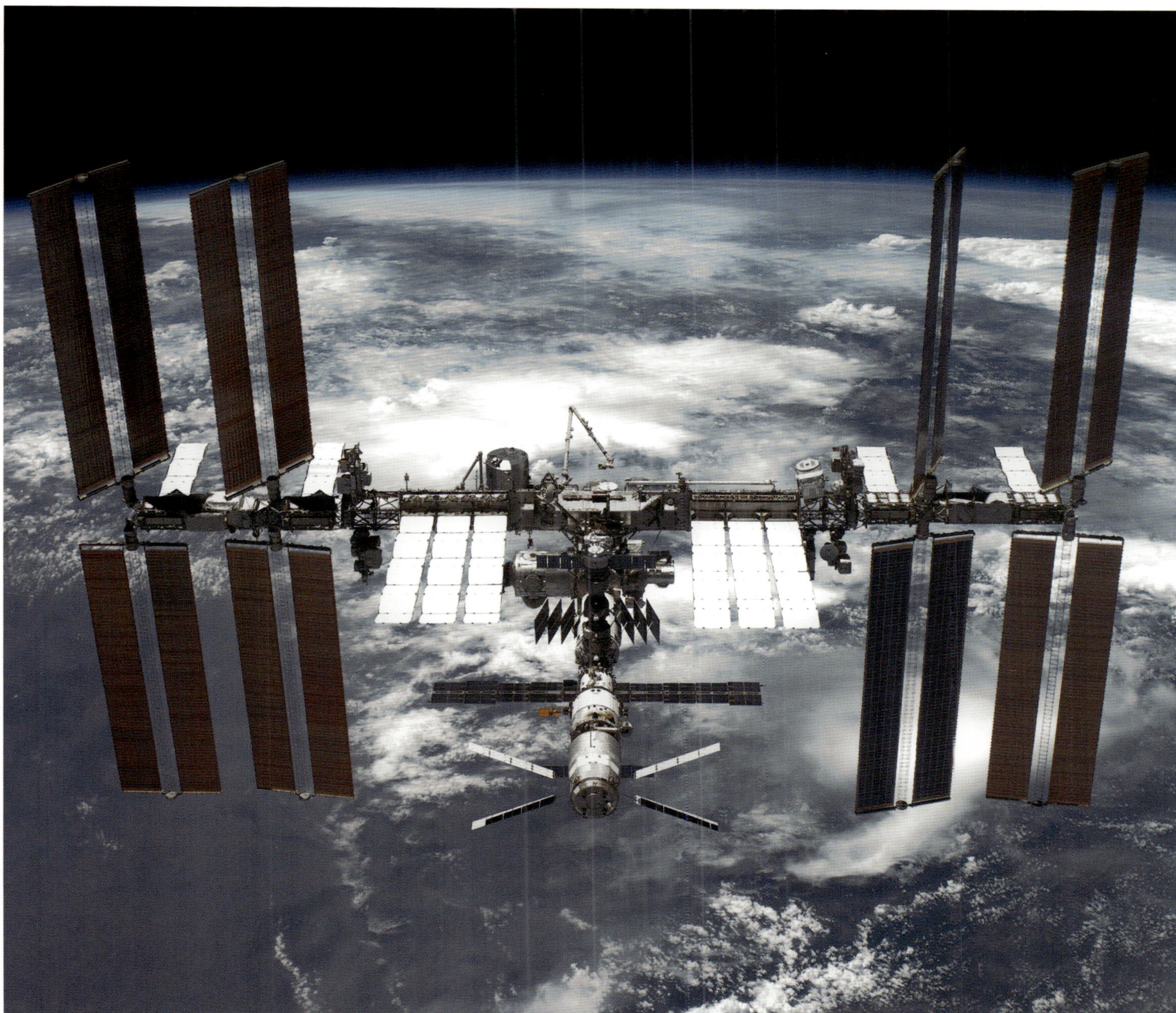

2018: Angénieux Aboard the *Dawn* Space Probe

Dawn was the ninth mission of NASA's *Discovery* program. Launched in the early 1990s, the program was designed to study the asteroid Vesta and the dwarf planet Ceres, two protoplanets that carry evidence from the birth of the solar system, located in the asteroid belt between Mars and Jupiter. Their characteristics have not changed substantially since their formation 4.6 billion years ago. The *Dawn* probe was launched in September, 2007, flew past Mars in February, 2009, reached Vesta in July, 2011, and studied it for one year, and then arrived at Ceres in February, 2015. During this astonishing odyssey, *Dawn* sent back incredibly sharp photos. These photos were the result of the Angénieux 150mm prime optics supplied in 2005 to the German company Kayser-Threde on behalf of the Max Planck Society, which was in charge of the probe's "framing camera" upper subassembly. There were four of these lenses in total, two to be integrated into the system and two used in the test and preparation phases. They had to meet very demanding specifications: to last more than ten years in extreme temperature conditions ranging from -50°C (-58°F) to 40°C (104°F), solar radiation, the vacuum of space, and vibrations, especially during takeoff. Thanks to a passive athermalization process, the Angénieux lenses were able to ensure perfect image sharpness whatever the temperature conditions. Specially treated specific lens elements were used to protect against gamma rays. The titanium lens body ensured the best resistance to vibrations.

Angénieux optics were thus able to bring the world new and unexpected details from this protoplanet of such great interest to us. NASA researchers believed its subsoil contained ice and organic molecules, the ideal conditions for preserving traces of life. *Dawn* will remain in that orbit until the hydrazine reserves of the electric propulsion system are exhausted. Its mission officially ended on November 1, 2018. After an odyssey of eleven years in space, the probe was to remain in orbit around Ceres for decades.

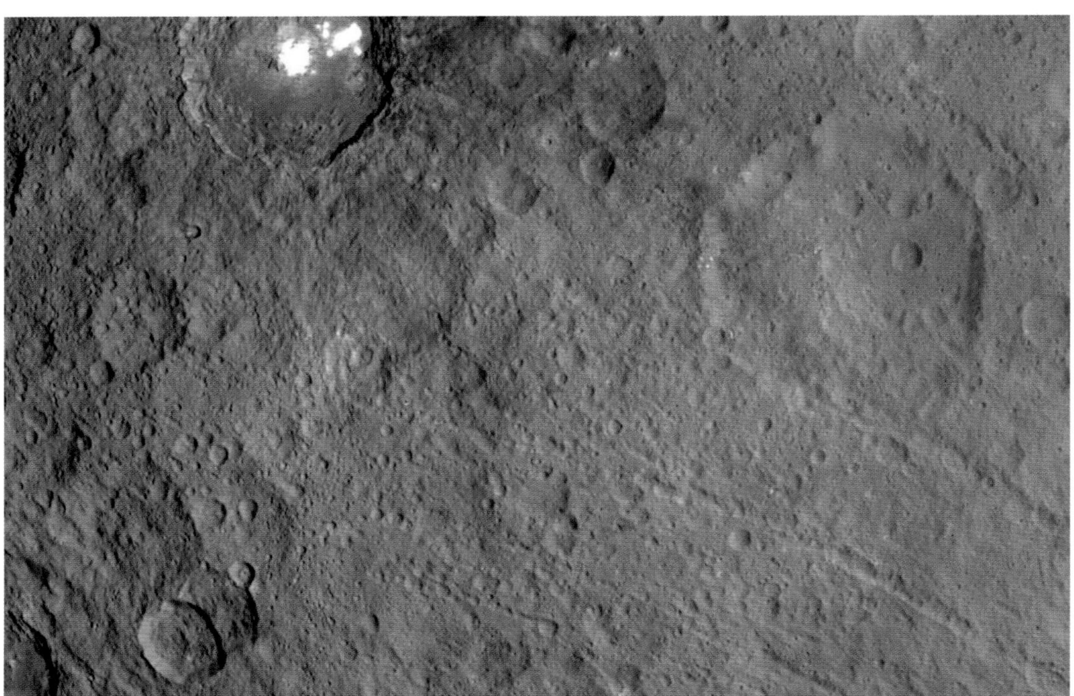

The images taken by the *Dawn* space probe draw attention to two strange bright spots on the surface of Ceres. Image taken on June 25, 2015 (© NASA/JPL-Caltech/UCLA/MPS/DLR/IDA).

Angénieux's Work in Space was not limited to NASA

In February, 1986, the SPOT 1 satellite for terrestrial observation, developed by French National Centre for Space Studies (CNES), was sent into orbit. Angénieux had designed the beam splitter.

Angénieux supplied all the beam splitters for the satellites SPOT 1 to SPOT 5. SPOT 5 remained active until 2015. The images taken by the SPOT satellites provided information mainly useful for agriculture and cartography.

During the time of the SPOT 4 and SPOT 5 programs, Angénieux technology was also on the four French reconnaissance satellites of the Helios 1 and Helios 2 programs, launched in 1995, 1999, 2004, and 2009. As of 2018, the two Helios 2 satellites were still operational.

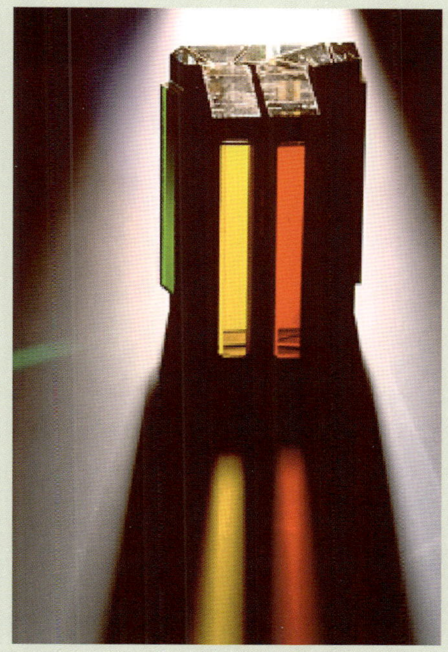

Angénieux beam splitter from the SPOT satellites (Angénieux archives).

When Reality and Fiction Meet ...

The world of film was gripped by the subject of space exploration from its very beginnings. Its ability to create images of as-yet unreal universes helped us to imagine our future. Here too, Angénieux was at the heart of the adventure. As far back as 1968, in Stanley Kubrick's *2001: A Space Odyssey*, considered one of the greatest films of all time, director of photography Geoffrey Unsworth used Angénieux zooms. In *The Martian*, released in 2015, Ridley Scott, with help from NASA, imagined what the beginning of mankind's presence on Mars could look like. It was filmed in 3D, the compact Optimo 45-120 and 28-76 zooms widely used on set. Released at the end of 2018, Damien Chazelle's *First Man* recounts the incredible feat of astronaut Neil Armstrong commanding the *Apollo 11* and becoming – fifty years ago – the first man to take a step on the Moon. It was a 6x25 zoom on board the *Apollo11*, and 24-290 and 17-80 Optimo zooms would be among the equipment selected for shooting the film on its fiftieth anniversary.

Medical Applications
Mastering Cold Light

The brand's reputation extends to the medical world, where its methodical and structured approach, similar to that of users including surgeons and healthcare teams, would allow it to offer a revolutionary surgical lighting solution. This innovation initiative, the legacy of Pierre Angénieux, supported today by the Thales Group, remains at the heart of the concerns of Saint-Héand's teams.

In Close Contact with the Medical Teams

To tackle this new and still unknown world, Pierre Angénieux and his teams would apply the same methods already tested in the areas in which the company had already taken an interest. At the beginning of the 1970s, in close collaboration with Professor Joubert in Saint-Étienne, Pierre Angénieux worked very methodically to become familiar with and gain understand of the setting for surgical operations. He thoroughly examined the lighting conditions. He carefully studied the work of hospital teams in the operating room. Interviews were conducted.

The commonly used "*scialytiques*[1]," or surgery lights, had one defect: they overheated the operating area. The increased temperature promoted dehydration of living tissue and added a risk factor for the procedure. In addition, the lights were not easy to manipulate.

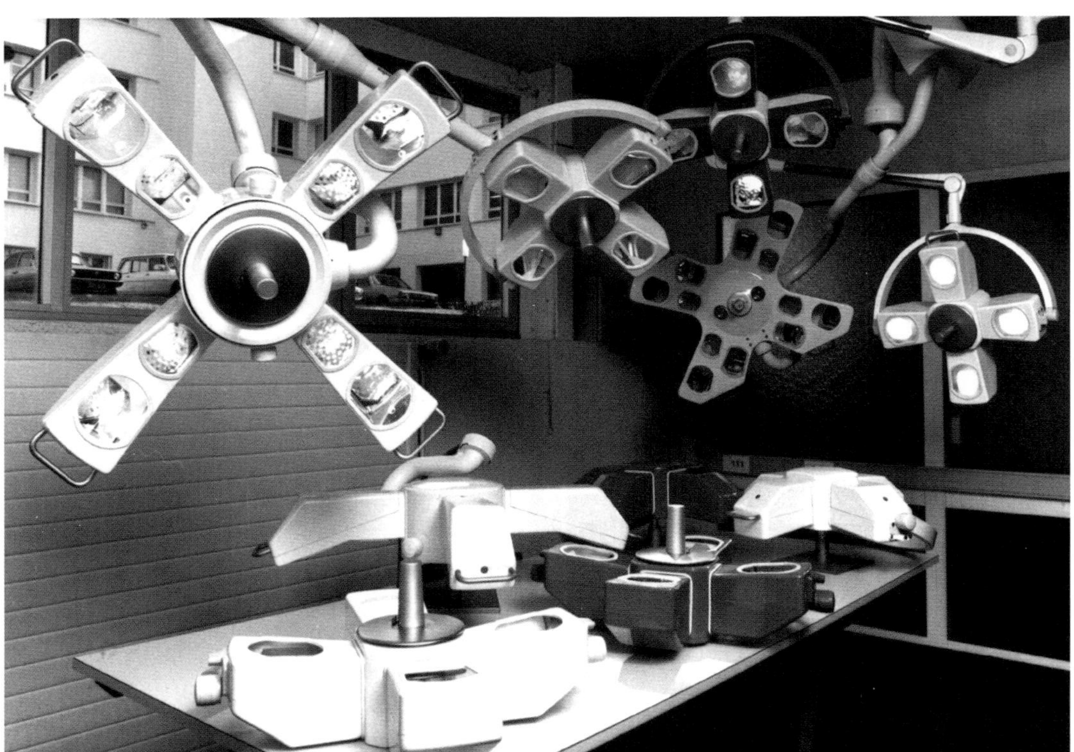

[1] Scialytique is actually a brand. The word has become a generic term for a lighting system used in operating rooms. The brand was registered in 1919 by Professor Louis Verain of the Faculty of Medicine of Algiers. Early surgical lighting replaced the old operating room lights, which were simply ceiling lights.

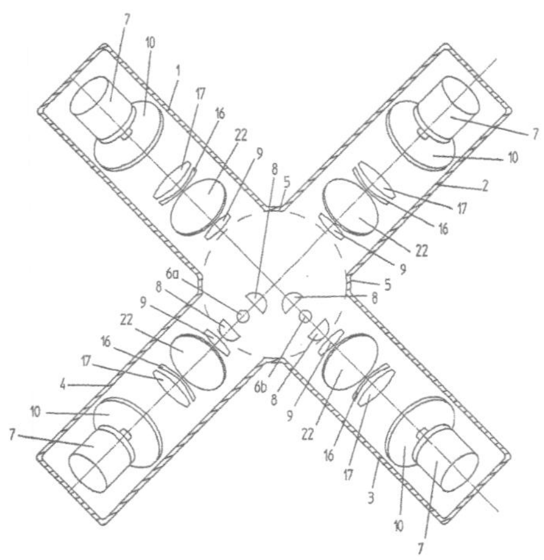

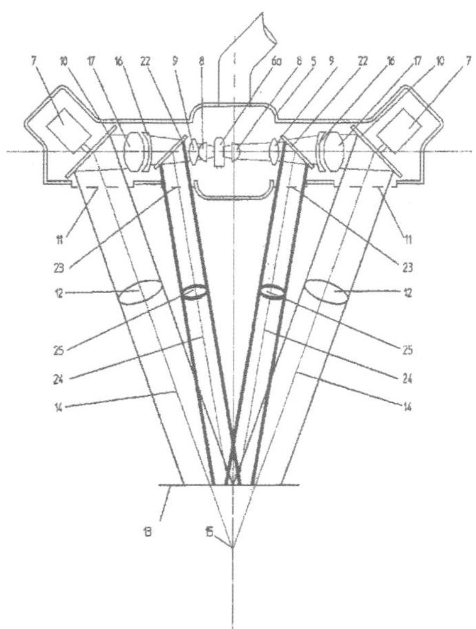

Capitalization and Innovation

Angénieux's teams built on the technological foundation they had acquired in the years preceding and made use of their experience with infrared and their expertise in treating dichroic mirrors with thin layers.
A new technology was born – cold light – and it would prove as revolutionary for surgical teams as the zoom lens had been for photographers and filmmakers.

In addition to visible light, the halogen lamps used in surgery at the time emitted considerable energy in the infrared spectrum. This energy is wholly transformed into heat. The innovative concept which Angénieux established, and which it would patent, consisted of directing the light onto specially treated dichroic mirrors (thirty layers were required), to reflect only the visible rays towards the operating area. The infrared rays passed through the mirror and did not reach the sensitive area.
The rise in temperature was thus reduced to 14°C (25°F, in contrast to the 45°F to 54°F with traditional operating lights) while maintaining the luminous intensity necessary to operate at an ideal color temperature of 4100 K without the nuisance of overpowering light changing the apparent color of tissues and even the diagnosis.

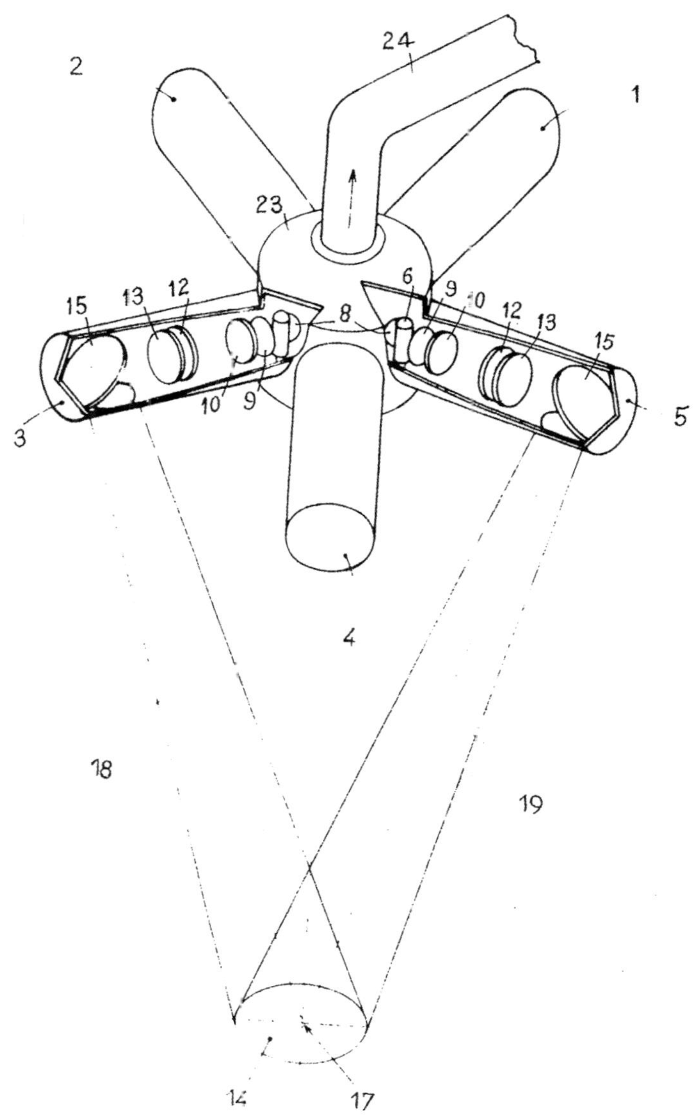

June 26, 1973. Pierre Angénieux's firm filed an initial patent for a "projecteur d'éclairage" using cold light for medical lighting. This patent will be followed by several later additions (Angénieux archives).

Angénieux's solution also faithfully delivered colors. The colors of organs range from yellows to reds. This is precisely the range of wavelengths of the visible spectrum in which the Angénieux lights develop the greatest energy. They thus allowed differentiation of the tissues in spite of the weak reflecting power of reds. All other surgical lights used heat-reducing filters. If blue dominated, they offered little energy in red and thus necessitated an increase the level of illumination, and therefore the glare experienced by users. If red dominated, they produced more infrared and therefore more heat.

Cold light technology was immediately adopted by hospitals. It contributed to the success of surgeries and to the comfort of surgeons and their teams.

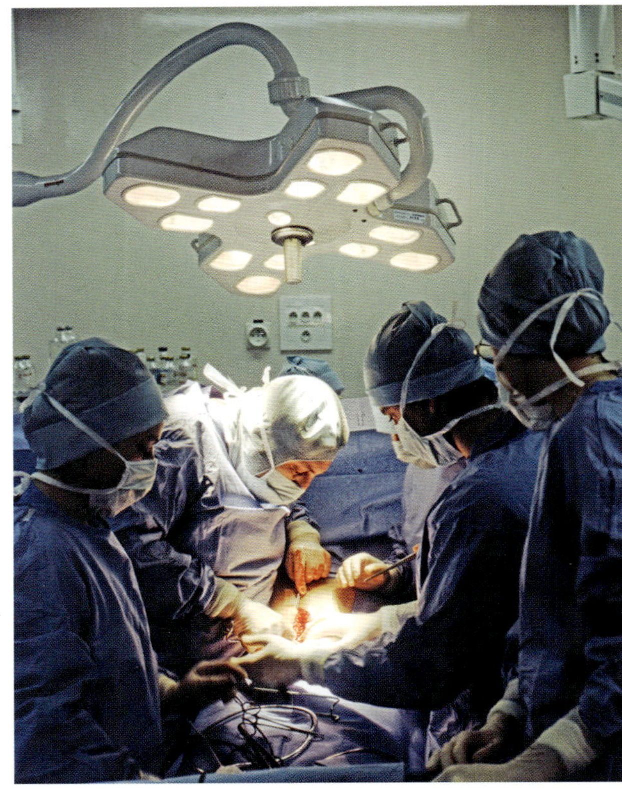

An incredibly Well-Designed Solution

In addition to the direct benefits of cold light technology, the ergonomics of the spotlights were also given attention.

Besides by the drop in temperature, the comfort of the medical team was further increased by the cross shape of Angénieux's spotlights. The surgeon's head, neck, and shoulders were thus much less of an obstacle to the light, and took the minimum possible heat. And, in operating rooms using laminar air flow[2], this cross shape also promoted the flow of filtered air over the patient. Increasing the number of light sources reduced the risk of shadows over the area being operated on, while the light, previously manipulated by the surgeon's assistants, could be moved with a sterile handle by surgeons themselves to obtain the best lighting they could wish for. These lights were also extremely flexible and could be jointed using a completely original system, the "spherical articulation", a symbol of the practical ingenuity of the Angénieux spirit, they employed a bronze ball as used in the sport boule lyonnaise.

Angénieux's range of AX2, AX4, AX8, AX10, and AX14 operating lights, in various sizes, allowed for multiple configurations adapted to each type of surgery, from the most complex cardiothoracic, deep abdominal, or multiple traumatic operations, to the most delicate pediatric or ocular surgery. A special version was designed for the French army.

In 1985, Angénieux was the world's third largest manufacturer of surgical lighting. Angénieux technology has equipped several thousand operating rooms around the world.

above
Operating room lit by an Angénieux AX 14 (collection of the Musée d'Art et d'Industrie, Saint-Étienne, photo Ito Josué).

below
February 24, 1993. Angénieux filed a patent for an optical device for dental radiography (Angénieux archives).

FIG. 1

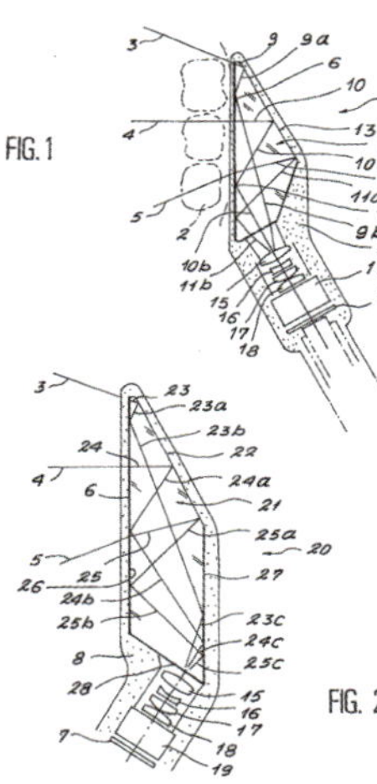

FIG. 2

In other areas of medicine, Angénieux developed a 20x8.5 parfocal[3] zoom lens for ophthalmological lasers in 1986 and, on February 24, 1993, filed a patent for an "optical device for dental radiology."

With the acquisition of Angénieux by Thomson-CSF in 1993, new strategies were put in place. Surgery lighting did not directly concern medical imaging. This activity was sold to ALM (Air Liquide Médical), with Angénieux remaining as a subcontractor. Angénieux continued to manufacture mirrors for MACQUET, the ALM subsidiary that handled the manufacture and marketing of surgical lighting equipment. Angénieux also fully completed development for ALM of the entire Prismalix 400, 600 and 800 operating lighting range. Beginning in 1993, Angénieux supplied TTE (Thomson Tubes Electroniques) with radiological optics and, collaborating with Thomson on the camera and CCD, created the ATC 3000 digital endoscopy system.

[2] In an operating room, laminar flow is a filtered, unidirectional airflow directed toward the operating table to prevent external contamination.

[3] All Angénieux zooms are parfocal zooms. They do not go out of focus; rather, focus remains constant with variation in the magnification of the image.

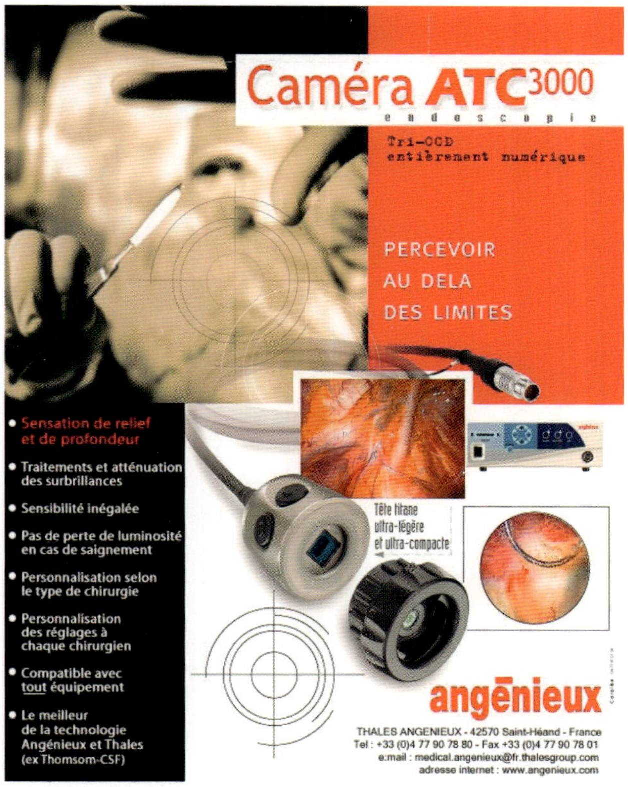

The Age of Artificial Intelligence

In 2017, the Saint-Étienne start-up Keranova asked the teams at Saint-Héand for information on ophthalmology for the design and manufacture of revolutionary optics intended for a surgical robot.

The young company was working to put to use a discovery made by a consortium of local academic laboratories, which made it possible for surgical lasers to become extremely fast. It was now possible to imagine a robotic machine, equipped with intelligence and vision, that could automate a large part of cataract surgery.

This is the most widely practiced surgery in the world. More than twenty-five million patients are operated on each year. Keranova was positioning itself as a future leader in the area of specialization with their revolutionary machine. The surgical robot would perform precise, fast, and nearly perfect surgeries better than the human hand. In this equipment, the most critical part was the design of very innovative optics.

« Our revolutionary robotic laser surgical equipment deserved something besides the same 'standard' optics as the others. It was necessary to design the best optics in the world to make this the best-performing machine possible! We found that excellence in the teams at Saint-Héand, and worked closely with them for over a year to achieve astonishing performance. This innovation should mark a shift in cataract surgery that will have a foothold in the future. »

Fabrice Romano

DVM, Founder, Chairman and CEO of Keranova

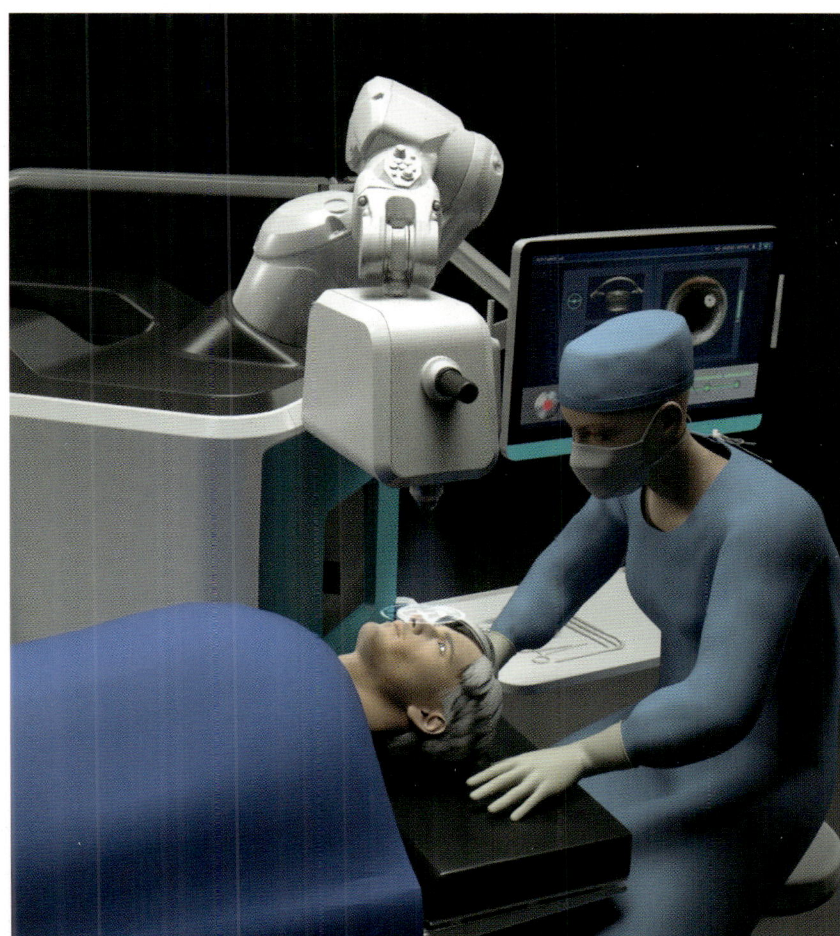

Optronics

Beyond the Visible

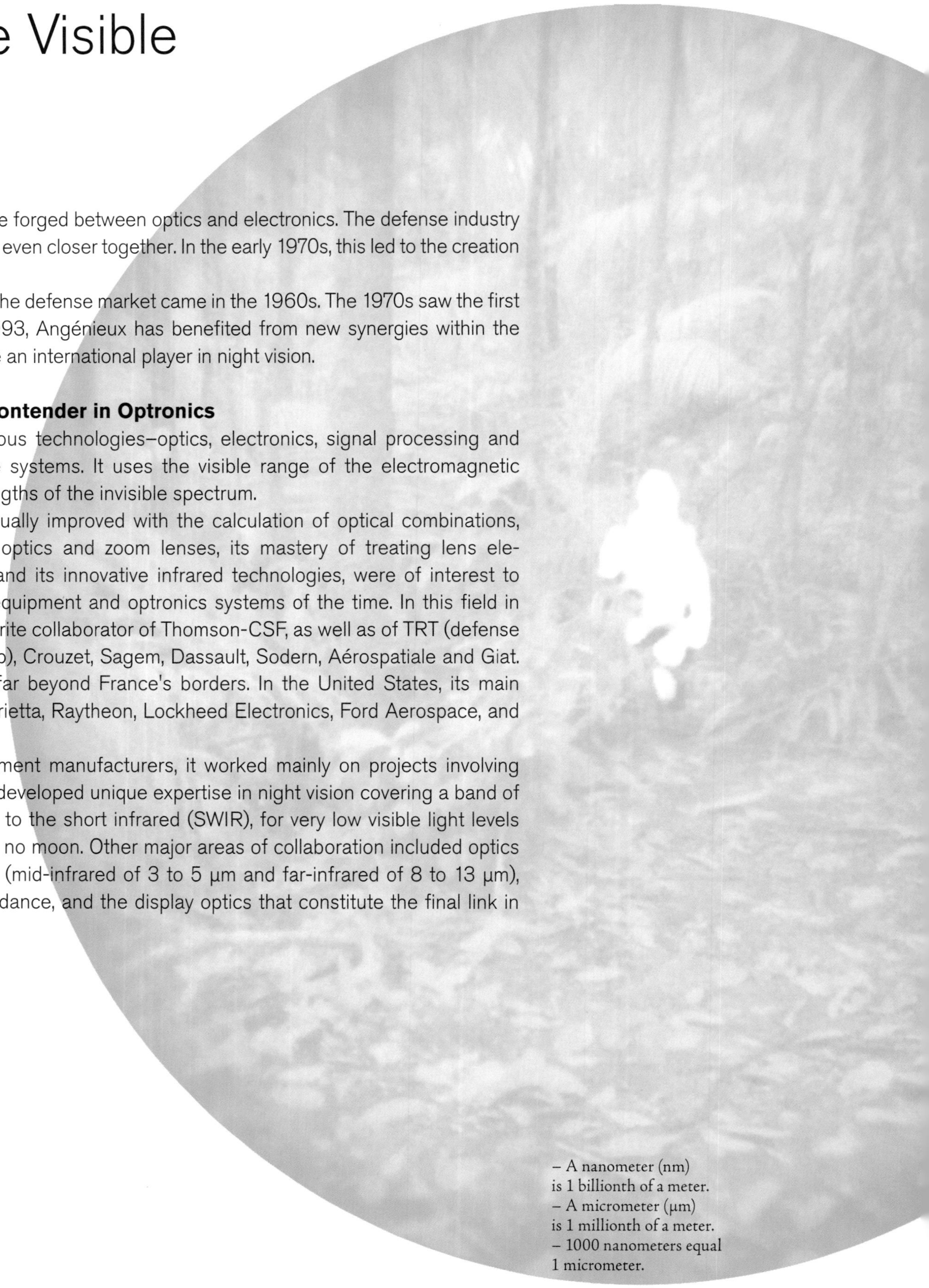

With television, strong ties were forged between optics and electronics. The defense industry would bring the two disciplines even closer together. In the early 1970s, this led to the creation of a new discipline, optronics.
Angénieux's first products for the defense market came in the 1960s. The 1970s saw the first major achievements. Since 1993, Angénieux has benefited from new synergies within the Thales Group and has become an international player in night vision.

Angénieux, avant-garde contender in Optronics

Optronics makes use of various technologies–optics, electronics, signal processing and mechanics–to form complete systems. It uses the visible range of the electromagnetic spectrum and certain wavelengths of the invisible spectrum.
Angénieux's expertise, continually improved with the calculation of optical combinations, especially for wide-aperture optics and zoom lenses, its mastery of treating lens elements with very thin layers, and its innovative infrared technologies, were of interest to the major manufacturers of equipment and optronics systems of the time. In this field in France, Angénieux was a favorite collaborator of Thomson-CSF, as well as of TRT (defense subsidiary of the Philips group), Crouzet, Sagem, Dassault, Sodern, Aérospatiale and Giat.
Angénieux's work extended far beyond France's borders. In the United States, its main collaborators were Martin Marietta, Raytheon, Lockheed Electronics, Ford Aerospace, and Boeing.
Alongside these major equipment manufacturers, it worked mainly on projects involving vision in low light. Angénieux developed unique expertise in night vision covering a band of the spectrum from the visible to the short infrared (SWIR), for very low visible light levels equivalent to a starry sky with no moon. Other major areas of collaboration included optics systems for infrared cameras (mid-infrared of 3 to 5 μm and far-infrared of 8 to 13 μm), laser designation weapon guidance, and the display optics that constitute the final link in an optronics chain.

– A nanometer (nm)
is 1 billionth of a meter.
– A micrometer (μm)
is 1 millionth of a meter.
– 1000 nanometers equal
1 micrometer.

The Spectrum of Light in Optronics

Light is made of electromagnetic waves. Some of these waves are visible to the human eye (the spectrum of visible light). Others cannot be seen: gamma rays, X-rays, infrared and radio waves. Optronics systems use visible light and infrared light. The infrared spectrum includes:

– near-infrared (NIR), near the visible spectrum, with a wavelength between 0.7 and 1 µm

– short-wavelength infrared (SWIR) with a wavelength between 1 and 1.4 µm. Night-vision devices use these rays as well as ambient light, such as from stars.

– mid-wavelength infrared (MWIR), in the center of the infrared band, with wavelengths between 3 and 5 µm, which makes it possible to detect bodies at an average temperature of 600K (620°F), such as aircraft.

– long-wavelength infrared (LWIR), with wavelengths between 8 and 13 µm, which can detect bodies around 300K (80°F) and can be used for night vision.

MWIR and LWIR are rays emitted by the heat of an object. NIR and SWIR are rays emitted by light reflected or absorbed by an object.

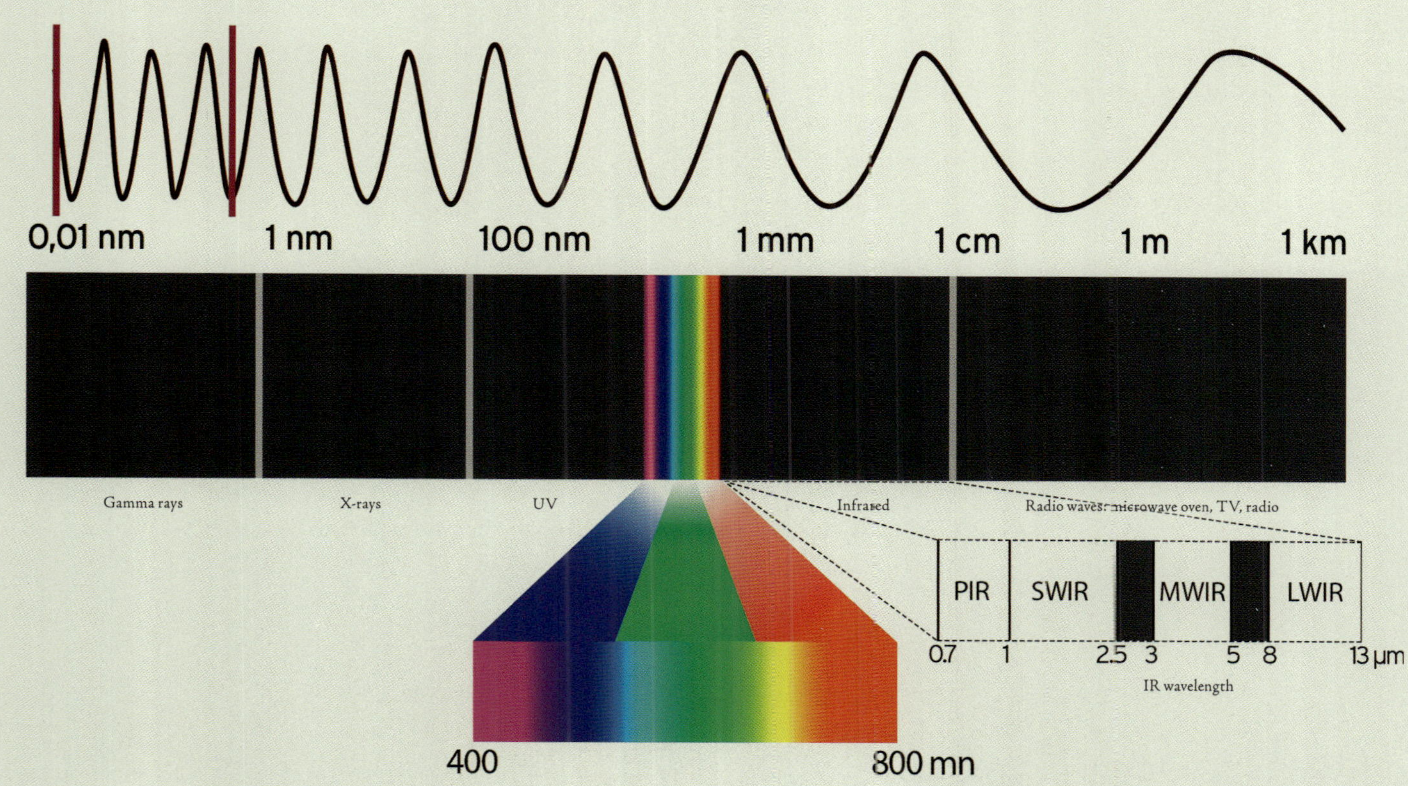

Image Intensification and Thermal Vision

Night vision relies on two different technologies: light amplification and thermal (or infrared) vision. Night-vision devices collect tiny amounts of ambient light (visible light and some infrared rays). This faint light may come from the moon, stars, and other smaller sources. The device will then amplify it (up to 50,000 times) to produce a visible image on the screen.

Thermographic cameras, on the other hand, detect and create video images from infrared radiation (heat waves) emitted by objects as a function of their temperature. LWIR is the band of the spectrum most used in this area. At the core of a night-vision device is a photomultiplier tube that collects visible and near-infrared light to amplify it and make it visible to the naked eye. A lens concentrates this light. The first end of the photomultiplier tube converts the photons (light) into electrons (electricity). The second end of the tube is a phosphor screen that the electrons strike, reproducing the original image seen through the lens. It is the phosphor in the screen that gives this image the green color characteristic of night-vision devices.

In a thermographic camera, the lens concentrates infrared light. This light is scanned by infrared detectors in the device which determine the temperature of the zones as a function of infrared wavelength. The thermal image thus created is converted into electrical pulses.

In the civilian world, the image is generally represented in color according to the temperature of the zones observed.

Red traditionally represents hot areas, and blue, cold areas.

Sophie thermographic cameras have applications in defense and area surveillance. They are also used in rescuing victims during earthquakes, for example, or in the fight against poaching networks.

below
Left, intensified image: Lucie night-vision goggles.

right
Thermal image: Sophie camera (Thales photos).

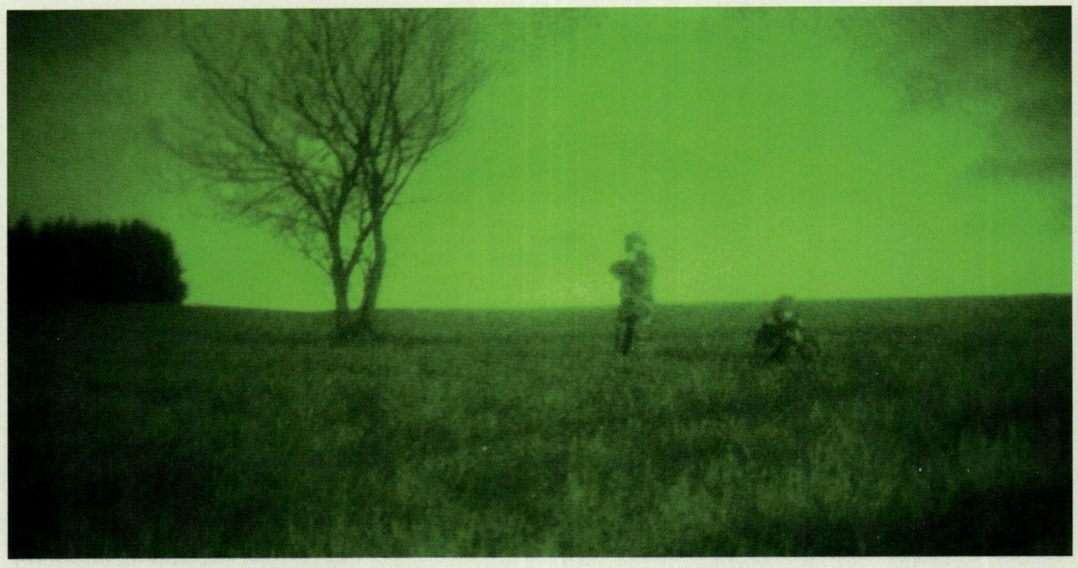

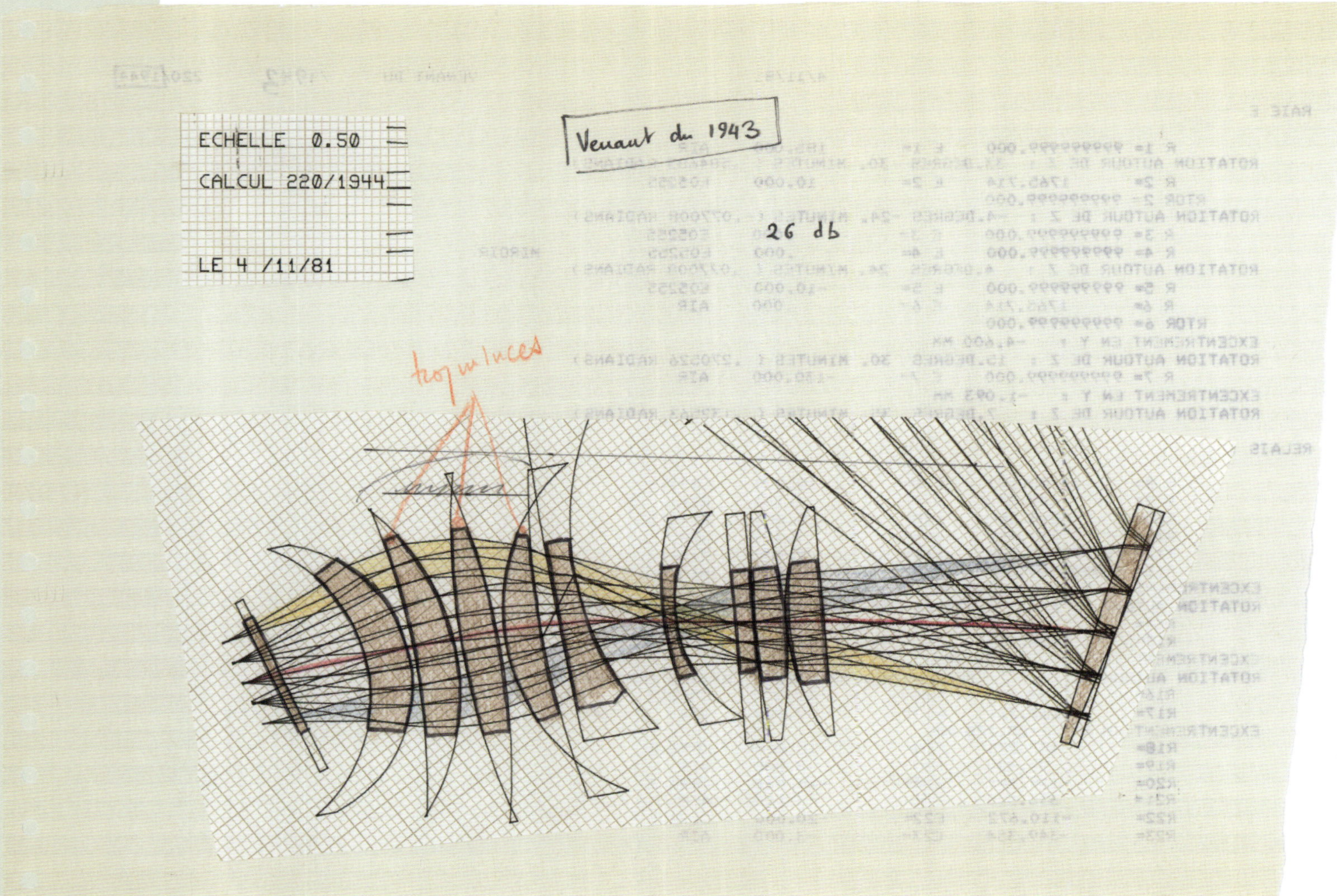

above
Notebook page with optics calculations for the design of the Mirage VH4 holographic display, Jacques Angénieux, 1981. (Angénieux archives).

In the displays sector, Angénieux participated throughout the 1970s in Thomson-CSF's work on Head-Up. This display allows pilots to see flight information and/or aim at a target without having to turn their gaze towards instruments inside the cockpit. In 1979, Thomson-CSF turned to Angénieux for its optical and mechanical expertise, considering the concept of a holographic display, at the beginning of the NVH series (Nouveau Viseur Holographique–New Holographic Display) of avionics displays for Mirage and, later, Rafale fighter jets.

In the early 1970s, Angénieux also supplied Thomson-CSF with infrared lenses with F=210mm and F=280mm focal lengths for observation and night shooting. They were used on AMX-10 and AMX-30 B2 tanks. Angénieux then offered TRT optics systems for its thermographic cameras. Thermidor (1975), Mephisto (1981), and then Castor (1985) would be employed in the AMX30 tank, for which Angénieux also provided an infrared zoom lens, F=229 at 515mm, and three bifocal lenses.
In 1973, Angénieux began work with TRT's night vision equipment programs using residual image intensification. Angénieux supplied various optics subassemblies (lenses, eyepieces, prisms) for OB31 episcopes for driving tanks at night and for OB41 and OB42 night-vision devices.

Night-Vision Optics

In night vision, as with optics that operate in the visible band of the spectrum, images are generated by optical arrangements comprised of lenses, prisms and mirrors, but the materials and processes are completely different.
The lenses are silicon, germanium, zinc sulphide, or zinc selenide, which are transparent for infrared radiation, and sometimes opaque for visible light.
The reflective layers are gold.
For these new systems, the materials were machinable. Making aspherical surfaces was easier. Aspherical lenses can correct aberrations, simplify optical combinations, and thus reduce the cost and weight of equipment. In 1975 Angénieux developed an original process of aspherization (action that transforms a spherical lens into an aspherical one) by vacuum deposition modulation.

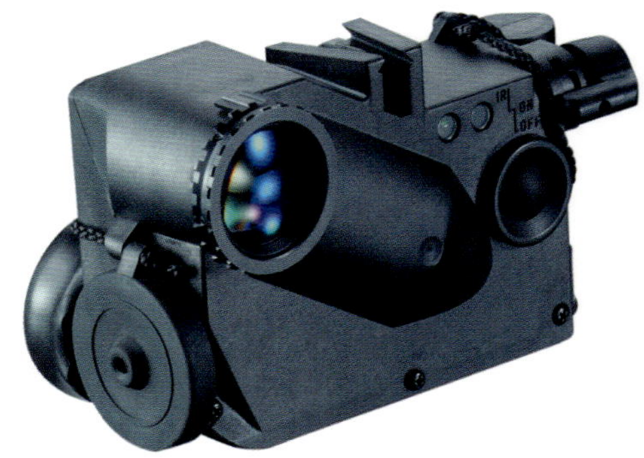

above and opposite
Lucie night-vision device for infantry. More than 70,000 of these goggles have been sold worldwide (photo by Thomas Goisque).

The first military use for a zoom lens was presented in 1975 at the Paris Air Show by the Swiss company Contraves for the surface-to-air Skyguard defense system. It was an Angénieux 10x30 zoom. Angénieux would later develop other lenses, including an F300mm and a 3x200 zoom with F/6.4 aperture for other defense systems.
In 1985, Angénieux's optronics department delivered an infrared simulator to the American company Raytheon for the Maverick missile test. The system was based on a 100x zoom, 3 meters long (10 feet), and weighing 40 kg (90 pounds), the largest zoom lens ever made.
From 1979 to 1985, Angénieux took part in the production of optics components for the NOVA laser from the Lawrence Livermore Laboratory in San Francisco, which was then the world's largest laser-based fusion laboratory.
In the early 1980s, at the height of the Cold War, the American company Martin Marietta ordered optics subassemblies from Angénieux to equip the new ADATS (Air Defense Anti-Tank System) weapon system of the United States and Canada, unique in its dual surface-to-surface and surface-to-air function. This system required implementation of three infrared zooms delivering a coded laser beam to ensure temporal and spatial guidance of the missile. Martin Marietta explains why Angénieux was chosen: "Zoom systems are typically avoided in military instruments because of the difficulties of maintaining boresight through the zoom range as well as through the operational environmental conditions. This problem is further compounded in ADATS by the fact that these zoom optics define line of sight for the entire ADATS system. Therefore, achieving and maintaining the system boresight is perhaps the most challenging aspect of GBA (Guidance Beam Assembly) design. It was for this reason that Martin Marietta sought out and contracted a world leader in the zoom lenses design. The Etablissements Angénieux, located in Saint-Héand, France, reviewed the requirements and submitted a winning proposal for the design of the ZPO (Zoom Projection Optics) and Command Optics (MTZ - Missile Tracker Zoom, and TMLO-Temporal Mode Laser Optic)".
With this program, the stakes were high for Angénieux. An American employee was posted to Saint-Héand for several years. Angénieux would provide more than a hundred zooms and optics systems. But the breadth of the ADATS program was not as expected. The end of the Cold War, still symbolized today by the fall of the Berlin Wall in 1989 and the dissolution of the USSR in 1991, marked the improvement in relations between the United States and Russia.

Optoelectronics
The term optoelectronic refers primarily to components that transform incident photons into electrons (detectors) or electrons into photons (emitting diodes, lasers, etc.).

Thales gives Angénieux's Optronics Work New Momentum

In the 1980s, the Angénieux family recognized that in order to ensure lasting durability for its company–which then counted over 600 employees–it had to bring on board a major investor. In 1986, a share was first sold to Essilor. It was acquired in 1993 by Thomson-CSF, which became Thales in 2000 with the merger of the military business of Alcatel, Thomson-CSF and Dassault Electronique. The group truly gave the company new momentum, and did not cease to develop it, in both civil and optronics work, while respecting the company culture, tied to its employees' exceptional and unique expertise.

In the early 1990s, the Gulf War brought to light the urgency of replacing a great deal of obsolete equipment. Angenieux's optronics work now focused on the development of night-vision devices and complex optical subsystems for other units in the group.

Lucie, the First Night Vision Goggles for Infantry

Drawing on its own expertise in opto-mechanics, as well as Thomson-CSF's optronics experience in image intensification, Angénieux developed a new night-vision goggle in response to a French call for tenders from the Direction Générale de l'Armement (DGA) for infantry equipment.

Based on an original structure patented by Angénieux, this goggle, called Lucie, differed from the competition in being lightweight, having a flat shape that rested close against the face, and offering an expanded field of vision. It won an initial DGA contract for 10,000 Lucies in 1996. Many other contracts in France and abroad followed, bringing the number of Lucies currently in use worldwide to more than 70,000, in over 30 countries.

This success made for lasting work at Saint-Héand. Angénieux's products soon included almost a dozen models of night-vision devices, covering all applications ranging from foot soldiers to modern infantry connected to a combat system, Special Forces, driving vehicles, or piloting helicopters and fighter planes.

In just a few years, Angénieux's expertise in night vision has earned international renown.

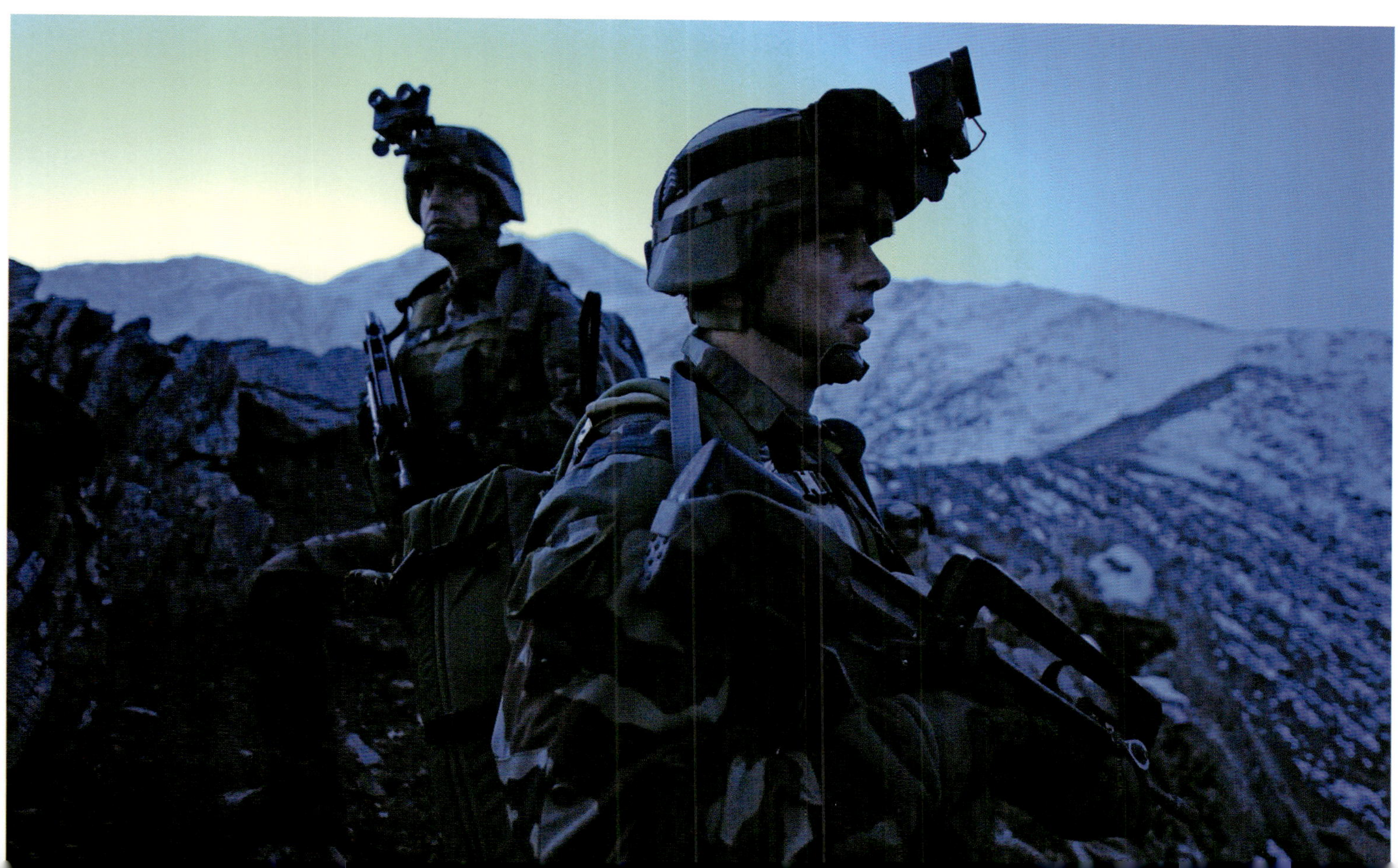

Biocular and Binoculars

Angenieux night-vision goggles are either biocular or binocular. With a single lens and a single tube, the biocular have the main advantage of cost and compactness, but do not offer 3D vision equivalent to that of the eye. Binoculars have two lenses and therefore two tubes. Their optical qualities are exceptional.
Today, the Angénieux catalog of night-vision devices includes bioculars: Lucie, Minie (lighter version of Lucie), Minie-D and Minie-Dir. The binocular types are Helie, NVD, Bonie HP (High Performance) and Bonie LW (Low Weight).
In 2009, the Minie-D connected device, which integrates image intensification and augmented reality, was selected by the DGA through a contract with Sagem (now Safran Electronics and Defense) for the future French soldier system Félin. This contract earned Saint-Héand a visit from French Minister of Defense Hervé Morin.
The Minie-D/infrared variant, which fuses image intensification, thermal imaging and augmented reality, was then selected by the German Forces for their IDZ (Infanterist der Zukunft) system.
Today, the Special Forces use Bonie binoculars in two versions, High Performance with f/0.95 aperture and Low Weight with f/1.2 aperture, versions designed specifically for them.

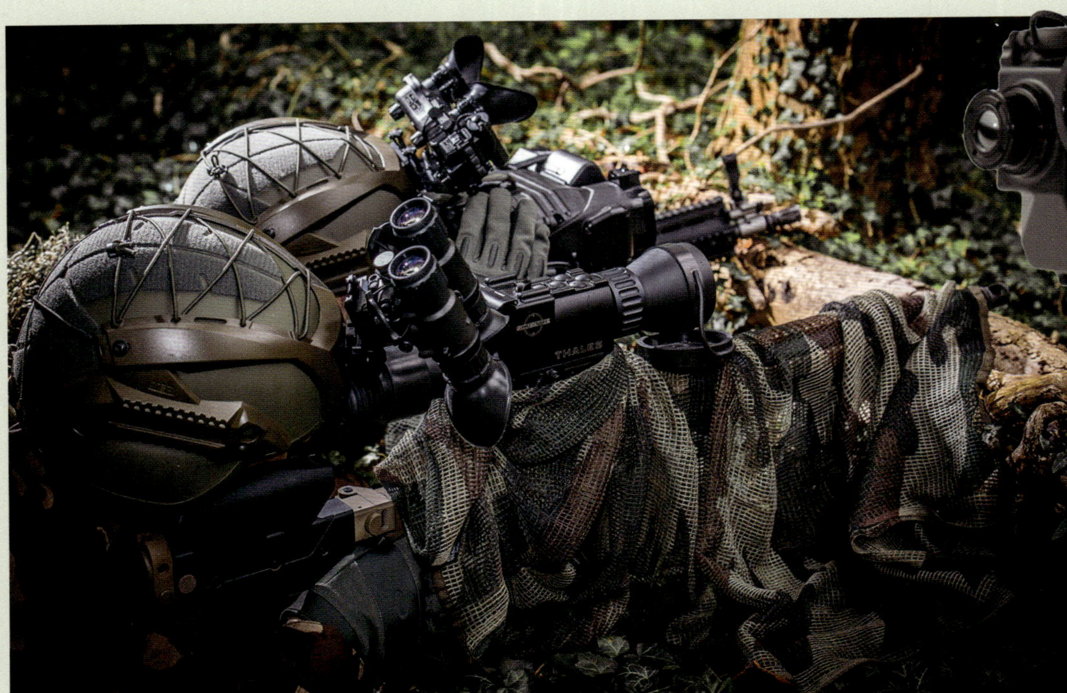

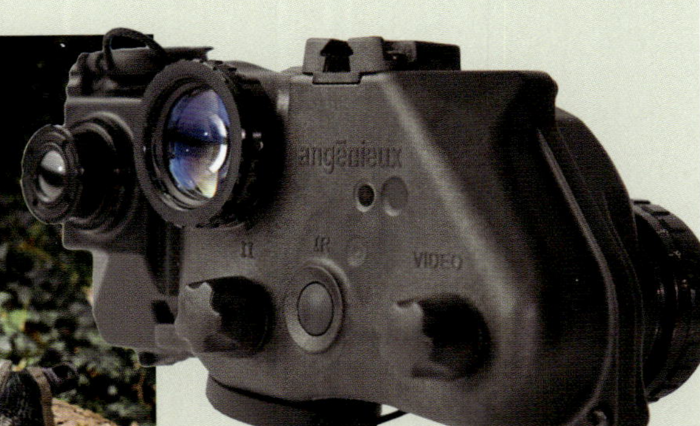

above
Minie-Dir night-vision device for infantry.

left
Bonie and Minie-Dir devices in position on soldiers' helmets.

top
Intensified image, and intensified image combined with infrared.

Angénieux and the Aeronautics Market

In 2007, under the leadership of Philippe Parain, Angénieux aimed to enter the aeronautics market. Initial research work was begun in view of an upcoming call for tenders from the DGA to update equipment for French helicopter pilots. An initial prototype was presented at the CEV flight test center in Istres a few months later. The presentation of this first prototype made it possible to understand very precisely the operational needs and to offer a product adapted to them. In flight situations, the homogeneity of an image is crucial. The quality from the margins of the field to the center have to be as uniform as possible. With the movement of a landscape, any drop in optical quality would be a disturbance for the pilot, whose gaze was focused on the image as a whole. Pilots must be able to fly at "level 5" darkness, to detect, for example, a power line on a very dark night with no moon. In addition, the devices had to be as light as possible to reduce pilots' neck fatigue.

Angénieux put to use its expertise in ultra-bright film lenses, and offered a new binocular design, Helie, which thoroughly met the operational needs of the French Army. These binoculars had an f/0.95 aperture without compromising optical quality, far beyond the f/1.2 aperture offered up to that time. Their mass was equivalent to that of competing goggles and their optical performance was exceptional for night flights. In 2009, the French Army began using Helie for helicopter pilots, a first contract that opened export possibilities for the binoculars.

Meanwhile, Angénieux also marketed a night-vision device (NVD for battle) for fighter pilots, to provide an economical and very light solution, compatible with unplanned ejection of a pilot in full flight, and also fire-resistant. Beginning in 2011, the first NVDs were provided to pilots of the French fighter jet Rafale.

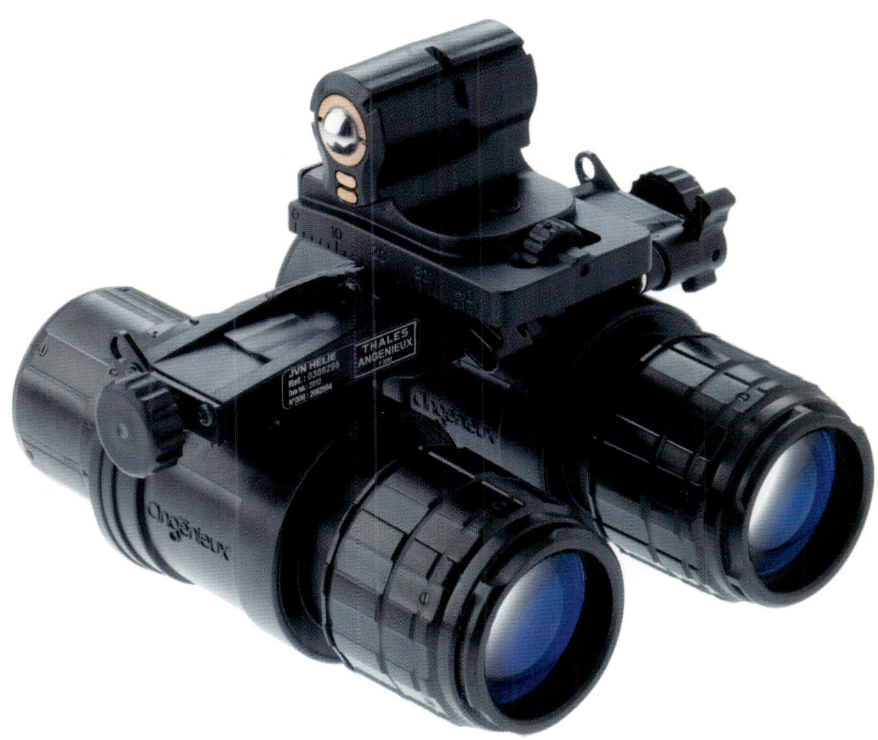

Helie night-vision binoculars for helicopter pilots.

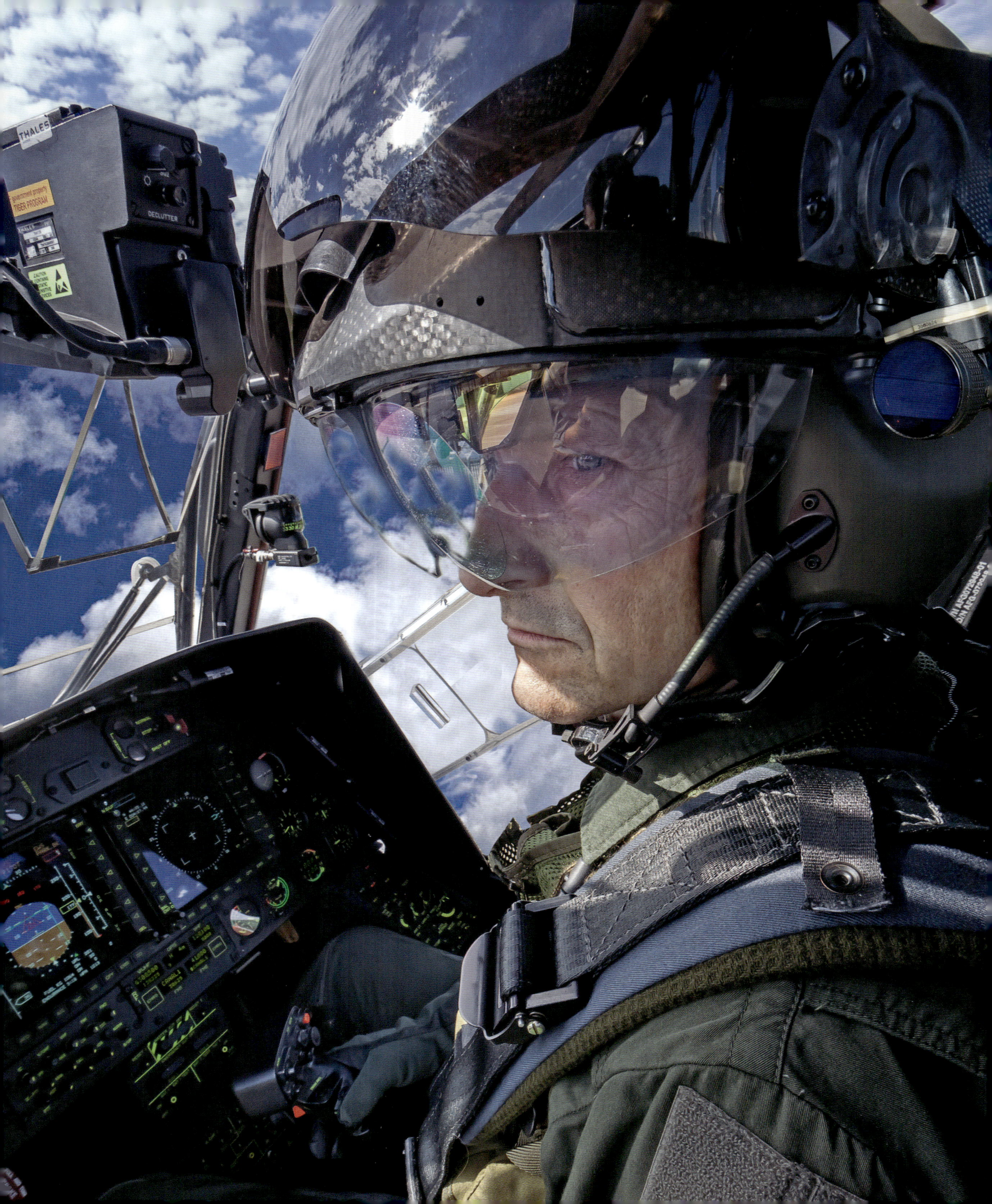
THALES
DECLUTTER

opposite
Pilot with TopOwl helmet in the cockpit of a Tiger helicopter (photo by E. Faz).

Complex Optics Subassemblies

Within the Thales group, the recognized expertise of the Saint-Héand site was regularly solicited when it came to supplying complex optics subassemblies for complete optronic solutions.

Given flying speeds, fighter pilots had to have access to information for piloting and aiming without having to turn their heads They had HUD (Head-Up Display) systems. A transparent glass screen, originally flat but now curved, is placed at the height of the plane's windshield, in front of the pilot. All the information necessary for piloting or aiming is projected via this screen, while maintaining the exterior view by day or intensified image night vision. Saint-Héand took care of fabrication and/or design of HUD subassemblies (optical projection system and glass display screen) for Rafale in particular.

By contrast, helicopter pilots, who are always moving their heads during flight situations, have HMD (Helmet Mounted Display) systems with flight and targeting information projected onto a visor integrated into the helmet, along with direct daytime vision or image-intensified night vision. Angénieux supplied the entire optical subassembly of Thales's TopOwl helmets. Since 1995, more than 2,000 of these helmets have been delivered. They equip, among others, pilots of Tiger and NH90 helicopters from Airbus Helicopters.

Internationally Recognized Expertise

Backed by unique expertise in optics and a strong culture of technological innovation and close contact with users and operators, the solutions developed at Saint-Héand offer an unprecedented agreement between performance, ergonomics and robustness.

In Saint-Héand, an original approach–the company spirit continues–wins the appreciation of clients and users for the performance of products for night missions. Special Forces, DGA delegations and other military authorities are regularly invited to binocular demonstrations in a nearby forest without the nuisance of light. In night vision, the competition is tough, the battle is fierce, and innovation remains the watchword. The next night-vision devices will surely see further improvements in ergonomics and optical performance. But the real challenge is the transition to digital and leaving electronic tubes behind, a technological transition already experienced by television, still to come in the world of night vision.

Security and Surveillance
Optics at the Service of a Safer World

Optics for security or surveillance must meet a number of new requirements. The Saint-Héand teams are capable of offering custom solutions tailored to the specific needs of each use, whether civilian or military: aerial photography and mapping; surveillance at sea, on coastlines and borders; surveillance of critical structures (nuclear power stations, soccer stadiums, etc.); one-off events (G7 meetings, Olympic Games, etc.); surveillance by air (police/customs) and of borders and coasts; search and rescue (SAR); trajectory tracking; etc. Angenieux zooms incorporate excellent visible image quality in near-infrared (NIR) for detection, reconnaissance or identification. They are robust enough to tolerate integration in systems subjected to extreme conditions, such as in gyrostabilized helicopter and drone ball mounts, or surveillance turrets and poles.

The teams, in close contact with the clients their solutions are intended for, have been able to develop optical processes that allow the best transmission of light in near infrared and thus optimize low-light performance. With their mechanical performance, these zooms offer the very precise tracking essential for following a target.

The processes of athermalization[1] and apressurization[2] ensure these zooms the same image quality regardless of temperature and altitude.

Mastery of large zoom ratios for television enabled the company to offer different versions of 60x (60x15, 60x30, and 60x50) and 40x (40x15HD and 40x34) zooms for long-range surveillance applications such as border surveillance or maritime surveillance.

With the advent of drones for a broad spectrum of applications–inspection of high-voltage lines, solar panels or various structures; search or rescue missions; surveillance for agriculture or to fight fires–the line of long-range zoom lenses has been expanded to include an ultra-compact, robust, and lightweight zoom, the 10x20 HD D/N. The 10x20 benefits from the latest technological innovations to best meet these new demands. In this ever-changing world, full of opportunities, these surveillance and security solutions from the Thales Group contribute to a better and safer environment. The Thales Group works alongside governments, institutions, and businesses to help protect countries, cities, and critical infrastructure in an increasingly unpredictable environment.

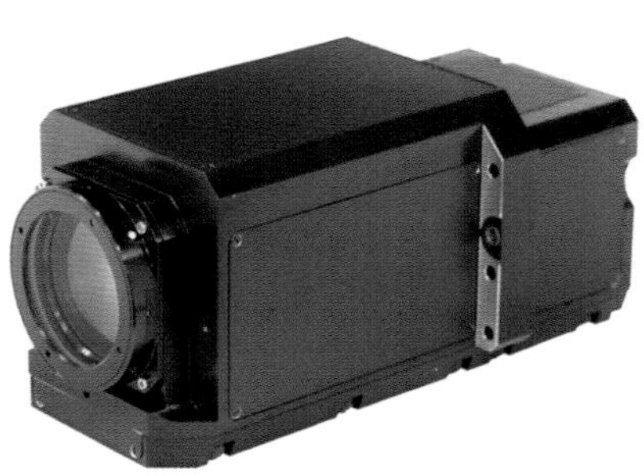

10x20 HD lens.

[1] Athermalization means rendering an optical system's performance insensitive to temperature. The image plane must remain perfectly focused on the sensor without additional focusing. This can be done in the following ways:
• active: for example, via a motor that will move an element or a lens within the optics according to a predefined rule or using a servo (autofocus). An external energy source is required. The optics and mechanics alone do not suffice.
• passive: materials are chosen so that the image plane remains on the sensor. No additional movement is needed. The optics and mechanics are sufficient on their own. There is no need to touch the MAP (focus) or the back-focus. The important parameters for the materials are the expansion coefficients of the mechanical and optical parts and the index variation of the lens elements as a function of temperature. Correct selection of these parameters is what makes it possible to achieve the desired results. This is obviously a constraint that limits the choice of lenses, but must not alter the performance of the optics: all in the art of good optical design.

[2] Appressurization is based on the same principle; in this case it applies to variations in atmospheric pressure.
A change in pressure means a change in the refractive index of air and a change in the focal length of the air lenses (space between the lenses), which changes the position of the image plane. Passive apressurization requires finding the right combination of air lenses so that the image plane does not shift with atmospheric pressure. Another constraint... which only concerns optics in flight.

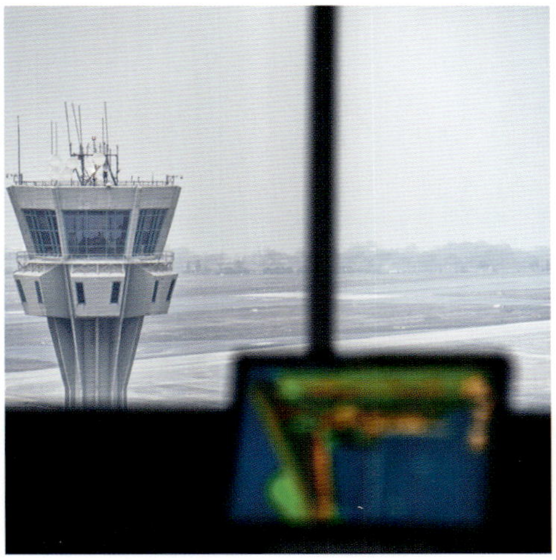

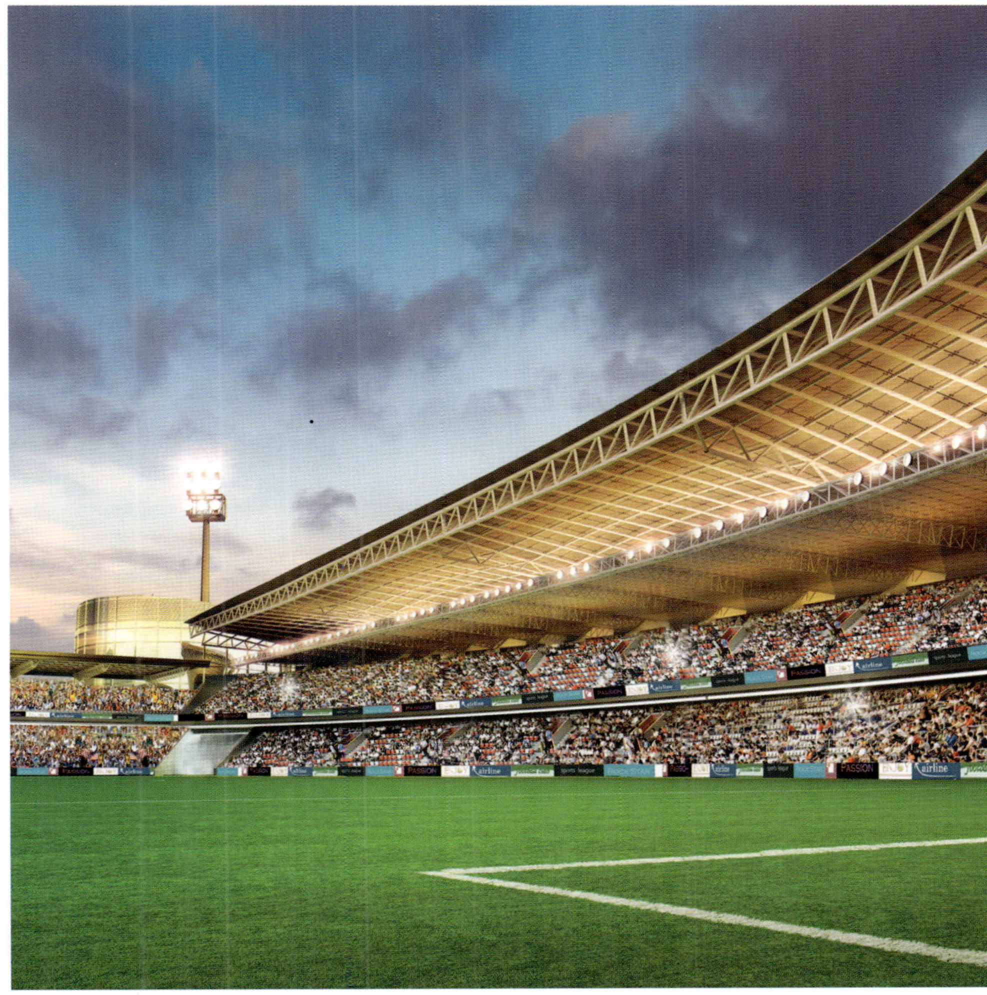

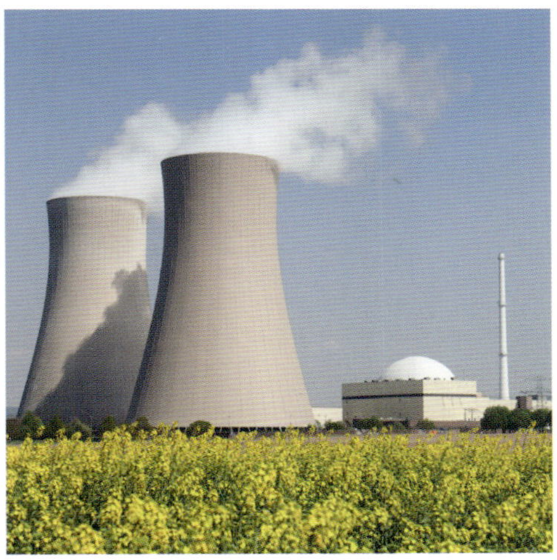

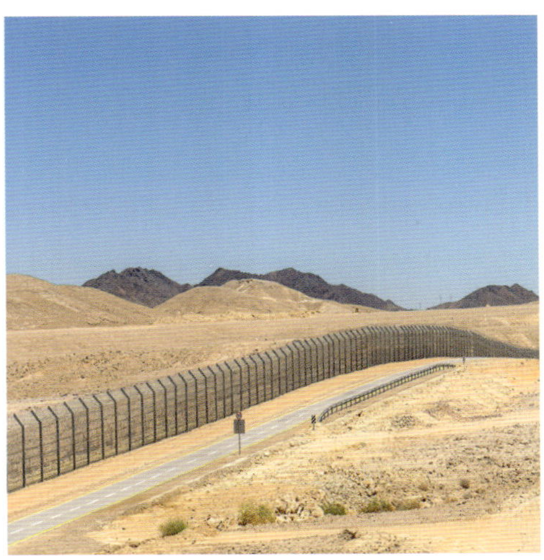

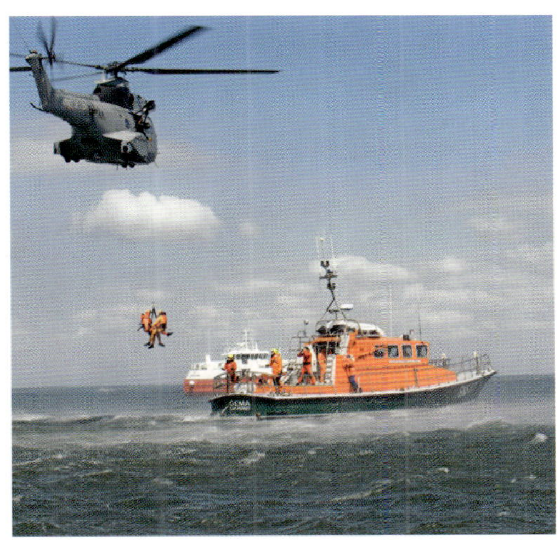

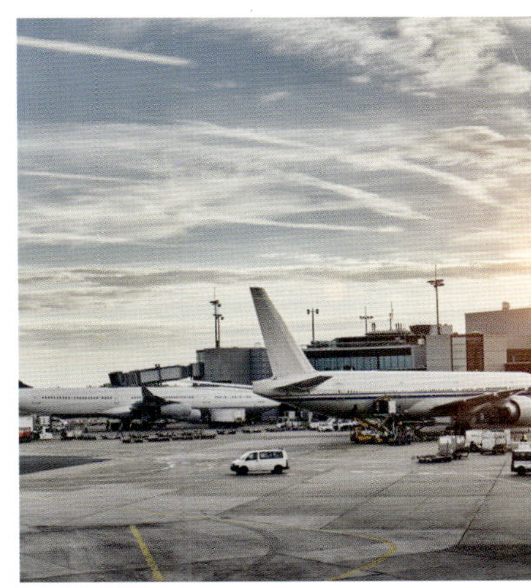

Security and surveillance solutions contribute to a better and safer world. photo by Baran Özdemir / Getty Images, photo by RelaxFoto.de / Getty Images; photo by Nick Brundle / Getty Images, photo by ViewStock / Getty Images, photo by Eric Beracassat / Gamma-Rapho via Getty Images, photo by Westend61 / Getty Images.

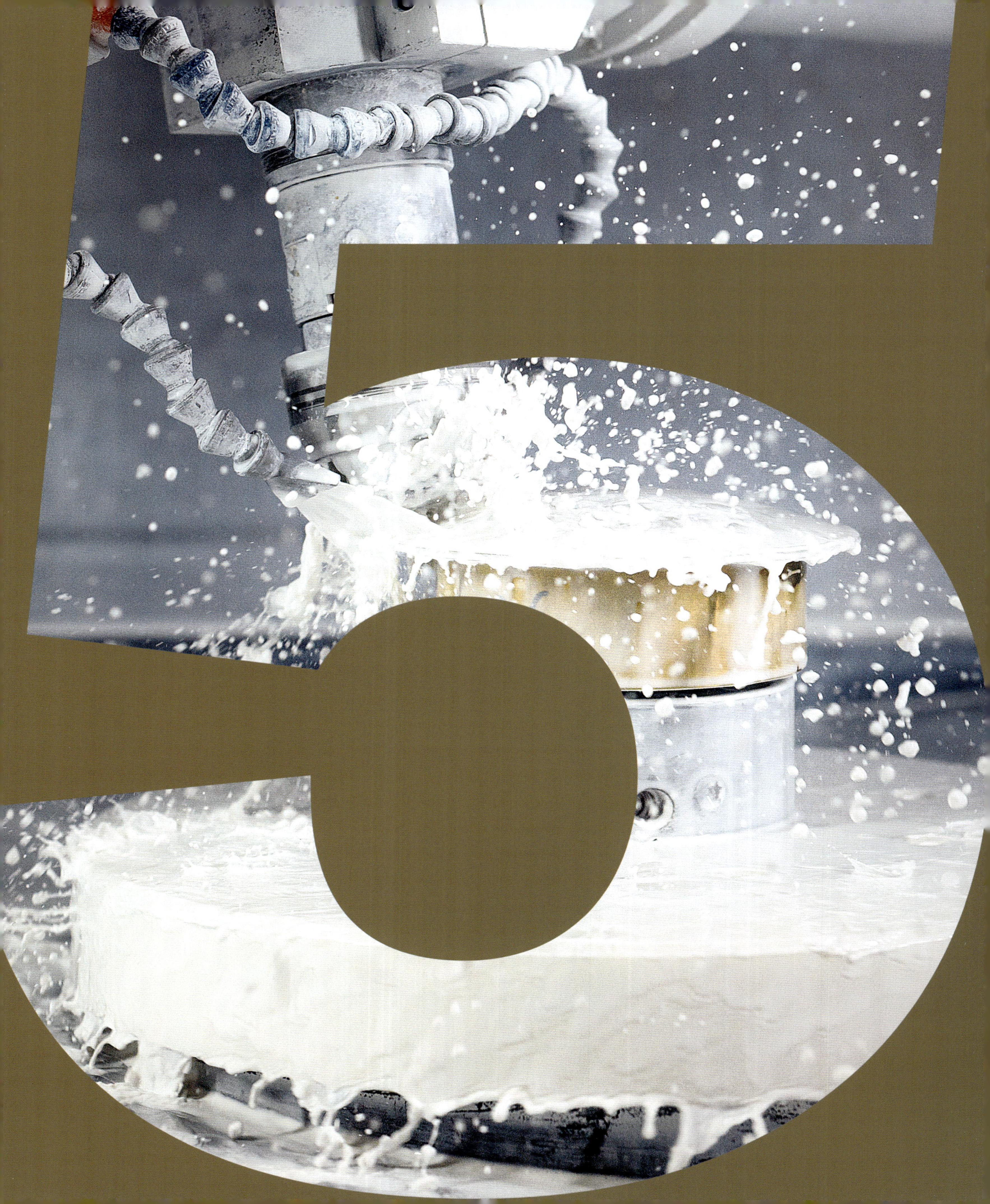
5

MADE IN FRANCE
EXCELLENCE
KNOW-HOW

The specificity of a manufacturer of precision lenses lies in the dialogue established between a **world of tradition**, where the know-how of the experts often requires years of training and of experience, and a world where **state-of-the-art digital technologies** are employed on a vast and constantly evolving scale. The **complexity** and **performance** of Angénieux lenses increases all the time and their production requires extremely high levels of **precision**.

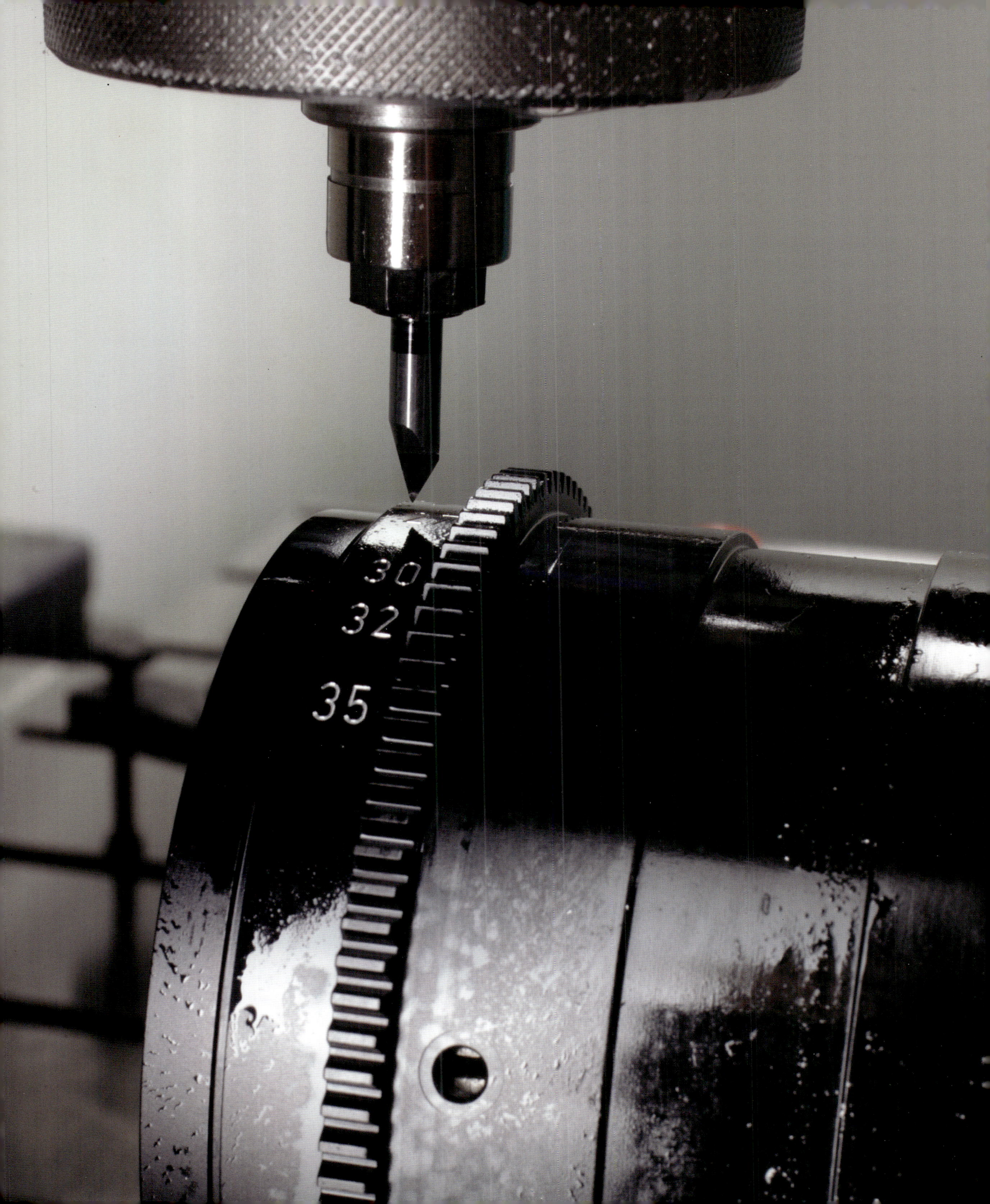
30
32
35

The Zoom: a Complex Optical and Mechanical Architecture

The introduction of every new lens is prompted by the needs of their users, namely directors of photography. As the paramount users, cinematographers can best indicate the new tools capable of meeting their needs in artistic and ergonomic terms.
These requirements are also juxtaposed with the economic constraints of Angénieux customers, primarily film equipment rental companies, production companies and TV channels. The solutions put forward must help ensure the profitability of their investment.
The role of a lens designer is to translate all of the needs expressed in terms for technical specifications such as focal length, ratio, aperture, weight, size and acceptable or desired degree of aberration[1].
The final choice of specifications for a new lens is therefore always the result of a long process of study to find the best possible balance and compromise between technical characteristics, optical performance, ergonomics and the constraints of weight, volume and cost.

In a zoom each lens—traditionally concave, convex or cylindrical in shape, convergent or divergent—will involve its set of aberrations. The success of a good optical architecture hinges entirely on the various aberrations compensating for one another in the end to obtain a perfectly correct image.
The first step in the mechanical design of a zoom is the construction of an architecture compatible with the optical solution envisaged as regards the size and position of lenses, the number, position and arrangement of moving parts, adjustment mechanisms, and type of mounting. Numerous exchanges of views then take place between the optical and mechanical teams to hone this design and arrive at a high-performance article that is viable in terms of production. Once the optical combination has been finalized, all the mechanical components are specified very precisely so as to develop plans for their manufacture.
The different lens elements are housed in barrels, which can be fixed or fitted into sleeves. The focusing mechanisms are located in the heart of the zoom. The spirally grooved cam shaft ensures that rotation of the outer guide slides the lens groups in the structure of the zoom. This function still employs the basic principle developed by Angénieux. Performance has been greatly improved by the use of new materials and new technologies for the machining of aluminum alloys. The iris or diaphragm is located at the rear of zoom as well as a mount for attachment to the camera. The other mechanical parts include the casing elements that incorporate the marks engraved on the moving parts, namely the focusing and aperture rings.

[1] See the glossary, pp. 178-179.

Sectional diagram of the Optimo 17-80 zoom. The Optimo 17-80 has 23 lenses.

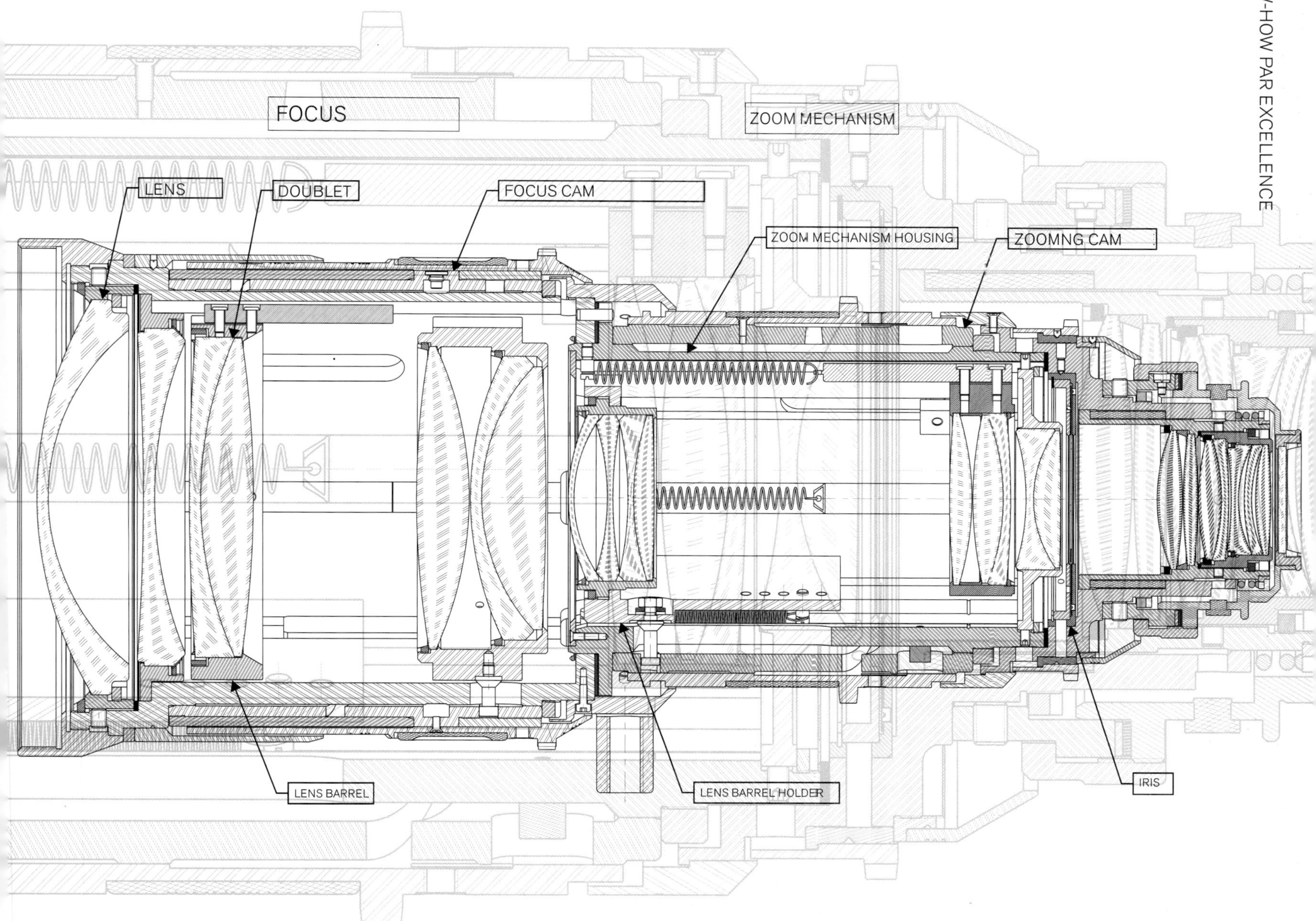

Innovative Engineers and Software: a Winning Combination

It takes many years of training in the Angénieux design teams before an optical engineer is capable of calculating a zoom. And then, despite the advances in information technology, it will take another calendar year before you can hope to design a technically flawless and manufacturable zoom.

As Angénieux has been producing lenses for over eighty years now, it has built up numerous records of calculation and each new development has a basis to start from. The starting point of the Optimo 28-76 in 2006, for example, was the architecture of the 28-70 f/2.6 for 24x36 photography developed in 1986 by Pierre Angénieux. The power of information technology provides undeniable assistance. There are today optimization and simulation programs that make it possible to calculate and visualize different configurational options billions of times faster than Pierre Angénieux's handbooks of calculations.

But the role of the engineer remains crucial. The analysis and selection of sound parameters of calculation still depend on the intuition developed by engineers over years of experience. If software can make it possible to simulate and estimate the rendering of an image through a given architecture in advance, only an experienced engineer can analyze the simulations and indicate the parameters to be modified.

Thanks to the power of these tools of digital modeling, it is very rare today for the choice of an optical combination to require revision. While it was once necessary to produce a prototype in order to assess the effective quality of an optical solution, the mass production of a zoom today is launched on the basis of simulated images with no need for prototypes.

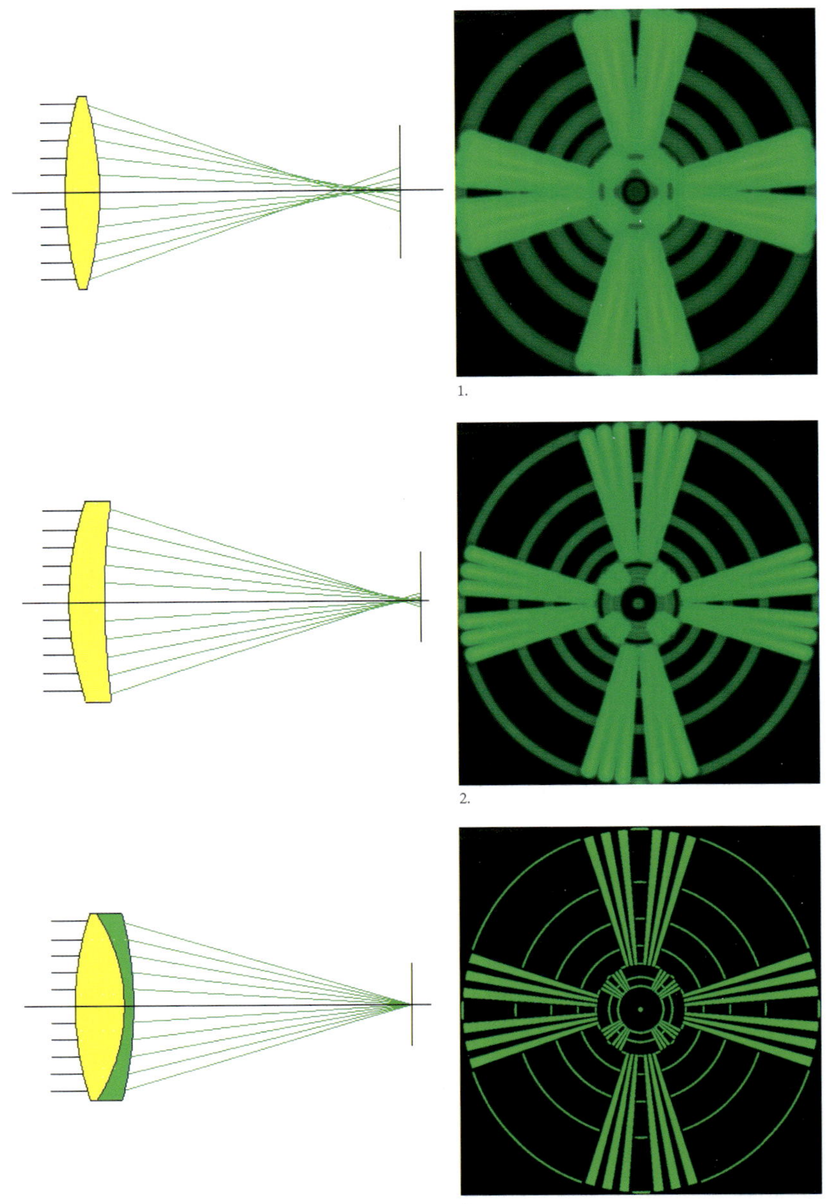

Example of the correction of spherical aberration by means of simulation software.

1. Spherical aberration.
2. Change of curvature and increase in the index of the lens. Spherical aberration partly corrected.
3. Application of a doublet. The remaining spherical aberration is corrected.

Enhancing the Performance of a Zoom

With the introduction of the new Digital Camera Initiatives[2] (DCI) specifications, images can no longer tolerate even the slightest optical flaw. If aberrations are to be invisible to the naked eye, they must theoretically be no greater than the pixel size of the cameras on which the zooms are used. This requirement is continually changing as the size of the pixels decreases in proportion to the ever-greater number of pixels of a camera sensor.

The 25-250 HR zoom developed in 1991 thus never had more than **20% of light around the image** a level of performance unthinkable in the era of digital projection.

In order to enhance the quality of Angénieux zooms still further, use is made of increasingly high-tech components and especially of high-precision aspheric[3] lenses. The selection of types of glass is obviously a parameter of paramount importance in the manufacture of these components. There are today more than two hundred types of optical glass. Each one has its own characteristics as regards refractive index, transmission dispersion, secondary dispersion, variation in relation to temperature, price, and manufacturing viability, and is subjected to a very precise selection process.

An Optimo 24-290, for example, has twenty-four lenses.

The use of aspheric lens elements, whose machining requires extremely high levels of precision, has two advantages. It makes it possible not only to correct defects like spherical aberration and coma by directly adjusting the shape of the lens element but also to reduce the size and weight of a zoom by minimizing the number of lens elements. Now used for years in photography and night-vision binoculars, where the level of complexity is not so high, these aspheric lenses are crucial to the design of the Angénieux range of compact zooms: Optimo 15-40, Optimo 28-76 and Optimo 45-120, Optimo 56-152 A2S and Optimo 30-72 A2S A2S for CinemaScope.

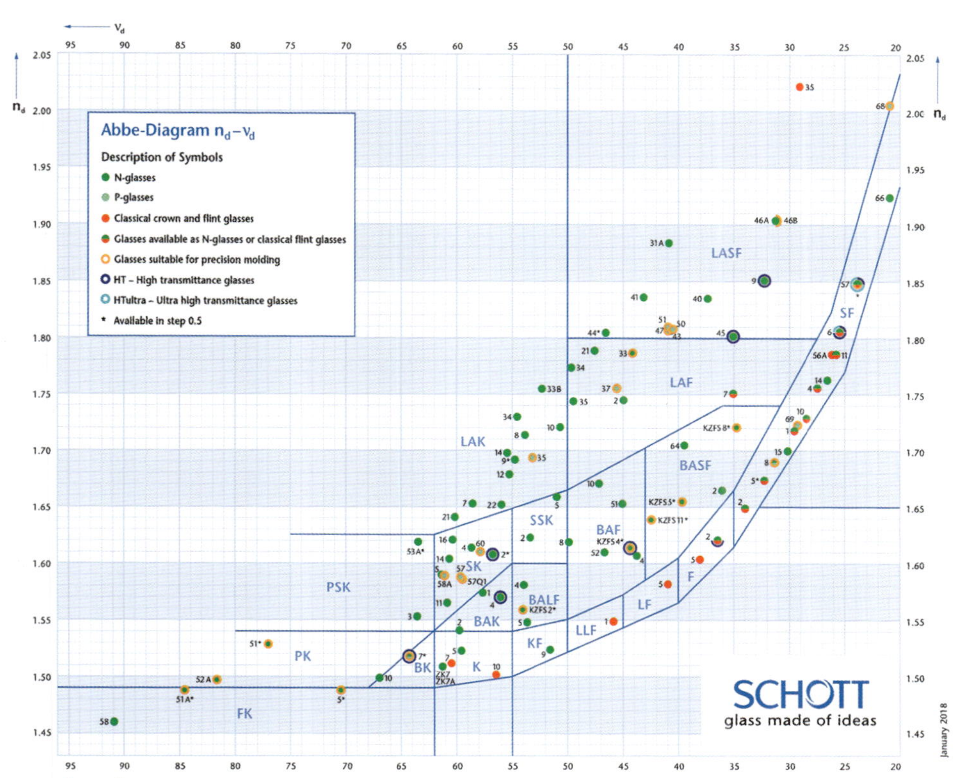

Example of the qualitative diversity offered by a well-known supplier of the industry. Each point corresponds to a different kind of glass.

[2] See the glossary, pp. 178-179.
[3] *Id.*

Items of the Optical Vocabulary

Convergent lens / Divergent lens

A convergent lens is thinner at the edges than in the center. It is the opposite for a divergent lens, where the edges are thicker than the center.
When a ray of light strikes the surface of a convergent lens outside the optical axis, the laws of the refraction cause it to deviate towards the axis. The ray *converges* on the axis. In the case of a divergent lens, an incident ray is deviated away from the optical axis. It *diverges.*

Spherical lens / Aspheric lens

The lenses traditionally used are spherical. An aspheric lens is one whose shape is close to a portion of a sphere but not strictly spherical. The interest of the aspheric lens lies in its capacity to improve optical performance.
Due to the complexity of the manufacturing process, optical systems were limited for a long time to the use of spherical or parabolic elements.
Spherical lenses introduce aberrations specific to spherical elements.
The addition of aspheric elements makes it possible to compensate for the defects of spherical aberration and coma. Due to the high cost of manufacturing aspheric elements, most optical systems use only one or two aspheric lenses, which are enough to correct a large number of aberrations. The large-scale use of aspheric lenses has led to a spectacular increase in the performance of economical photographic lenses. The machining of aspheric lens elements for cinema lenses requires extremely high levels of precision (in the region of one tenth of a micron).

Aberration

An aberration is a defect of an optical system that impairs the quality of the image. Aberrations can be divided into two groups. One consists of chromatic aberrations caused by wavelength, which cause errors in the coloring of images, and the other of geometric aberrations, dependent on geometrical parameters (especially digital aperture and angle of field as well as position of the pupil) which have an impact on the system's degree of resolution.

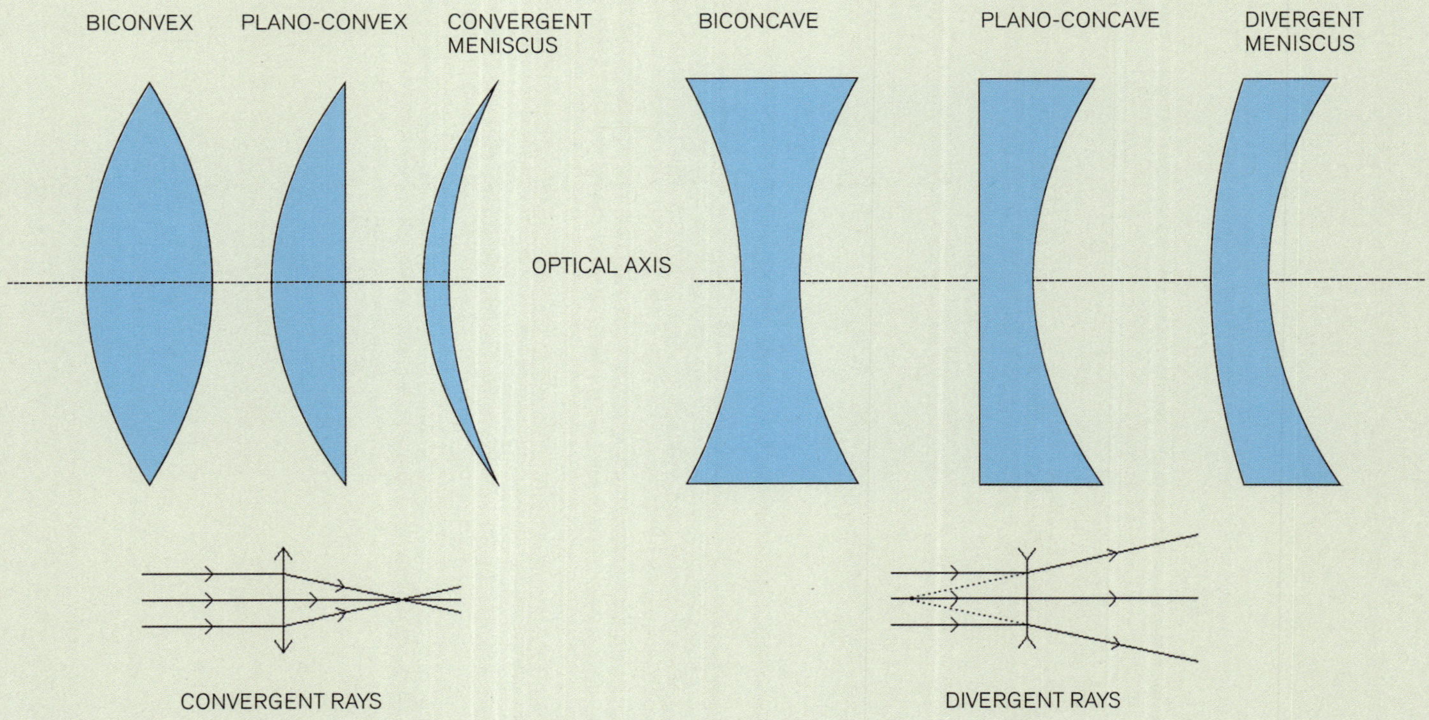

A successful optical design is one that combines the lenses in a system in such a way as to compensate for their aberrations. Chromatic aberration produces differences in focusing in relation to wavelength. It arises from the decomposition of white light into bands of colors and results in a blurred image with an iridescent outline.

Spherical aberration, coma, astigmatism, curvature of field, and distortion are the major geometrical aberrations.
Spherical aberration causes loss of focus. The rays from the edge and center of the optical system are no longer focused on the same point and produce a caustic in which the expected image point is replaced by a more or less blurred surface. Spherical aberration can be minimized by carefully choosing a particular curvature of the lens surface. It manifests itself in photography as loss of definition in areas of strong contrast.
Coma is an aberration that depends both on the field and on the aperture, i.e. on the distance of the point of impact of a ray on the pupil in relation to its center. This is called a *field aberration* because it does not exist on the optical axis. Its effect is to produce a comet-shaped spot in a plane of observation. Like spherical aberration, coma can be reduced by selecting a particular curvature of the lens.

An optical lens element necessarily has axial symmetry. As this symmetry is never perfect in practice, the focal distance from the objective is not constant around the optical axis.
An element corrected for defects of astigmatism by making its radius of curvature regular is called *anastigmatic*. Modern lenses, being calculated and manufactured by means of processes of the utmost precision, are almost entirely free from astigmatism. The correction of aberrations is effected by combining the individual lens elements in such a way that their flaws compensate for one another.
The distortion manifests itself most on straight lines.
The terms *barrel* and *pincushion* distortion are used respectively when the image of a grid appears to curve outwards or inwards.

DCI Specifications
Digital Cinema Initiatives (DCI) was created in March 2002 by the seven majors of American cinema (Disney, Fox, MGM, Paramount, Sony Pictures Entertainment, Universal and Warner Bros. Studios).
Its principal goal was to define the set of technical specifications of digital projection equipment so as to offer top-of-the-range image quality (equal at least to 35mm), to ensure a high level of protection against piracy, and to remain open to the broadest possible range of technologies and providers.
In July 2005, the DCI published a document presenting a set of specifications. These have since been subjected to technical testing in the USA and other countries, including Germany, France, and Japan, and undergone repeated revision and updating.

The Production of a Zoom: a Combination of Know-how and Digital Technologies

The manufacturing of an Angénieux zoom has undergone radical change over the last decade. The production of the optical and mechanical components of a variable-focus lens has become an incredibly sophisticated process. There is a world of difference between the Optimo 24-290 of the early 2000s and an Optimo Ultra 12x of 2017. The Angénieux teams are now able to create large-diameter aspheric lenses by polishing on digitally controlled machinery. Mechanical guidance systems have improved greatly since the first patents registered by Pierre Angénieux and performance standards have evolved so that lenses now function perfectly in extreme situations of temperature (from -20° to +40°C).
Digital technology is also at the service of the skilled workers involved directly in operations of polishing, assembly, and fitting, such as the cementing of doublets (two lenses joined together) and applying varnish to the rims of lenses by brush, where the human hand is still more precise than a machine.
Another specificity of the Saint-Héand plant lies in the dual role it has played in the civilian and military sphere since the 1970s. The teams in Saint-Héand develop and produce not only zooms for cinema but also binoculars and other complex optical systems of night vision like the HUD (head-up display) and HMD (head-mounted display) systems installed respectively in the cockpit for the pilots of fighter planes and on the helmet for helicopter pilots.
If developments in cinema prompted the development of new activities in the history of Angénieux, including the fields of television, spaceflight, medicine and optronics, interaction has now become more frequent with the increasing complexity of night-vision systems and the different spheres of activities fuel one another.

In 2013, Angénieux received the Entreprise du Patrimoine Vivant (EPV) label, awarded by the French government to firms distinguished by excellence in terms of artisanal and industrial know-how.

opposite
Above, inspection of a lens element.
Below, polishing of a lens element with magnetic fluid.

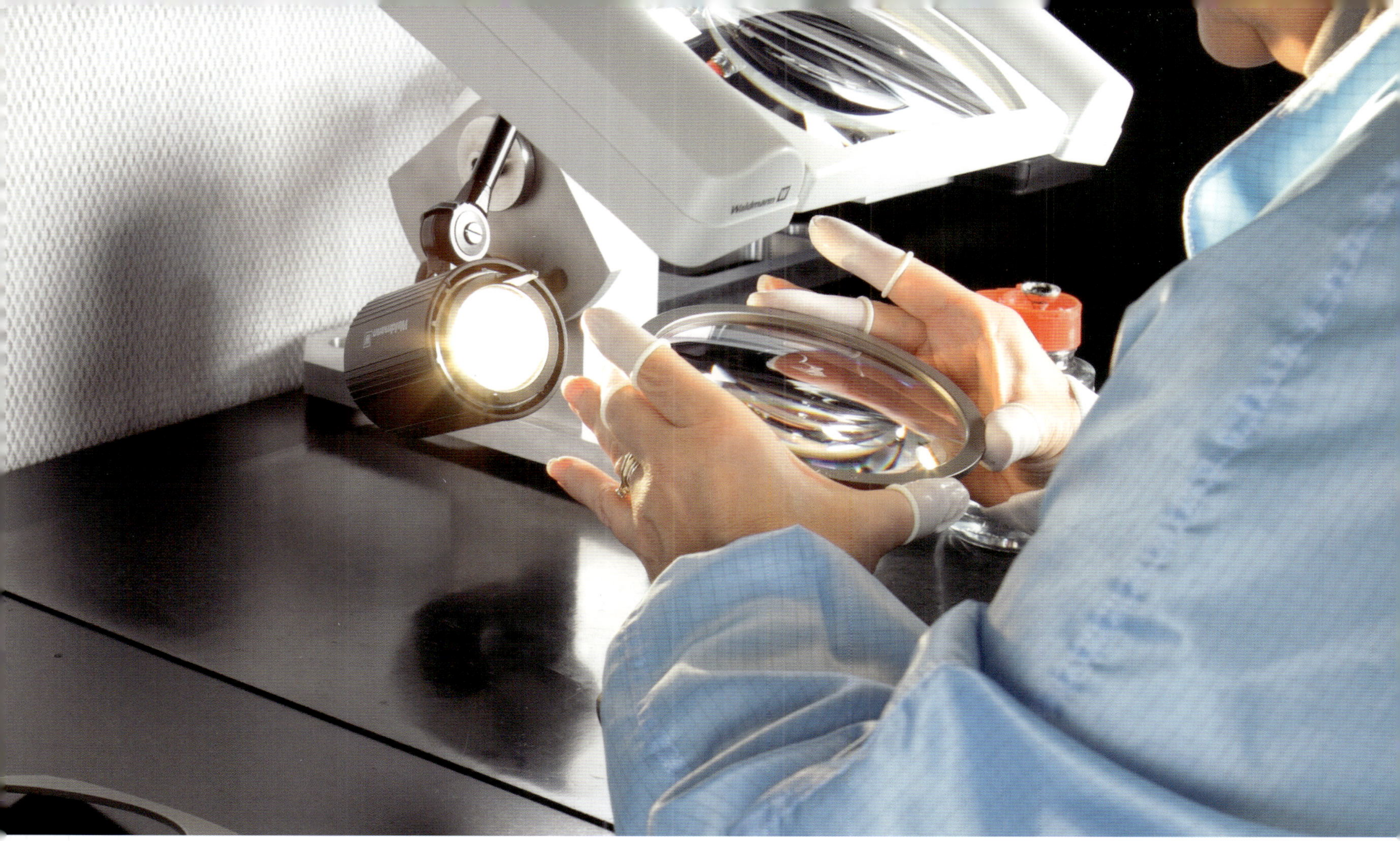
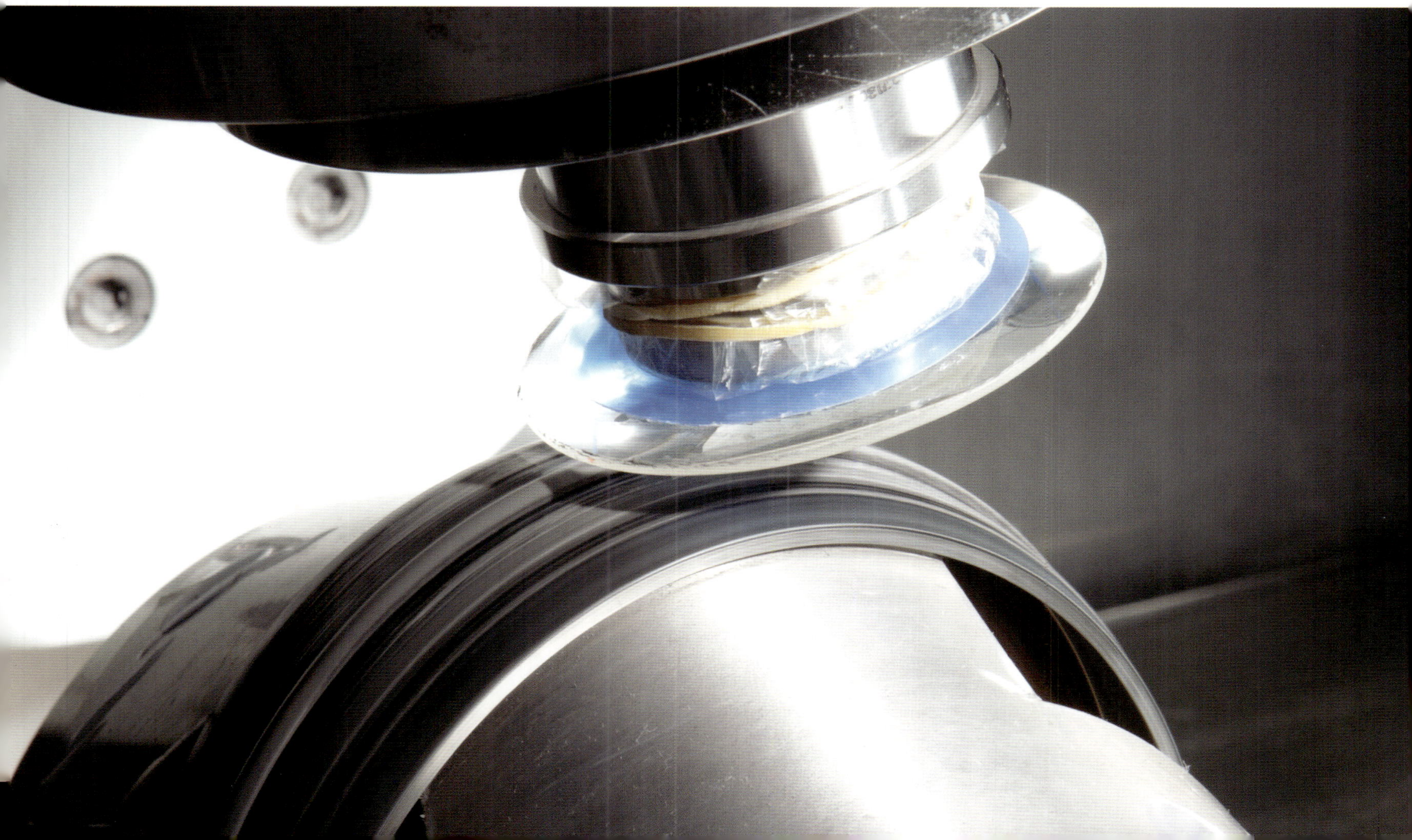

The Development of Mechanical Components

Over the years, Angénieux developed unique expertise in the production of the numerous mechanical parts that make up a zoom, including the sleeves for the different optical groups and the cam shafts for focusing and focal length. The machining of these strategic parts is carried out on digitally-controlled machinery and usually requires a precision of about ten microns.

After the machining of the elementary components, the second key stage of the mechanical process is assembly of the components of the variator and focus groups. The variator group makes it possible to achieve the zoom function, i.e. to change focal length with no loss of focus. This is a crucial element of the zoom. Fitted into a cam, it ensures the guidance of the optical groups. This stage is crucial to ensure regular functioning and obtain the "firm and soft touch" that is the hallmark of Angénieux zooms.

Production of a strategic mechanical component (Angénieux archives).

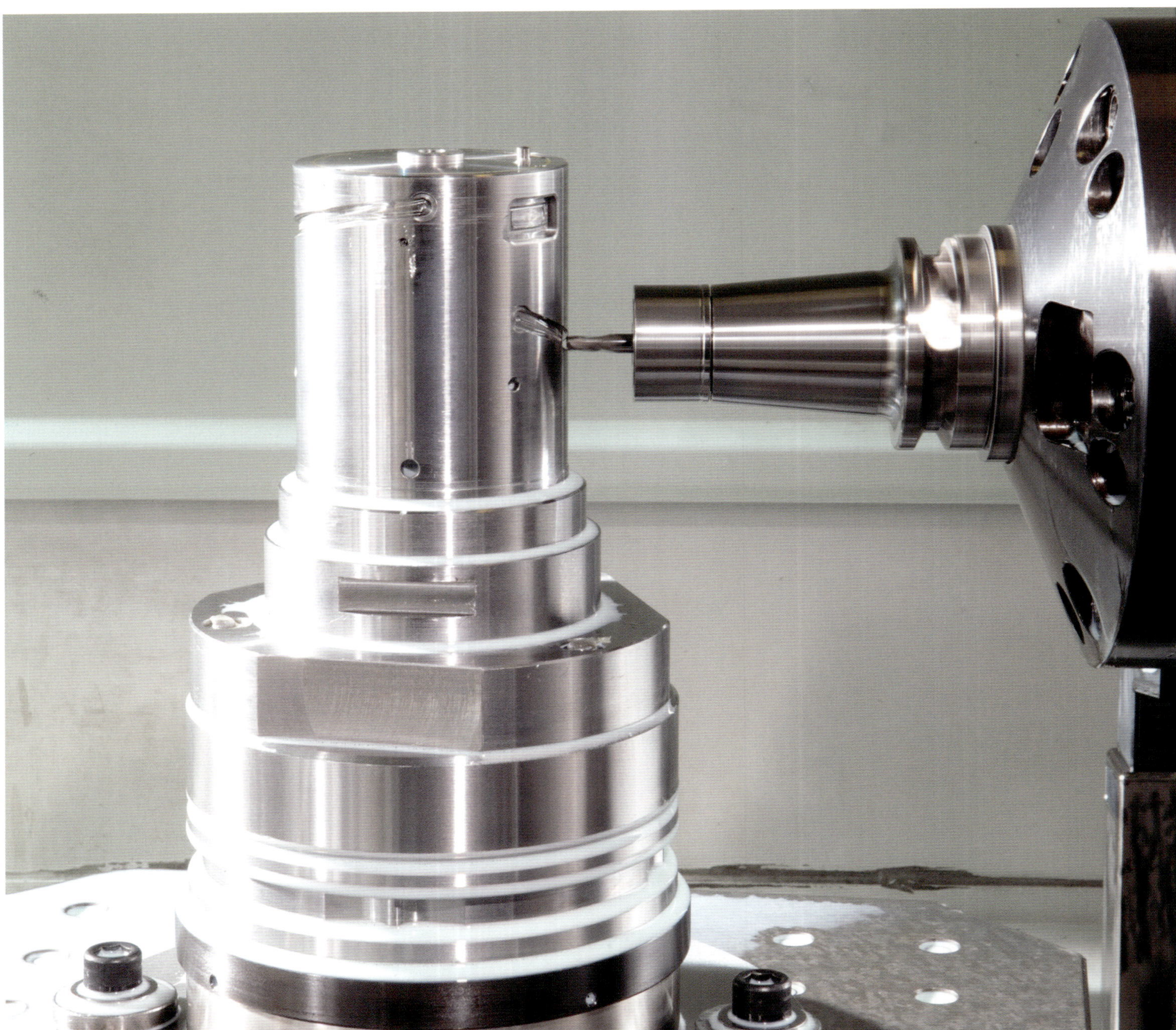

The Development of Optical Components

Starting from blocks of glass, Angénieux is able to produce lenses of the biconvex, biconcave and meniscus types[1] for spherical, aspheric, cylindrical, and even free-form surfaces with diameters from 20 to 320 mm. Once again, it is digital machinery controlled by programs developed by Angénieux's engineers that handle the different stages of lens production, from grinding the blank to polishing and edging with precision in the order of about ten nanometers (1nm = 1 billionth of a meter). In this field, **metrology** (see p. 185) is of indispensable assistance in developing and ensuring the reproducibility of these manufacturing processes on a day-to-day basis.

[1] See the glossary, pp. 178-179.

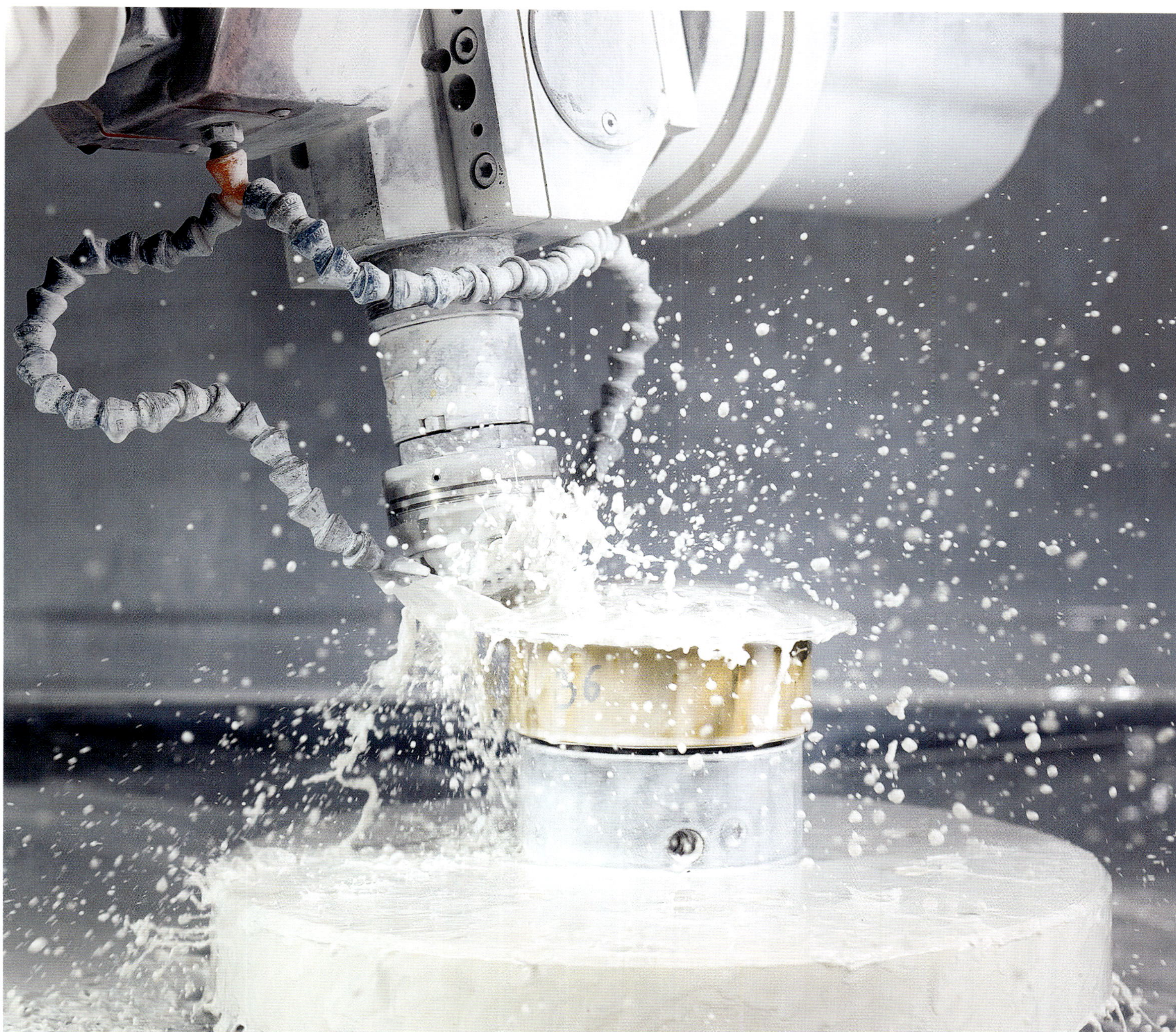

Lens polishing (Angénieux archives, photo by Nicolas Grognet).

The Coating of Optical Components

The coating of lens elements is a crucial stage in the production of an optical system. Without coating, a zoom lens would transmit no more than 8% of the light of the object filmed. If the lens elements were not coated, the final image would be completely dark and impaired by ghosting or the addition of unwanted objects and severely reduced contrast. After coating, the total transmission of a zoom with more than 20 lens elements is 90% over the entire visible spectrum of light. The optical properties of a lens can be modified through vacuum coating with one or more (up to around 300) thin layers (from 10nm to 500nm) of one or more materials. By varying to the number and nature of these layers, it is possible to increase reflection (HR or high-reflection coating) or decrease it (AR or antireflective coating). It is also possible to combine high reflection for red light with antireflection for blue. The coatings regularly used are effective across the spectrum from visible light to short-wave infrared (SWIR). The application is carried out in a cleanroom environment and requires extremely sophisticated machinery using vacuum processes, plasma generators, and electron guns to evaporate the materials and deposit them on the surface of the lenses.

below
The coating of lenses with fine layers in a cleanroom environment (photo Angénieux archives).

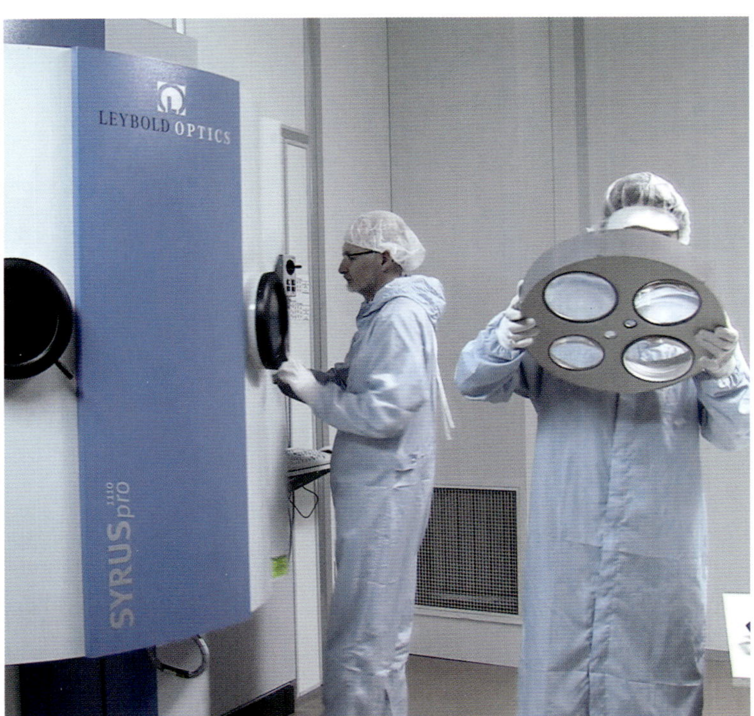

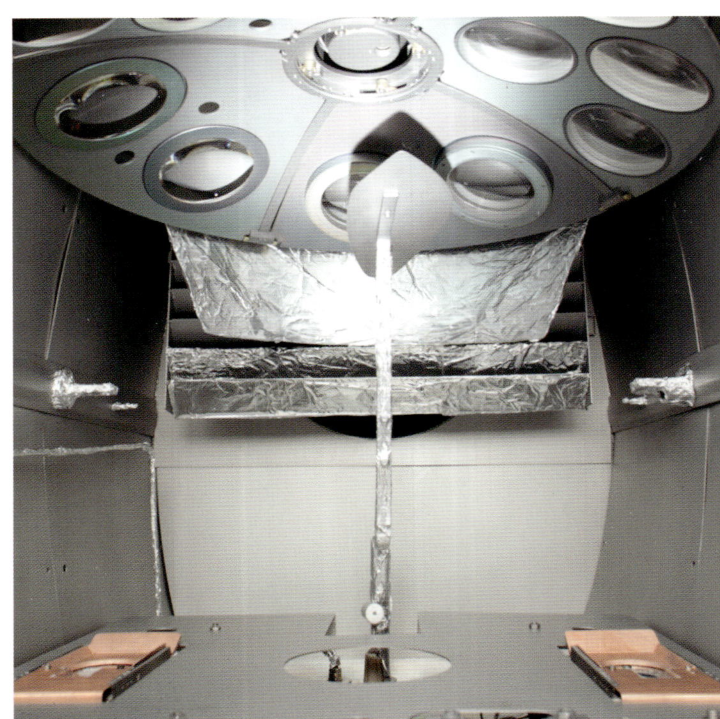

Metrology
Metrology is the set of techniques for taking measurements, interpreting them, and ensuring their accuracy. For industrialists, ensuring the traceability and reliability of measurements is not only essential to control over manufacturing processes and product quality but also makes it possible to achieve interchangeability.

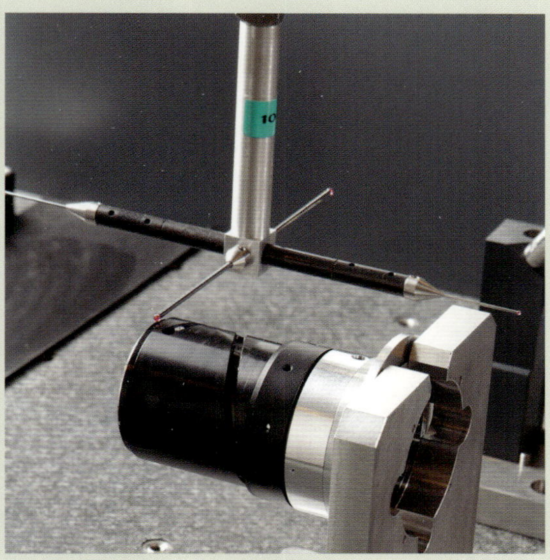

Cementing Lenses

Doublets are not infrequently used in an optical system, as cementing two lenses together by means of a transparent adhesive makes it possible to reduce chromatic aberrations. This operation entails the greatest care and manual dexterity, as the degree of precision required is about ten microns.

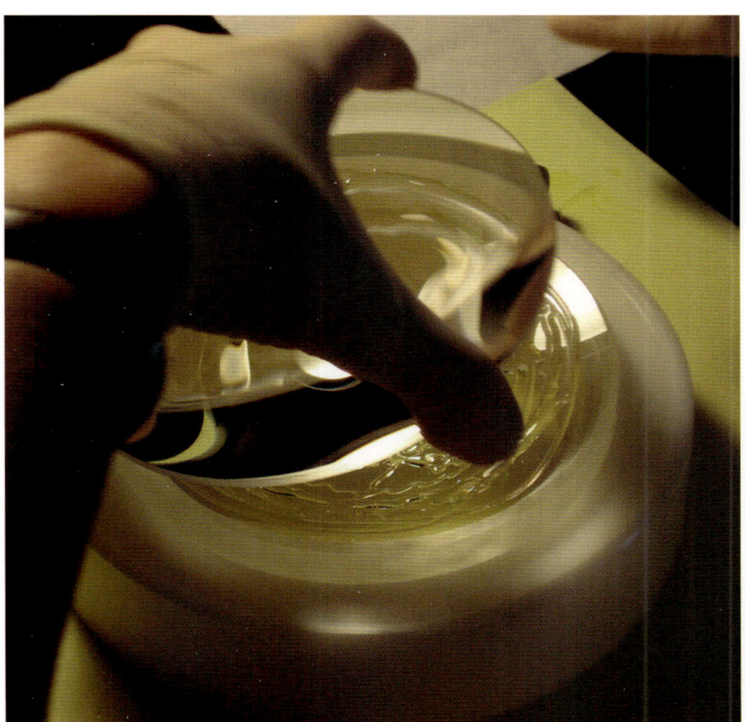

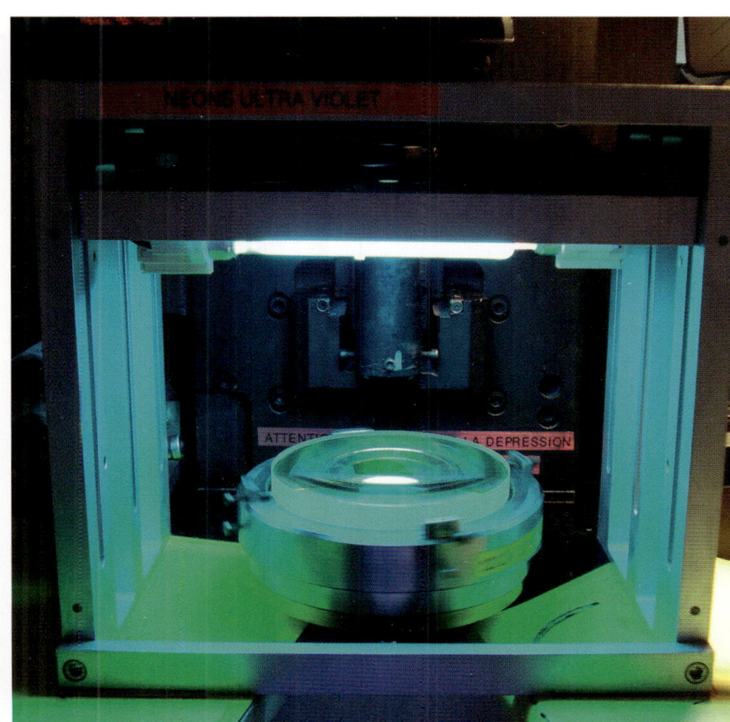

The operations of cementing a doublet (a pair of simple lenses) and polymerization of the adhesive.

Varnishing the Rims of Lenses

The rim of each lens element is coated with black varnish to limit ghosting and block unwanted rays of light. This manual operation is again highly demanding as it requires the utmost dexterity and concentration.

Assembly of Optical and Mechanical Components

Once the lenses have been coated and varnished, and the necessary doublets or triplets have been made, the optical components of a zoom are fitted into the mechanical parts. This assembly stage can be purely mechanical, with the lenses positioned by means of nuts, or entail cementing of the optical component into its mechanical sleeve. It takes dozens of hours of work to assemble these optical-mechanical sub-units and all the successive phases of adjustment must be followed meticulously in order to attain the standards expected of an Angénieux zoom.

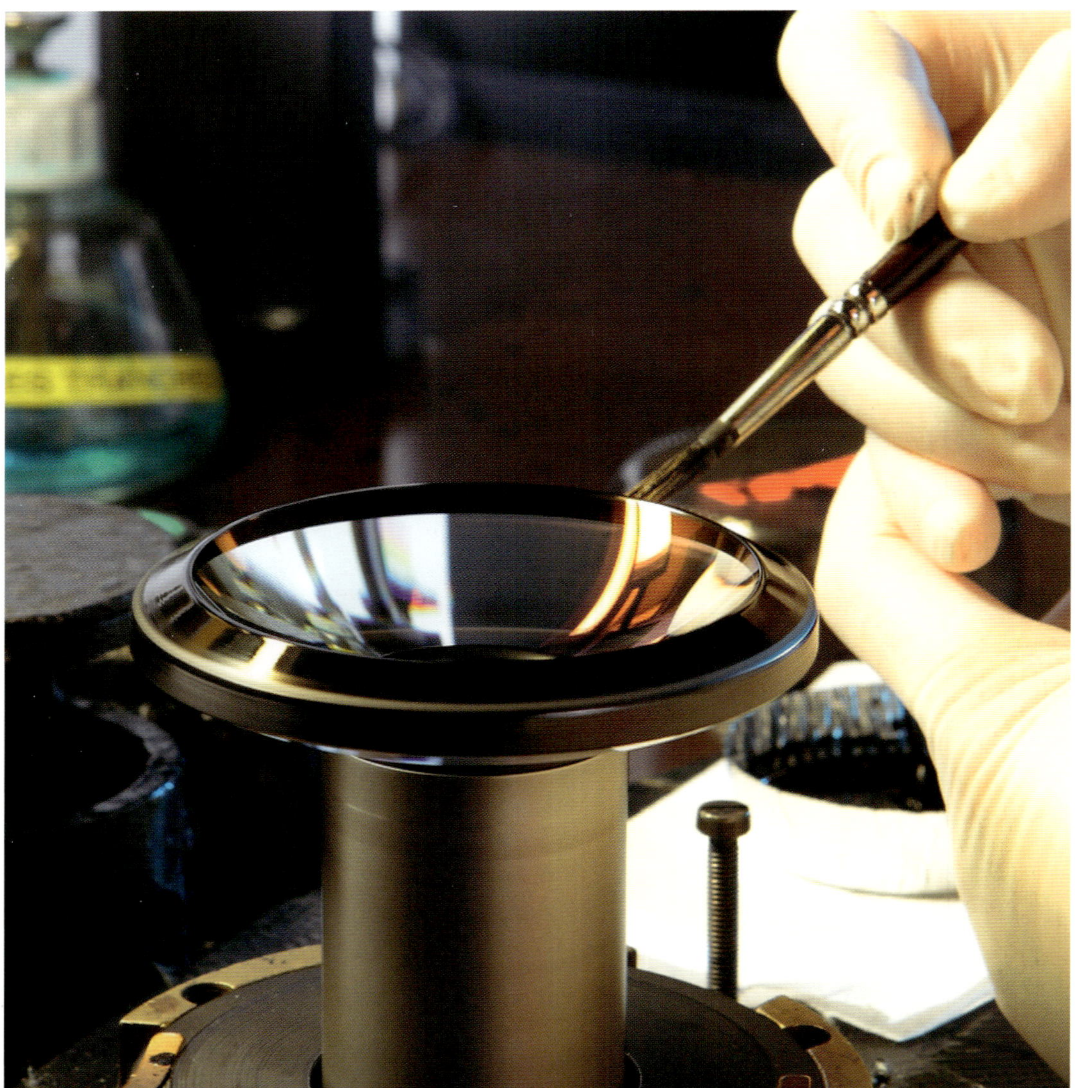

left
Varnishing the rim of a lens element.

opposite
Assembly of optical and mechanical components.

following pages
The testing of an Optimo 24-290 in the projection room.

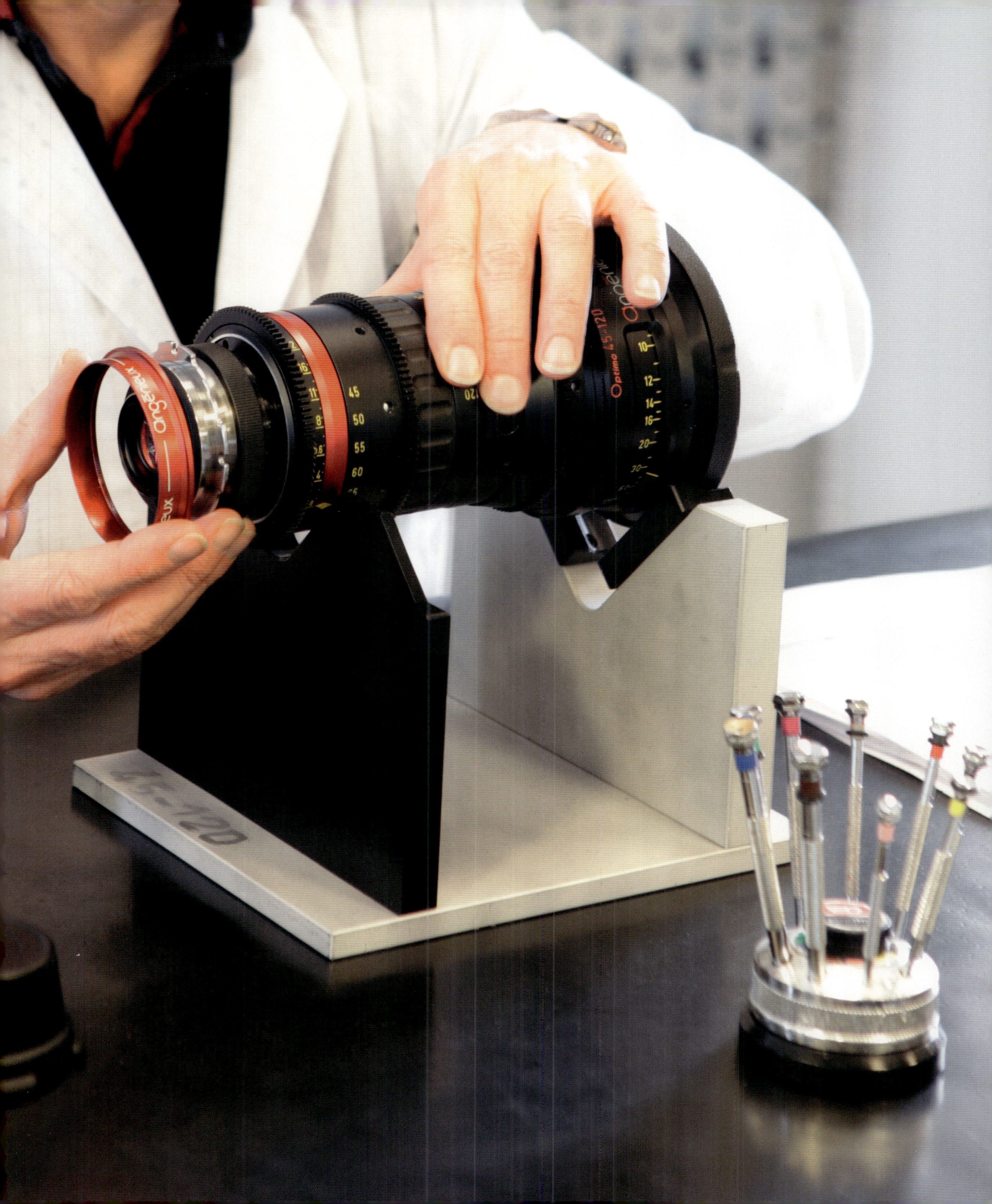
Optimo 45-120

Final Check

The last stage is the final testing of a zoom before shipping.
The zoom is placed on a collimator to ensure the absence any residual defocusing (loss of focus in zooming). The resolution of the lens is also subjected to MTF testing.
This is followed by calibration and casing. The focus rings are engraved once the zoom has been calibrated. Each zoom is unique.
The zoom is finally tested at extreme temperatures (from -20° to +40°C) to ensure its stability regardless of the filming conditions.

And because a reputation for excellence is something you earn, almost all Angénieux zooms, once packed and ready for shipping to the other side of the world, is subjected to a drop test. After hours and hours of delicate, painstaking work, the zoom is dropped in its packaging from a height of one meter and checked again. This is the only way to ensure it will withstand the sometimes rough conditions of shipping by air or other forms of transport.

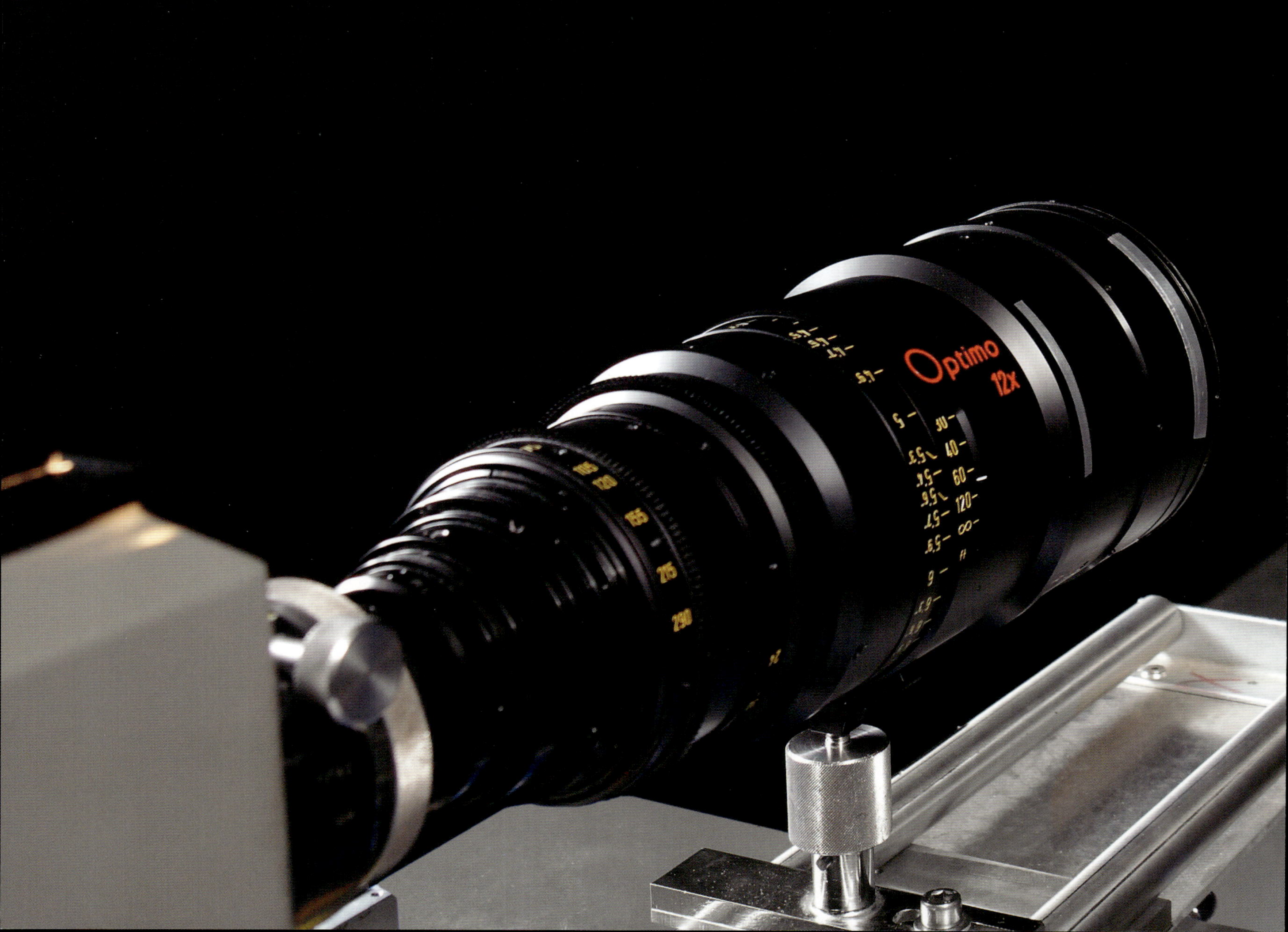

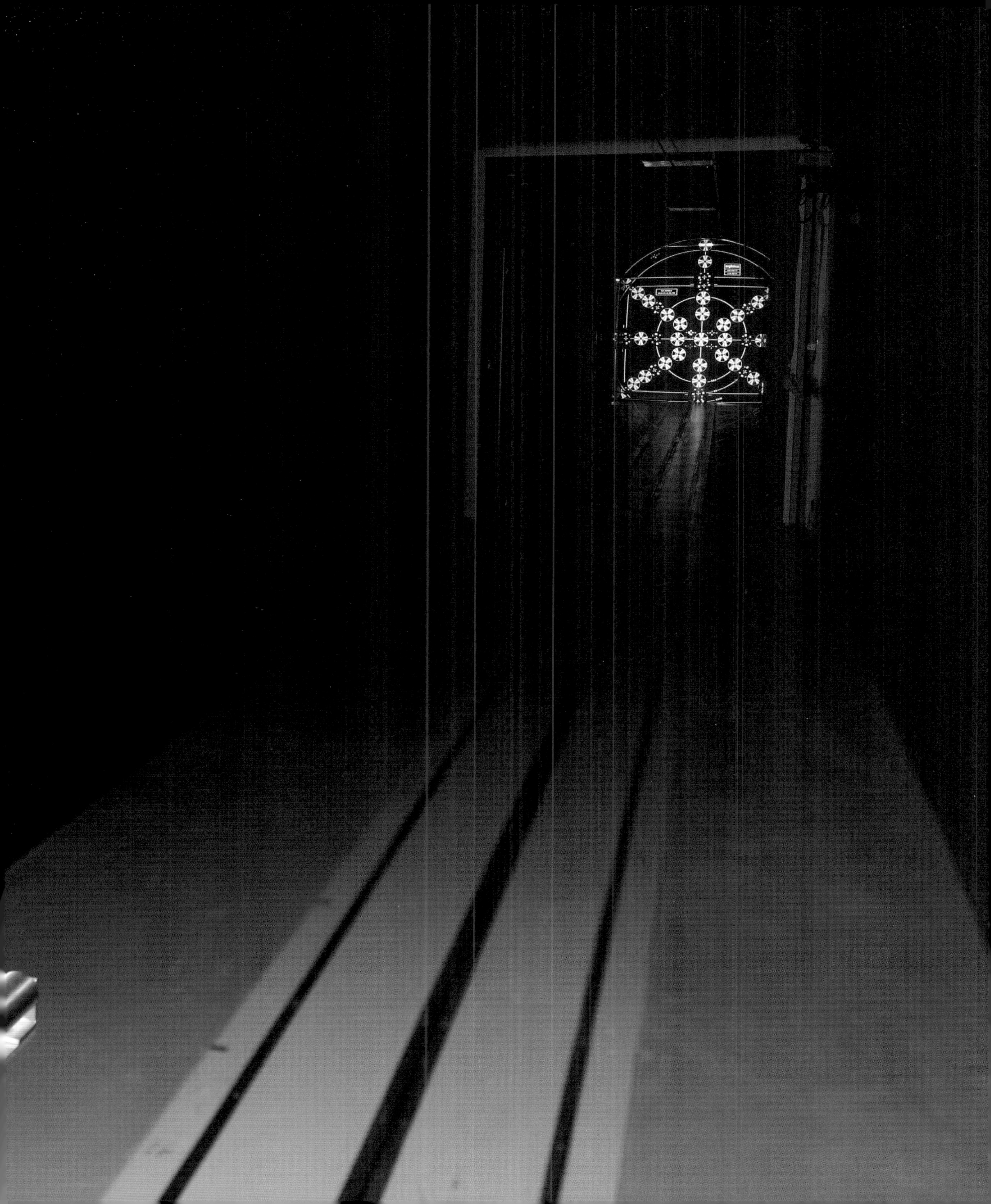

The Saint-Héand Plant often Welcomes Illustrious Visitors

« I didn't imagine the factory as it is at all. I said to myself, 'OK, they make lenses' but what struck me, apart from the splendid place where the company is based, is the people on the spot. They are involved in the mission to get the image right, and not only in technical and technological terms. They are concerned with image that the lens will provide. This has always been something marvelous for me, something I can count on. And I did not know where it came from ... Was it the water? Was it the air? No. ... It is all the people involved, the people from the different departments of the company aware of everything that goes to make up a beautiful image. »

Kees Van Oostrum

President of the ASC, American Society of Cinematographers, after visiting Angénieux in January 2017

« It was indeed a great honour and cause for joy to visit the Home of Angénieux and to experience the exemplary hospitality of Angénieux Family. Thank you to Pierre, Jean-Yves, Paulette, Edith and Dominique and other great souls of Angénieux family. Being at Angénieux also meant to be with the history of cinema and to witness the dedication and love you bestow upon creating the 'eye' of the cinematographer.
As a cinematographer I am proud to be part of the Angénieux family by using your marvellous creations.
I wish Angénieux family a great era of continuous success and prosperity.
Hope to meet you all again and share the joy of being a part of the magic of Cinema.
Cinema is Love 24 Times per Second.
Thanking you, Affectionately, »

Sunny Joseph

Director of photograpy
President of ISC - Indian Society of Cinematographers, and Life Member of SICA - Southern India Cinematographers' Association

« It was an honour to visit Angénieux and to relive the history of Angénieux unfold in front of our eyes. It is a beautiful and exemplary story to witness, the story of how Angénieux came into being and what it stands for. One could feel the passion, perfection and excellence which took the company from its foundation by Pierre Angénieux in the 1930s till today in 2017 to become a trusted friend of every cinematographer. Its humbling to see the kind of precision, research, engineering, meticulous human judgement along with the endless hours it takes to design and manufacture each and every lens. And to know that every Angénieux zoom I used for some of my most special films took birth here at Saint-Héand was a very special feeling. My relationship with Angénieux is a trusted one and this visit made it even deeper. As a cinematographer I am proud to be shooting with the supreme optics of Angénieux. I sincerely thank every member of Angénieux family for this beautiful visit to Angénieux. »

Savita Singh

Director of photography
Member of WICA - Western India Cinematographers' Association

« Thank you ever so much. Our trip to Angénieux factory was a very memorable one indeed!
I remember the day, some three decades ago, when I first saw a 25-250 Angénieux zoom lens ... and I also remember how excited I was when it was mounted on a camera and I looked through it. It opened up my imagination with huge possibilities in designing shots! Since that day Angénieux has remained very much a part of my journey as a DOP [director of photography].
It was very thrilling to see the most intricate science and engineering and precision that went into the design and manufacture of each part of the lens. I was deeply moved by the high degree of dedication and commitment that each member of the entire team Angénieux brought to their effort to achieve excellence ... each one of them seemed so proud to participate in creating such a world class product.
I returned from the visit with my faith firmly rooted in Angénieux's constant quest for excellence.
Thank you so much for the warmth and such gracious hospitality.
With warm regards »

Govind Nihalani

Director of photography
Member of ISC - Indian Society of Cinematographers and WICA - Western India Cinematographers' Association

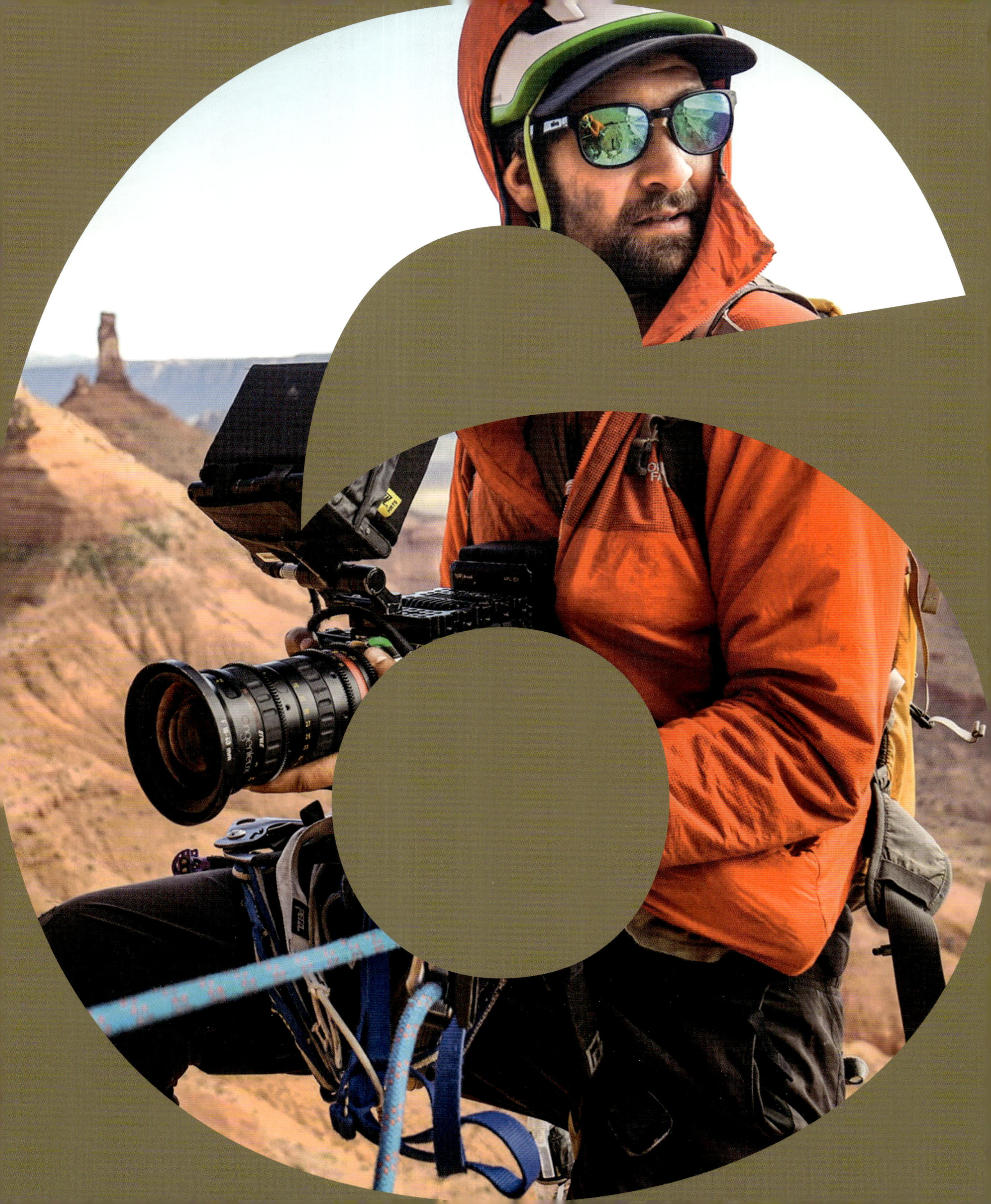

OPTIMO
FROM FILM TO DIGITAL

Thales imparted **fresh impetus** to the Angénieux brand. The combination of the flexibility of Angénieux, where design, production and sales all rub shoulders, and the strength of a major group contributed to the success of the **Optimo** range, a new line of cinema products. After the initial cinematographic experiments of the late 1990s, the **transition to digital technology** came into effective operation in the **mid-2000s**. Angénieux has successfully handled this technological advance by revolutionizing all aspects of the cinema world. The sudden increase in the number of movie cameras opened up new horizons for lens manufacturers in the digital sphere.

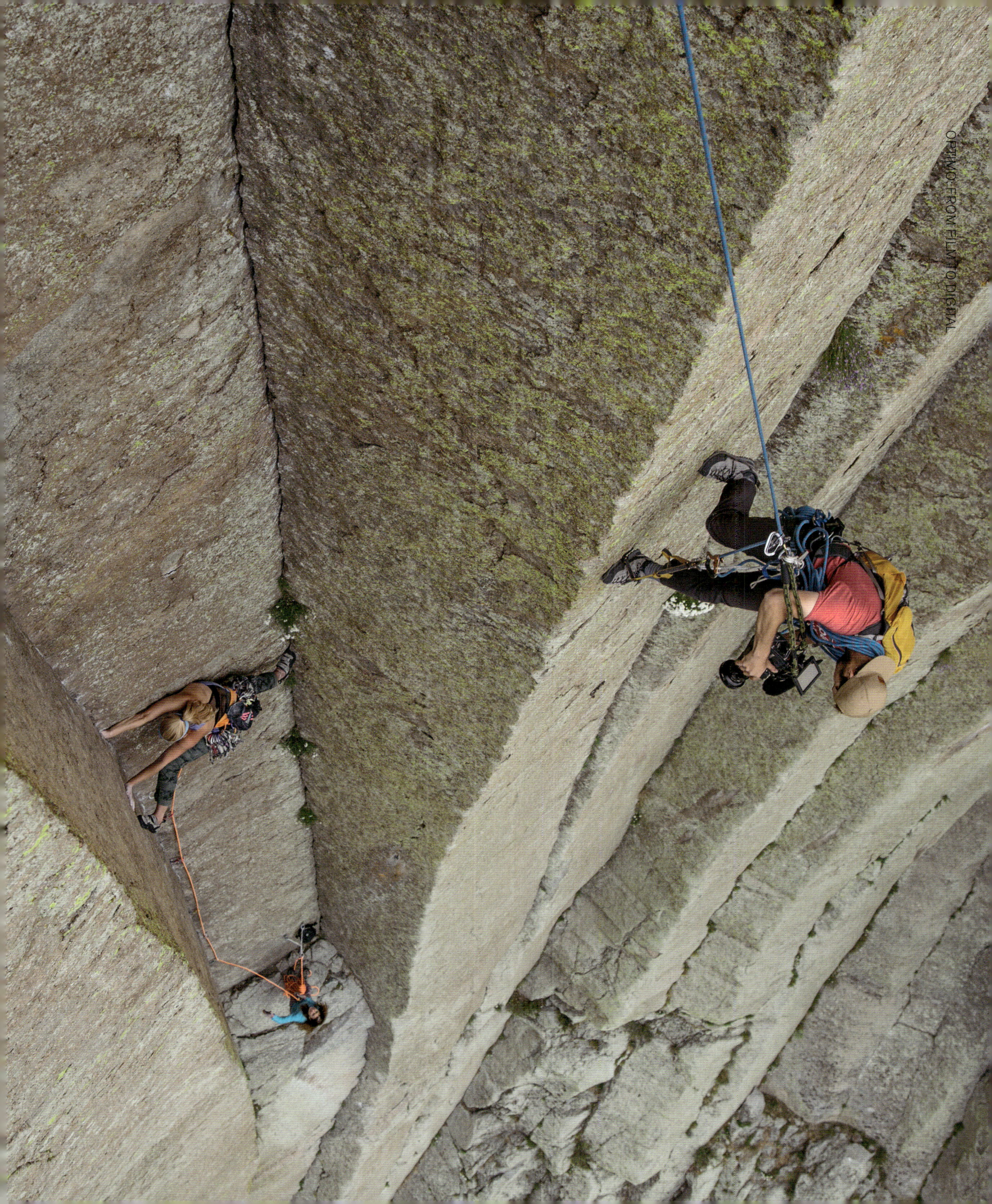

Cinematographers in Digital Times
Trends and Technology in Art and Cinema

Jon Fauer, ASC

"Film is like a colony and there are very few colonists. It is essential for every responsible artist to cultivate the ground that has been left fallow. I just want to occupy an unoccupied terrain and work there," Orson Welles said in a *Cahiers du Cinéma* interview. "I work with the [Angenieux] 18.5 mm lens only because other filmmakers haven't used it. The 18.5 lens is a new, important invention. Each time I give it to a Director of Photography, he is terrorized: but by the end of the film, it's his favorite lens."

Great Directors and Cinematographers have always wanted to be first in line for the latest equipment in their quest to stand out, to be unique, to be remembered. In this constant quest, cameras and lights and lenses sparked new styles, new looks. At the same time, these new trends pushed manufacturers to respond with more new tools. The 18.5 mm lens of Orson Welles's desire was introduced by Angénieux in 1951. It was the company's first lens made specifically for 35mm cinematography. It was fast, lightweight, small and had a unique look.

The appeal of cinema lenses certainly was apparent to 28-year-old Pierre Angénieux when he opened his first workshop in August 1935 at 7 rue Henri Murger in Paris. Two years later, he set up a second workshop in his hometown of Saint-Héand, where the company continues today. Pierre Angénieux had the credentials and the contacts. He had studied under Henri Chrétien, inventor of the anamorphic process, and received a double degree in optical and mechanical engineering from the École Supérieure d'Optique and from the École des Arts et Métiers. His first job was at Pathé, once the world's largest film equipment and production company. He met directors like Jean Renoir and Abel Gance and listened to their suggestions about lenses.

Fritz Gabriel Bauer famously said, "A camera is just a box onto which we put beautiful lenses." That's a modest statement from the man who invented the Moviecam and Arricam cameras. What he meant was that, above all, the lens defines the images we see on screen. It is the tool that translates ideas, suggests styles, provokes emotions and provides the look. Most important, the lens is the key to creating images that are unique—different from everyone else's.

Cinematography Trends and Pictural History

Let's take a quick detour back a century in time.

"Initially hailed with enthusiasm, photography, like all inventions, has stirred up a storm of criticism. Certain minds see this admirable discovery as posing a danger for art … lest the human hand might forget its skill. This fear is without foundation," Théophile Gautier, artist, art critic and photographer, commented in 1857.

It was a time when tools and taste were colliding at a prodigious rate. Photography was affecting art and the nascent Impressionist movement. Impressionism, in turn, would go on to influence technology and the birth of cinema. Tools shaped styles. Trends transformed equipment. The momentum of converging currents in painting, photography and

[1] See pp. 212-213

cinematography were bringing about spectacular societal changes in the last decade of the 19th century. Cinema, with images projected multiple times per second, soon became the medium destined to become the most widely-viewed art form of all time. It was a disruptive force, a revolution in technology that enthralled the largest audiences in history. Cinema was made possible by tools—cameras, lights, film, and lenses that changed, in successive waves, the way films looked. The synergy of style and invention in the art of cinema continues to this day—again, made possible by machines, and, as any Cinematographer will tell you, mostly by lenses.

When the migration of cinematography from silver to silicon gathered momentum in the past ten years, after a century of analog, the reciprocal relationship of technique and technology became even more apparent. Ineffable differences between photochemicals and pixels sparked the inevitable questions: "What was the film look? And why did digital look different?" Shooting with analog motion picture film introduced all kinds of "happy accidents." These serendipities helped distinguish one Cinematographer from another. There was the random texture of grain from silver halide crystals. Despite the best efforts of the best mechanical engineers, and no matter how tight the registration pins' tolerance, the film negative inexorably bounced around in the camera's gate. And the film continued to wobble all the way along the production chain, long before the word "workflow" was conceived. It bounced when the lab contact-printed the release prints. It bounced during projection. All these variables contributed to a film look that enraptured audiences and captivated cinematographers.

Cinematographers also had many choices of film stocks and labs and processing techniques. Additional idiosyncrasies came into play: the water, chemicals, temperature of the "baths," and the technicians who managed the magic in the lab late at night.

A few years into the digital age of film, Cinematographers were despairing at the idea of images looking the same or being "overly pristine." The cameras and image sensors were made by a handful of manufacturers and everyone was using the same ones. As Checco Varese, ASC said, "It was like everyone was shopping for the same vegetables at the same grocery store. The difference was how you cooked them."

Hang on. Had everyone suddenly forgotten how to light or use filters? The recipe for skillful lighting remained essential. In some ways, it required even more finesse. Cameras were now more sensitive. You didn't light for exposure. Often, it became a matter of skillfully taking light away. Lighting was still the paint. Lenses, as always, were the brushes, the instruments, the tools shaping the image.

And then a curious evolution took place. The mantra, for so many years, had been sharper, crisper, faster, flareless. Because the bouncing film chain's mechanical flow degraded the image with every step, an optimized lens was considered an essential ingredient in preserving the original integrity. In the new digital paradigm, something remarkable was happening. Lenses that had been retired as obsolete, considered too soft or too hazy, suddenly looked better in this new digital age than they had ever looked before. They often looked more interesting on high resolution digital sensors than they ever did on film. Why was this? Flares and flaws became objects of desire. The buzz was all about "look" and "character." Vintage lenses were "rediscovered."

And yet ...

The mandate of a Cinematographer is to tell the story with moving images, to illuminate the script. But, unlike monks illuminating manuscripts in their scriptoria, future success in the competitive freelance filmmaking arena often depends on creating memorable images unlike anything any other colleague would create. Do you follow trends or create new ones?

 Handheld or dolly, drone or Steadicam, epic style or documentary look?

The pendulum swings. Styles change. Soft and romantic one year might turn to hard realism the next. So, what is the lens designer to do in this new vortex of digital dreams? Do you build a lens with look baked in, with character preordained? And what of the cinematographers' aspirations to be unique, while still advancing the story without becoming too ostentatious.

Being a cinematographer was not even a job description until the 1890s. On December 28, 1895, Auguste and Louis Lumière projected their first short film on a screen in the Grand Café in Paris. Louis Lumière personally shot over 30 of his company's short films. They were documentaries with roots in Impressionism's subjects and styles: workers leaving the factory, a train arriving in a station, a family dining together, street scenes of Paris.

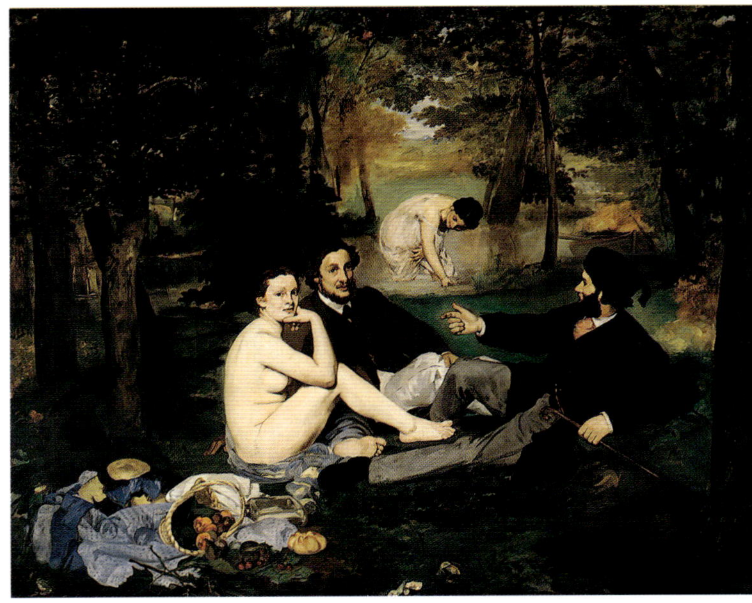

Édouard Manet, *Le Déjeuner sur l'herbe (Luncheon on the Grass)*, 1863
Paris, Musée d'Orsay.

The audiences at the Grand Café in 1895 attended the birth of the French film industry which would grow to become one of the most influential worldwide. The dialog and dialectic on matters of look and tools and taste continued to this day. The swing of styles in contemporary cinema, from pristine to distressed, vintage to pure, flared to clear, has precedents in the "-isms" of the art world after 1895: realism, classicism, romanticism, impressionism, pointillism, post-impressionism, fauvism, expressionism and so on.

These were the years when art was democratized along with the society it often depicted. Style was influenced by technology (railroads, paint in tubes, photography). Édouard Manet and his fellow Impressionists painted scenes of modern society in a way that critics at the time called "unfinished" or "sketched." We might call it documentary style.

In 1863, Manet exhibited three paintings, including the (then) scandalous *Déjeuner sur l'herbe*. The public professed to be shocked by the subject of a nude woman blithely enjoying a picnic in the company of two fully clothed men. Perhaps what upset critics most was the flat and contrasty front lighting, rescued only by the dark jackets worn by the men in the painting. And here's where technology came in. Manet was most likely influenced by the flat-lit photography of the day.

By this time, the Impressionists noticed how much faster photographers could produce images. While a classical realism masterpiece for the Academy might take years to complete, an Impressionist could dash off a painting with impressions of light and an atmospheric palette in one day on location. By painting the same subject repeatedly, at different times of day and under various lighting conditions, the artist could create paintings at an astonishing rate, almost as quickly as rival still photographers. Were these the ancestors of today's independent filmmakers?

1876 was another disruptive moment in art history. American department store magnate Alexander T. Stewart purchased Ernest Meissonier's *Friedland* painting, sight unseen, for the astronomical sum of $60,000. Meissonier was the highest-paid artist in France. He was decidedly not an Impressionist. He specialized in action, adventure and romanticized battle scenes. With an attention to detail that some called obsessive, and others would later see in Kubrick, Ernest Meissonier wored on *Friedland* for fourteen years. In his vast garden near the Seine, assistants pushed a dolly at breakneck speed with Meissonier perched on board, diligently sketching (handheld) the precise details of the horses galloping back and forth.

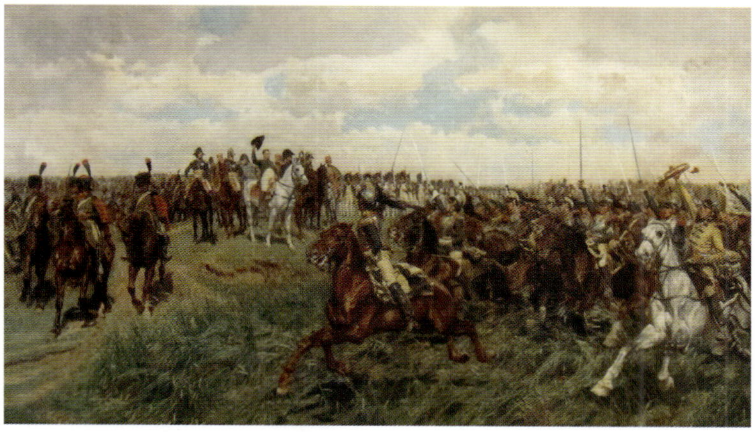

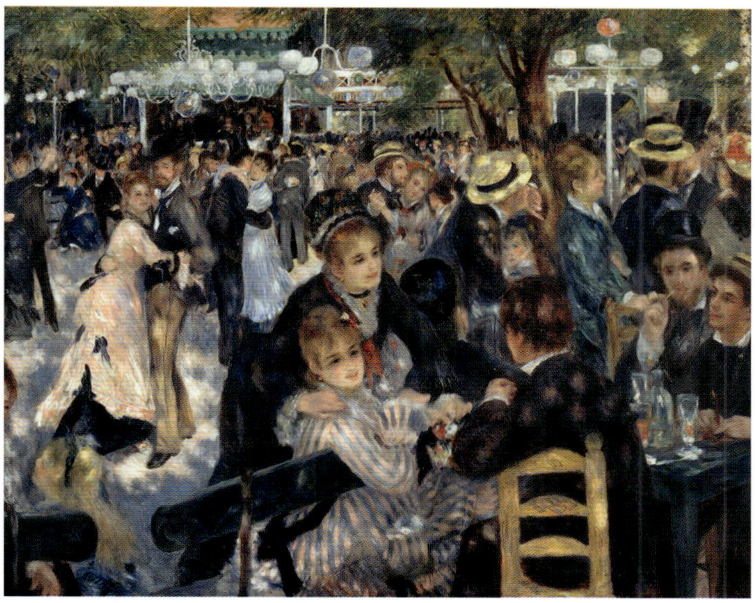

Ernest Meissonier, *1807, Friedland,* 1861-1875 New York, The Metropolitan Museum of Art.

Pierre-Auguste Renoir, *Bal du Moulin de la Galette (Dance at Le moulin de la Galette),* 1876 Paris, Musée d'Orsay.

In that same year, 1876, Pierre-Auguste Renoir declined a dinner invitation from Madame Marguerite-Louise Charpentier. He wrote, "Alas I shall very probably not be able to dine with you. I began a portrait this morning. I begin another this evening, and it is extremely likely that I shall have a third to do afterwards." Renoir wisely understood the eternal predicament of a freelance artist or Cinematographer: never turn down a job. He also put technology to work to keep up with the prodigious pace of production. Impressions of the moment and broad brush-strokes could almost keep up with the speed of a photographer, and certainly art works could be completed at a much more rapid rate than Meissonier's fourteen years on *Friedland*. Once again, technology influenced technique, tools shaped style, styles and tastes changed. As Renoir remarked "Without paint in tubes there would have been ... nothing of what the journalists were later to call Impressionists."

Were Impressionists troubled by the same dilemmas as Cinematographers in the digital age? Were they shopping for the same vegetables at the same grocery store—using the same canvases, pigments in tubes and brushes from the same manufacturer? Decidedly not. Each artist created a different look. Canvases could be distressed and aged, paint-brushes and palette knives blunted, fine lines softened into blurry blotches of dappled light.

Degas, in 1876, went darker: night, interior, off-center framing, a vertical aspect ratio. His painting was originally called *In a Café* and later renamed *Absinthe* when exhibited in London to horrified reviews. A critic wrote in *The Westminster Gazette* (1893), "The two sodden people at the café ... will never be a work of art, however 'incurable' your 'affliction and confusion.'"

Later, in 1895 (the same year the Lumières developed their Cinématographe), Edgar Degas produced a pastel and a photograph. This was the Post-Impressionist age of rapid and broader brush strokes, and less concern for available light. It was essentially a reaction to the very nature of photography and Impressionism itself.

In an ironic footnote of history, *Absinthe* is now what the Tate Museum calls, "One of Degas's most celebrated works." Meissonier, once the highest paid artist the world ever knew, is hardly remembered. Renoir's *Dance at Le moulin de la Galette* recently went for $82.5 million at auction. Styles change and the taste of the audience evolves.

Cinematographers and Digital Times

The role of Cinematographers has evolved as well. In analog film days, Cinematographers were like magicians. They were essentially the only ones on set who could conjure up images and know what they would look like. Everyone else on set would have to wait until the following day, after the ministrations of processing and printing ultimately revealed projected images while screening dailies. Video assist chipped away at the mystery a little bit. At first, it was a grainy, flickering black and white image that represented a hint of framing and not much more.

Today, the digital image is live, high resolution, often high dynamic range, and viewed on monitors everywhere. The magician is no longer a solitary explorer but the conductor of an on-set, real-time orchestra. The cinematographer has new tools, new methods and responsibilities more encompassing than ever before, with even more time devoted to previsualizing in preproduction, collaborating on set and grading in post.

Vittorio Storaro, ASC, AIC said, "The experience of digital cinematography was the end of my little nightmare of agonizing how the image would look on screen. With film dailies, we waited until the day after, sometimes the week after, depending on where the laboratory was. There were so many things that could change or go wrong, and no matter how vast your experience, knowledge and technical preparation, there was always the challenge of something going differently from the expectation, in the developing of the film or the printing or the color timing in the lab. As much as you could be knowledgeable or experienced, there was always doubt, followed by the great emotion of relief upon seeing the first image on screen."

Today, there are more distribution avenues than ever before. New technologies are changing the way we view motion pictures and how they are completed. Streaming, OTT (Over the Top), smartphones, tablets and laptops are competing with traditional broadcast, cable and cinema theaters for what the industry calls "eyeballs." Historic studios are competing with new providers like Amazon, Apple, Netflix, and many more.

The challenge for Cinematographers in this rapidly accelerating world of new technology is to maintain control of the process by embracing all aspects of the imaging process. Moore's Law is pushing us all into 24-month cycles of exponential innovation. And yet, there are benefits. If you cannot clear your schedule to sit with the colorist during the final grading sessions, the latest versions of Blackmagic DaVinci Resolve and other software let you collaborate from anyway in the world, in real time, with an internet connection. Small digital mirrorless cameras let you scout and previsualize scenes as stills or editable video. Software for tablets enable wireless control of lighting intensity and color. Lens metadata is advancing to the point where software can change the characteristics of lenses not only in post but also in real time, on set or location.

It has always been this way. Filmmaking has always involved evolving technique and technology. The role of the Cinematographer and the Artist has always been challenged from actual process to projection and ultimately preservation.

For example, Rembrandt Harmenszoon van Rijn completed *Night Watch* in 1642. It is neither Night (there are no torches or lanterns) nor are the characters Watchmen. They are actually "The Company of Frans Banning Cocq and Willem van Ruytenburch," one of the civilian militias that guarded Amsterdam. For much of its life, the painting was coated with a dark varnish that gave the incorrect impression it was a night scene and leading to the name by which it is now commonly known. The varnish was finally removed in the 1940s.

Night Watch was one of six group paintings installed wall-to-wall in the great hall of Amsterdam's civic guard building, Kloveniersdoelen, a kind of clubhouse. The *Night Watch* was "reformatted" in 1715 when it was moved from the Kloveniersdoelen to the Amsterdam Town Hall. The painting was shortened on all four sides to fit between two columns, cutting off the top of the arch and losing two characters and the railing on the left. Of course, the railing and void behind it were important elements that create a dynamic sense of depth and give the scene a sense of urgency and movement. A 17th century copy of the *Night Watch* by Gerrit Lundens, on display in the National Gallery, London shows how Rembrandt's painting looked originally.

There are a number of things conspiring against the success of any work of art or motion picture.

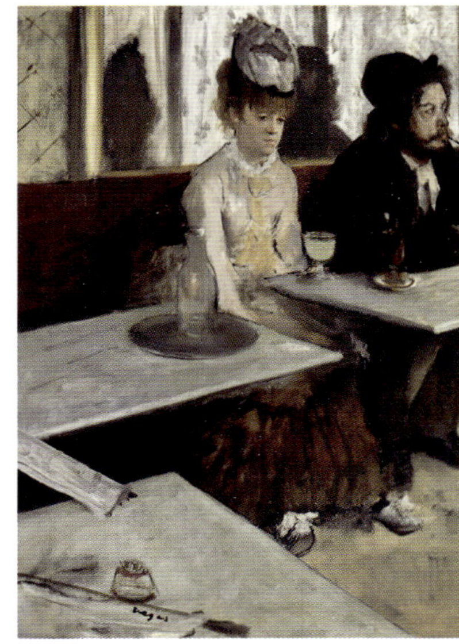

Edgar Degas, *Dans un café (In a Café)*, also called *L'Absinthe (Absinthe)*, 1875-1876 Paris, Musée d'Orsay.

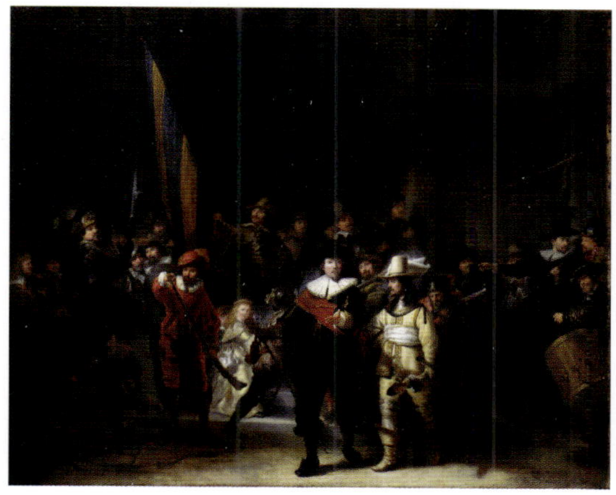

Rembrandt van Rijn, *Night Watch*, 1642, oil on canvas, 363 × 437cm Rijksmuseum, Amsterdam.

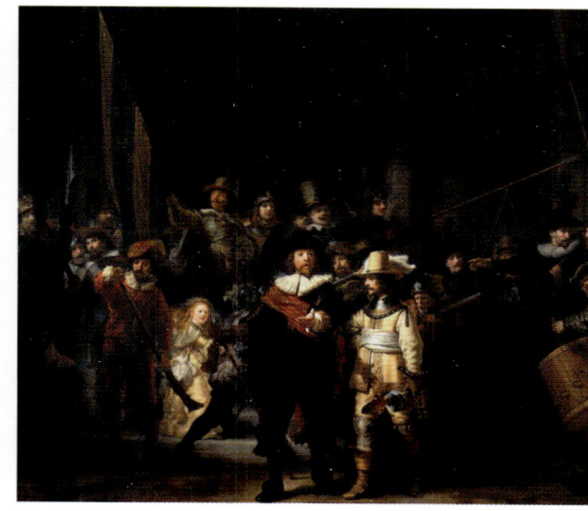

Gerrit Lundens (after Rembrandt) *The Company of Capt. Banning Cocq and Lt. Willem van Ruytenburch* later than 1642, oil on oak, 66.8 × 85.4 cm. On loan to Rijksmuseum, Amsterdam.

When Cinematographers and Directors begin work on a movie, inevitably there are lingering doubts. Opening night is the moment of high anxiety. Will the audience applaud? Will the critics complain? Will the producers demand a re-cut?
Lens designers face similar pressure. "When we begin work on the design of a new lens, we really don't know what the reaction will be from our customers until they try it. In the beginning, it's pure intuition. It's always a challenge. We have many gates to go through until we reach the end. It's quite complicated. It's somewhat like being a Cinematographer" use to say the optical engineers at Angénieux.

Optimo Range, the New World Reference

« I was in charge of developing the Optimo 24-290 with my teams in 2000. I had already been with Angénieux for 14 years and this was my first movie camera zoom. To be quite frank, our aim was basically to do as well as the Primo 24-275 T 2.8 just brought out by Panavision, which was establishing itself in the United States. With respect to our own 25-250 HR, the goal was only to improve the vignetting at short focal length, limit or even eliminate ramping at long focal length, and avoid any breathing. We did not confine ourselves to the specifications.
In order to achieve our goals, it was necessary to rethink the architecture of the zoom completely, as that of the 25-250 HR did not work. For a ratio of 12x and an aperture of T 2.8, the optical performance of our zoom was exceptional as regards colorimetry, resolution, ramping, breathing, and vignetting, far beyond those of the zoom Panavision and our own hopes. Especially as regards the minimum focal distance. The image quality of the 24-290 at 1.22 m (4ft) was far superior to that of the 25-250 HR at 1.7 m. »

Bruno Coumert

Optical engineer at Angénieux since 1986

New Competitive Context on the Dawn of the 1990s

In the early 1990s, the American motion picture equipment company Panavision began to offer its new range of Primo zoom lenses for rent, in particular the Primo 24-275 with a ratio of 11x and aperture of T2.8.
By request of Century Optics Precision, its distributor at the time in the United States, Angénieux sought to meet the needs of Panavision's competitors in the rental market, especially Clairmont Camera and Otto Nemenz International, in the early 2000s by developing a zoom of large focal range. The result was the Optimo 24-290.

The Optimo 24-290, aka "the Hollywood Lens"

The Optimo 24-290 was launched in 2001 and enjoyed immediate success on the market. It became the favorite zoom of numerous filmmakers for shooting full-length movies as well as advertisements and TV series. It was soon very popular in Hollywood for filming both inside and outside the studio. With a constant aperture of T2.8, exceptional optical performance, and a very limited degree of breathing, it was approved by directors of photography and used not only for zooming but also to replace a whole series of lenses with fixed focal length. The popularity of the Optimo 24-290 on sets in Los Angeles was such that it came to be called "the Hollywood lens."
Angénieux followed up this success in 2004 with the new Optimo 17-80 T2.2, a wide-angle studio zoom with a ratio of 4.7x. At 5.5 kg, this lens is much more manageable than the 24-290, which weighs 11 kg. Directors of fiction films were particularly pleased by its ability to cover the entire range of the most commonly used primes and adopted it very quickly. The Optimo 17-80 is also the ideal lens for shooting from a crane.

Sales of the Optimo 24-290 were to exceed all expectations. It was not until June 2018, seventeen years after its launch, that the last of these lenses left the production line in Saint-Héand, a highly emotional moment for all the firm's employees.

The Optimo 24-290 was used in the making of countless movies all over the world and remains a kind of talisman for many directors of photography.

The first zoom in the range, the 24-290, boosted the prestige of Angénieux at the dawn of the third millennium and made the name Optimo known all over the world. The reputation for excellence and reliability attached to the Angénieux Optimo showed no sign of waning in 2018 with the twelve zooms that now comprise the range. The 24-290 was followed by the Ultra 12x, a highly innovative, multipurpose zoom perfectly designed to cope with the evolution of digital cinema. So as not to cause despair among unconditional devotees of the Optimo 24-290, Angénieux made sure that the specifications, aperture and focal lengths of their favorite zoom were preserved in the S35 version.

Since its presentation in September 2017 at the IBC show in Amsterdam, this zoom has found favor with equipment rental companies and appears likely to enjoy all the success of its predecessor.

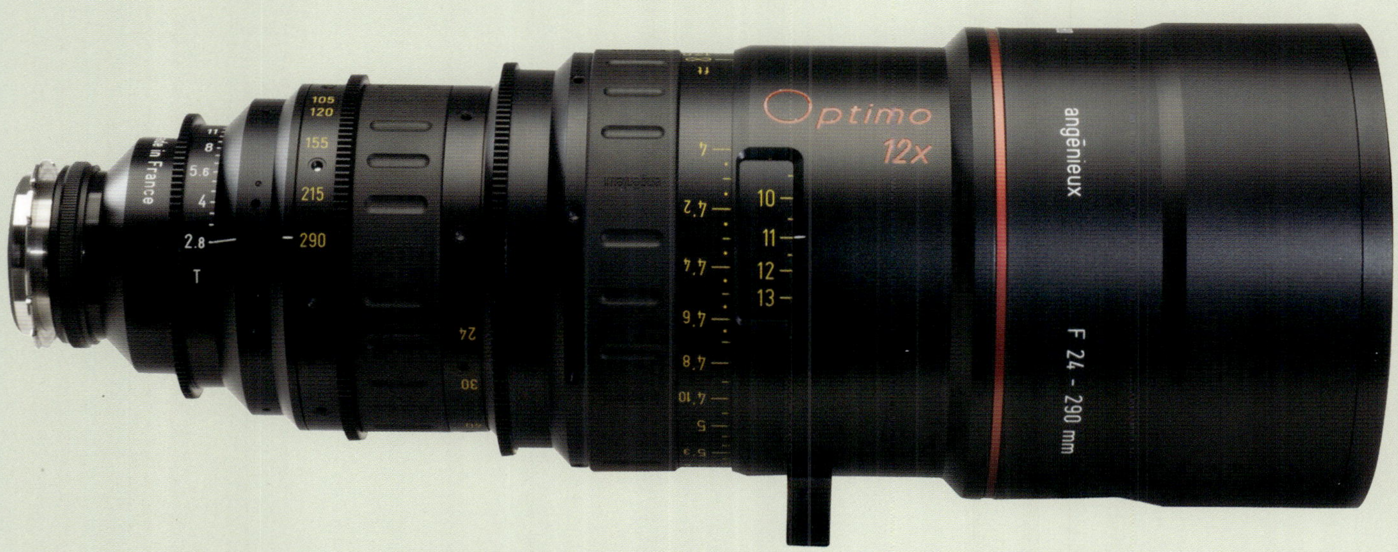

The Testimony of Equipment Rental Companies

« Our clients have been enamored with Angénieux glass and consistently request their zooms since we first opened our doors for business. Of the vast array of 35mm format zooms we carry, Angénieux's extremely high quality Optimo line far and away represents the lion's share of what we own. Their 24-290mm Optimo zoom has been the de facto workhorse in episodic television since the beginning of this century, and is the Angénieux lens we carry the most of - and it is still in such high demand to this day.
We're proud to offer our clients the best tools in the industry and Angénieux consistently delivers with excellent optical quality, mechanical consistency, and superior color rendition. Angénieux has definitely set the bar high. »

Brad Wilson

Vice President, Business Development at Keslow Camera - Rental House for cinema equipment in United States

« We have been providing our clients with Angénieux zooms for a long time now as that's what our customers have been requesting. The 24-290 revolutionised the possibilities with a long zoom. Its great T-stop, colour renditioning, contrast, sharpness and most important for a rental company, its build quality and reliability. With the addition of its compact "sisters" which despite their size and weight, you don't lose any of that quality, they are the go too zooms to match any prime for features, drama and commercials. »

Russell Allen

Director – Business Development
Camera & Grip Division ARRI Rental UK

« I've personally been involved with Angénieux lenses since 1964 when I was working with the 25-250 zooms at Panavision. When I was a working cameraman I did a lot of 16mm work and my favorite lens was the 9.5-57. Otto Nemenz International has always featured Angénieux zooms. We were the first in the U.S. with the 16-44 T1.3 zoom for 16mm as well as the 25-625 zoom for 35mm.
We invested heavily in the Optimo line starting with the 24-290. That lens was a game changer. It became the defacto zoom lens on any production be it commercials, television or features. Angénieux lenses have been a big part of my success in this industry and always will be. »

Otto Nemenz

Founder of Otto Nemenz International - Rental House for cinema equipment in United States

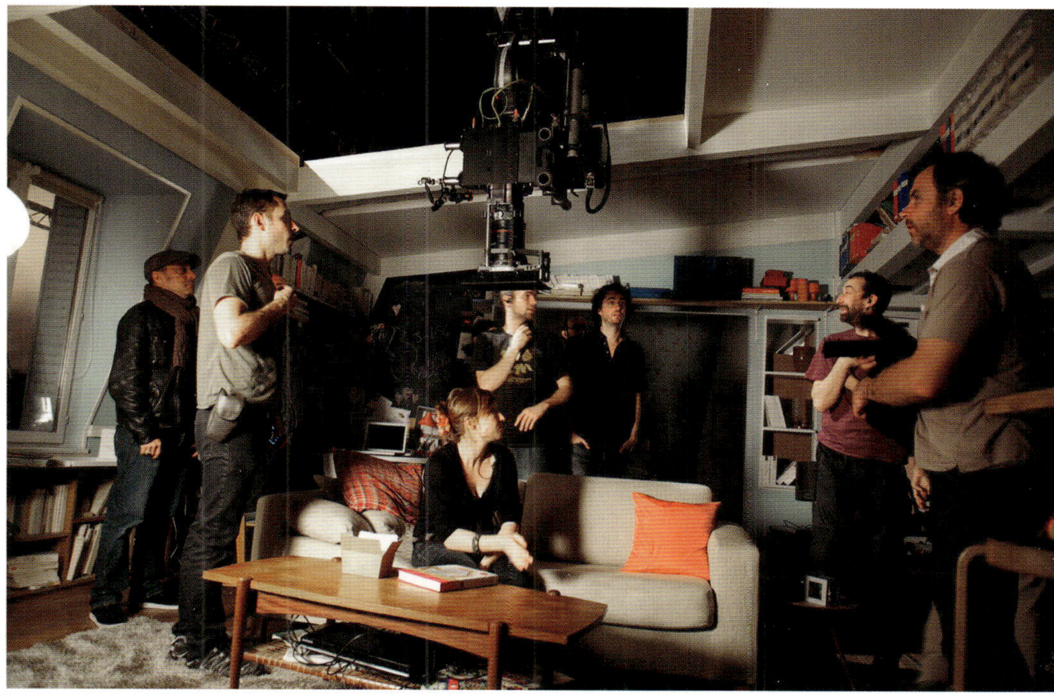

The Optimo 24-290 during the filming of *Un Heureux Évènement* (2011) by Rémi Bezançon DP and Antoine Monod AFC. From left to right: Maamar Ech-Cheick (production designer), Antoine Monod, Valérie Houdart (seated, 2nd assistant director), Paul Henri Belin (1st assistant director), Rémi Bezançon, Laurent Usse (key grip), and Pascal Roussel (unit production director), Mandarin Production (photo © Nicolas Schul).

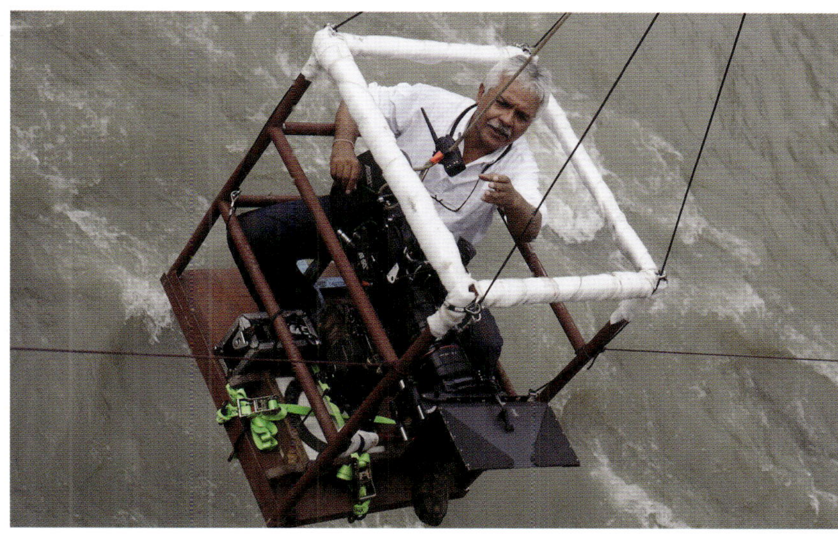

The director of photography Anil Mehta ISC on the set of *Jab Tak Hai Jaan* (2012). Mehta is also known for his work in *Beyond the Clouds* (2017), *Secret Superstar* (2017), and *Sui Dhaaga: Made in India* (2018) (photo provided by Anil Mehta).

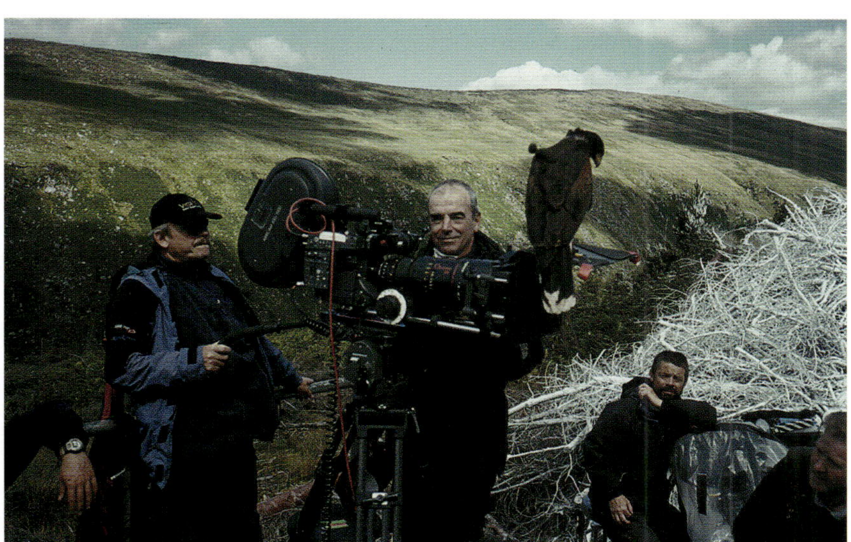

Optimo 24-290, on the set of *King Arthur* (2004), directed by Antoine Fuqua. DP Slawomir Idziak using the zoom (photo provided by Slawomir Idziak).

Slawomir Idziak, a director of photography with a staggering filmography, "always with Angénieux," as he puts it. His films include *Trois Couleurs: Bleu* (1993), *Proof of Life* (2000), *Black Hawk Down* (2001), *King Arthur* (2004), *Harry Potter and the Order of the Phoenix* (2007), and *A Tale of Love and Darkness* (2015). His work regularly wins prizes at the most illustrious festivals. He is also the founder of the Film Spring Open Festival in Poland.

The Optimo 15-40, 28-76, and 45-120, the First Ultralight, Compact Zooms

In the mid-2000s, in collaboration with the American distributors Ste-Man Inc., owned by the Manios family, Angénieux began to work on compact lenses for shoulder-mounted filming, Steadicam, and all forms of shooting in which weight is a limiting factor.

The result was the Optimo 15-40 and the Optimo 28-76, which appeared respectively in 2006 and 2007, two lightweight lenses of less than 2 kg, similar in size, with very good image quality and a constant aperture of T2.6. While innovative optical design ensures no change of field in focusing, their simultaneously precise and sturdy mechanical elements are perfectly suited to the requirements of shooting films. Their characteristics are ideal for the quick pace of filming TV series. Their success was immediate, especially in the filming of TV series, where their characteristics are perfectly geared to the quick, precise pace of production. These two zooms strengthened the international renown of Angénieux's compact Optimo range and won their engineer designers a Scientific and Technical award at the Oscars in 2009. The range was completed in 2011 by the Optimo 45-120, once again immediately acclaimed in the world of cinema.

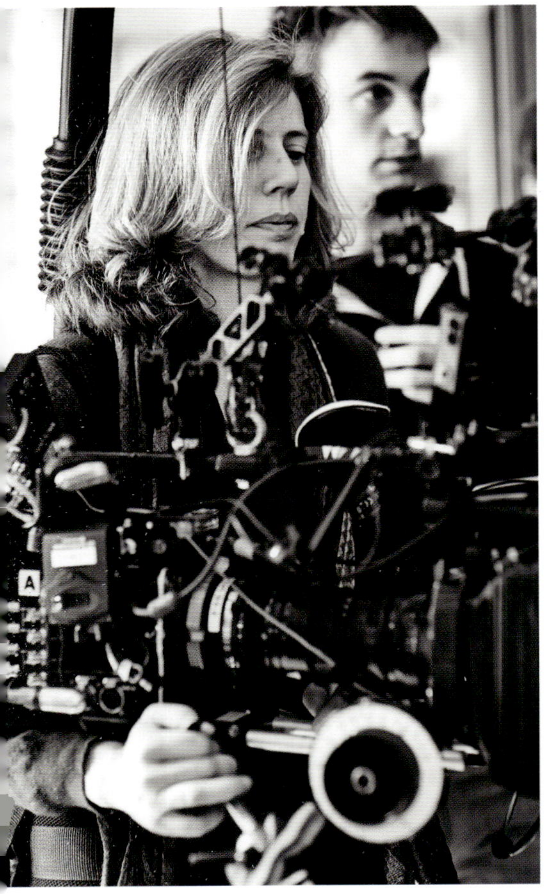

The director of photography Claire Mathon AFC during the shooting of *Mon Roi* by Maïwenn in 2014 (production: les productions du Trésor). An Optimo 28-76 on an Easyrig system. Claire Mathon is also known for her work on *Les Deux Amis* (2015), *L'Inconnu du lac* (2013), and *Une vie violente* (2017) (photo by Marcel Hartmann, provided by Claire Mathon).

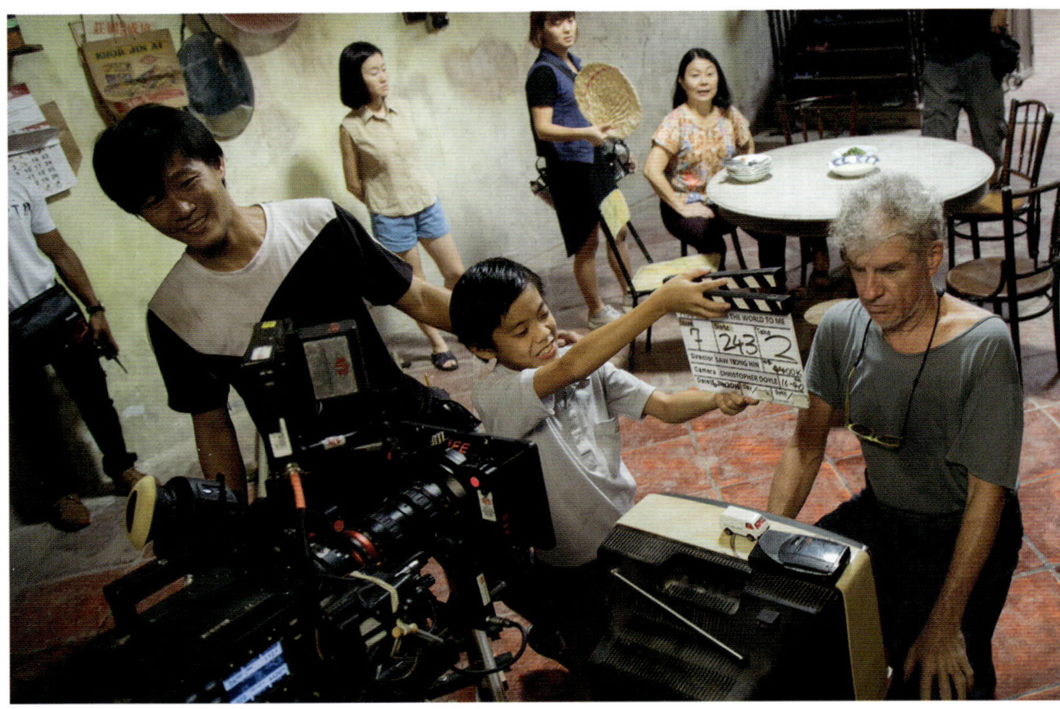

Christopher Doyle during the shooting of *You Mean the World To Me*. An Optimo 15-40 zoom on the camera (Astro Show and Real Films Prod. © Jack Cheung)

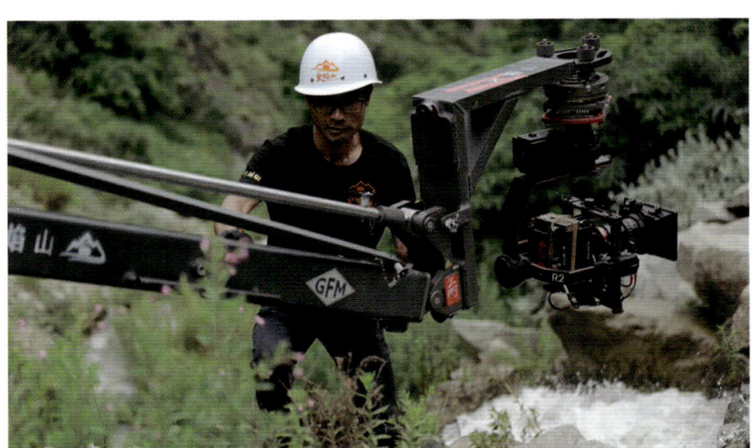

The Optimo 15-40 used to film *Da Lu Chao Tian* by Miao Yue (image provided by Flaming Mountains Co, Ltd).

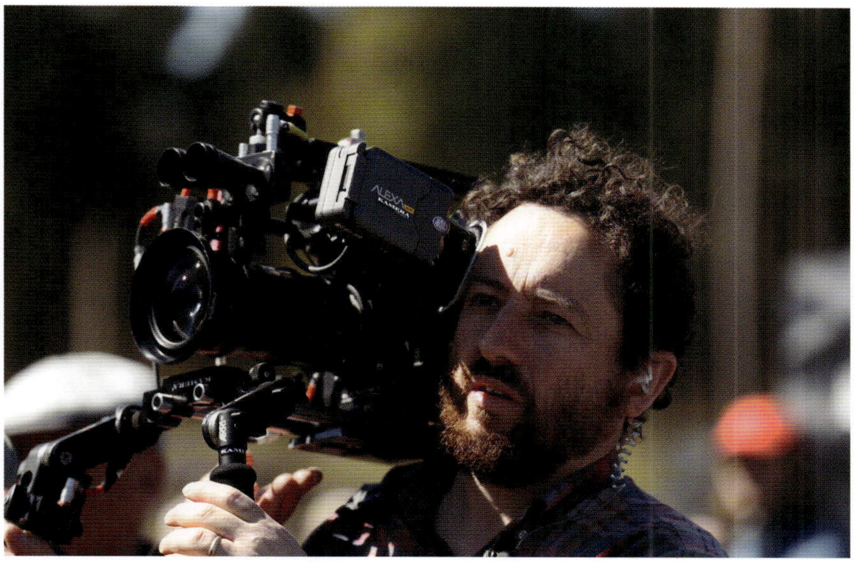

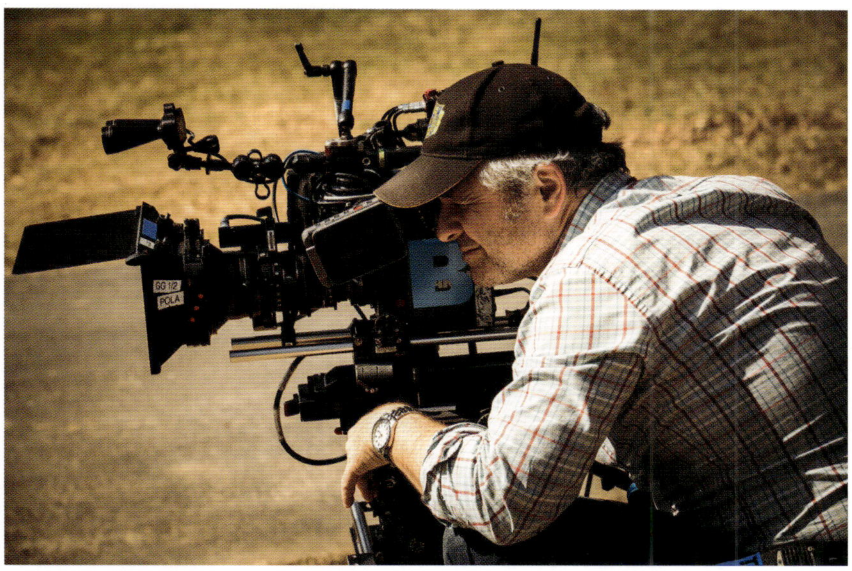

top
The director of photography Manuel Claro with an Optimo 28-76 on the set of Lars von Trier's *The House that Jack Built.*

above
Cinematographer Toby Oliver (photo by Justin Lubin).

Digital Transition

Digital cameras arrived on the cinema market in the late 1990s and gave rise to two very different creative trends.

Some directors started out with HD, 3-CCD, 2/3" digital cameras like the Sony F900 CineAlta, Thomson Viper, and Panasonic Varicam, which paved the way for special effects. Georges Lucas exemplifies this tendency with films like episodes 2 and 3 of the *Star Wars* saga, *Attack of the Clones* (2002) and *Revenge of the Sith* (2005). Angénieux developed two zooms for these HD 2/3" cameras, an 11.5x5.3 HD and then a 12x9.7 HD, as well as a cine video 35 adapter developed in collaboration with Zeiss making it possible to use 35mm zooms on these early digital cameras. Despite a certain degree of success, these failed to satisfy filmmakers completely, as the beam-splitter prism limits the aperture of the lens and hence the creative possibilities.

Other filmmakers working on lower budgets preferred to experiment initially with small, lightweight, compact cameras like the consumer camcorder (1/3" or 1/4" mono-sensor with digital recording on miniDV) for quick shooting with no need for large crews. Leaders in this field included the Danish directors Lars Von Trier and Thomas Vinterberg, the founders in 1995 of Dogme 95, an artistic movement of rebellion against mega-productions and in favor of a far more sober, realistic approach. The movement lasted no longer than about a decade.

The use of digital technology for cinema did not really get under way until the arrival of single-chip cameras.

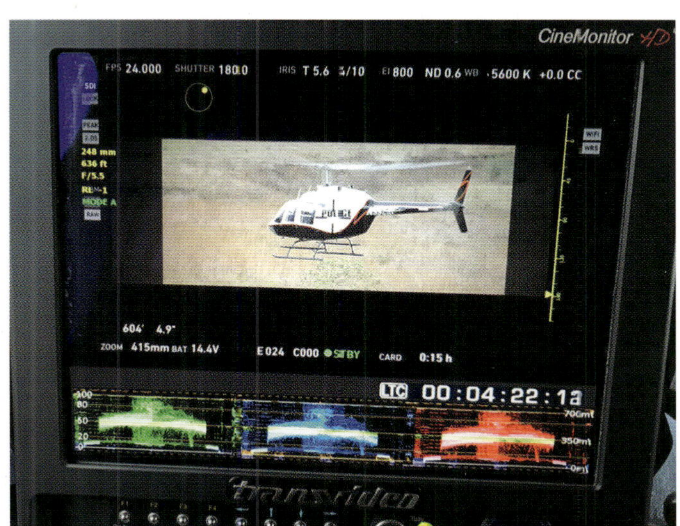

left
Digital display on a control monitor showing the primary filming parameters for cameras and lenses.

Single-chip digital cameras made their appearance in the mid-2000s with the Dalsa Origin, Arri D20, D21, and Panavision Genesis. The market was revolutionized in 2007 by the appearance of the Red One, produced by the American manufacturer Red Digital Cinema, a newcomer in the world of cinema owned and run by Jim Jannard, the legendary founder of Oakley. This camera marked a major technological breakthrough. Its price and its 35mm sensor (22mm x12.6mm) with 4K resolution, making it possible to use all 35mm film lenses with PL mounts, won over a large number of the undecided.

Filmmakers found familiar operating features, especially as regards depth of field. Video recording in the form of digital files became a reality. Special effects and postproduction work were all possible. Digital technology opened up new creative paths and gave rise at the same time to all kinds of fears. How to preserve the work of directors of photography when everything can be altered in the post-production stage? How to preserve the integrity of an image from its capture to its digital projection in the newly-equipped cinemas? What about the preservation of the works themselves? All crucial questions for the future of the image and memory through images.

The success of the smaller and incredibly affordable Red One prompted Angénieux to develop in 2008 and 2009 a new range of lenses, namely the Optimo DP (Digital Production) 16-42 and 30-80, easily recognizable on the set by their red rings, an obvious reference to the camera. Offering levels of performance particularly geared to digital cameras as regards chromaticism and telecentricity for superior illumination of the sensor, Optimo DP lenses proved a great success in the early digital productions. Blue rings were made available for users of Arri digital cameras. In 2014, the large-scale use of cameras with EF mounts prompted Angénieux to modify the rear sections of its Optimo DP for both PL and EF mounts. A revision of overall design led to the birth of the new-look Optimo Style 16-40 and 30-76. The range was completed the same year by the introduction of a 10x zoom, the Optimo Style 25-250. On the basis of a completely new design, this zoom offered the focal length and mechanical sturdiness of the legendary 25-250 HR while meeting the new needs of digital cinema: optimized optical performance, image coverage of 31.4mm diagonal, recording and display of data. The Optimo Style 25-250 is used in particular for shooting from helicopters, while the Optimo Style 16-40 and 30-76 are used above all for filming documentaries.

The first zooms of the Optimo range initially designed for film cameras cover the sizes of the early generations of digital camera sensors. The advent of digital technology, which had a major impact on many companies in the movie industry involved with film and film processing, came as an opportunity rather than an earthquake for lens manufacturers like Angénieux. The "film look" of Optimo zooms is still particularly prized today for productions in the digital era.

Single-sensor and 3-CCD Digital Cameras

Apart from a few exceptions, all digital cine cameras now have one sensor so as to avoid the separators of 2/3" broadcast cameras, which restrict aperture and depth of field. CMOS technology has made it possible to fit the photosensitive part (the sensor) and other processing circuits, especially for analog/digital conversion of the video signal, onto one chip. The sensors can have direct digital output, whereas conversion circuits had to be added externally in the case of CCD. This technological advance has made it possible to reduce the size, consumption, and cost of cameras considerably. While increased resolution (4K, 8K, etc.) is the goal of broadcast cameras, though more simply attained with single-sensor models, this is not so important to filmmakers, whose priorities rather comprise color, sensitivity in the areas of high and low luminosity, and repeatability (identical image quality from one camera to another for similar video parameters), which is a very important quality in the case of multi-camera productions.

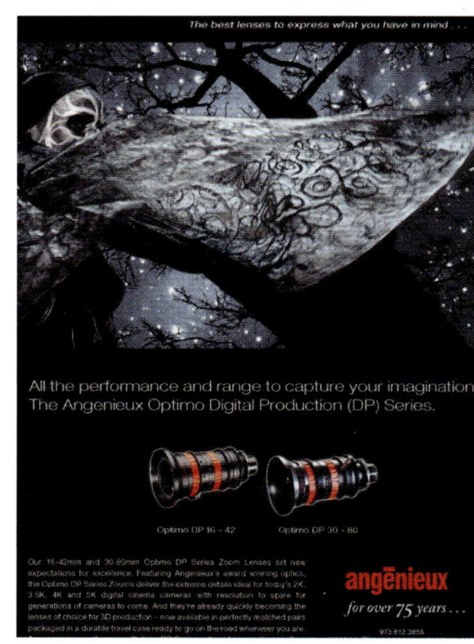

above
Advertisement for the Optimo DP, 2008.

below and opposite
The Optimo Style 25-250 used to film *What If You Fly*, produced by The North Face and Camp4Collective, directed by Renan Ozturk and Taylor Rees (© Renan Ozturk, photo by Taylor Rees).

The cinematographer Tom Bergmann and Optimo Style 16-40 on the set of *The King*, directed by Eugene Jarecki (photo by David Kuhn. Production Ghost In The Machine Films).

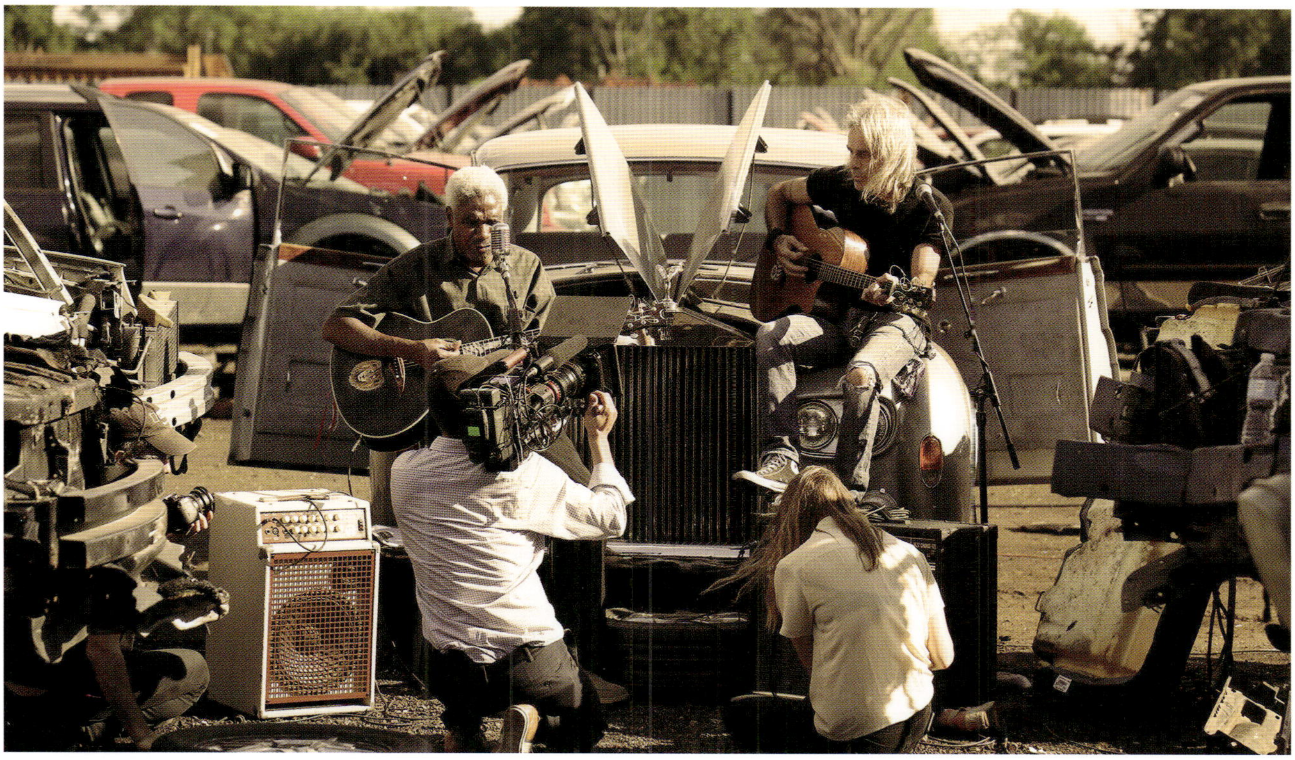

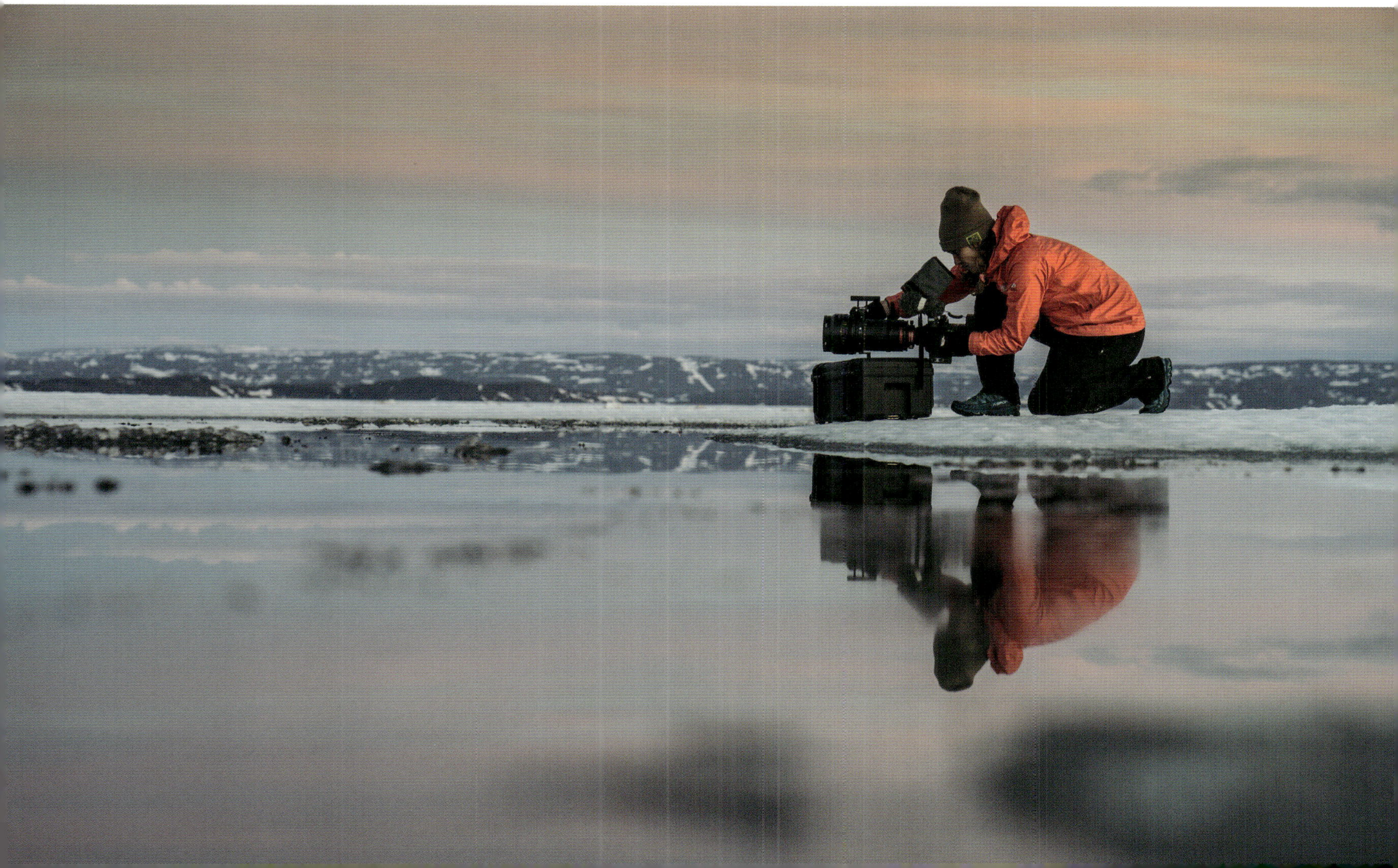

About the Transition to Digital Technology

« The digital transition came about brutally and gave rise to every kind of misunderstanding. A large part of French production tried to pass off as progress an evident technical decline in the quality of cinematographic images, those created for the big screen. Images with low definition, no definition, no contrast, and no brilliancy were passed off as 'modern'. Calls for quality equal or superior to 35mm met with derision. It was not the same, there was no comparison, even though all skill and knowledge derive from the comparison of what existed in the past with what exists in the present. (...) It was feared that this technical decline, supported and sustained by the bad production principle of 'faster and cheaper,' would be established in our cinema. These fears have still to be confirmed. Today's digital cameras have large sensors. They are produced with the aid and advice of cinematographers and cameramen by manufacturers that have realized the need for cameras of quality equal or superior to 35mm in order to make films of equal or superior quality for showing in suitably equipped cinemas. »

Pierre-William Glenn

Director of photography and director
Joint founder of the Association Française des directeurs de photographie Cinematographique (AFC) and president of the same for many years.
President of the Commission supérieure Technique de l'Image et du Son (CST) from 2002 to 2018 and currently joint director of the Image department of the Fémis.

« Everything has its place. Digital technology offers something new, possibilities that film does not offer, and vice versa. It seems to me that movies can still choose their support. It is once again a question of what is right with regard to the subject, to the sensation you seek to produce in relation to the images. I am not nostalgic about film. I love it and used it myself for my graduation film at the Fémis, exploited to the utmost with out-of-date, overexposed and superimposed stock. There is an immediate poetry with film that it is harder to find with digital, but I love the challenge of finding an identity in the digital realm, of recreating something material out of what is not 'physical'. It is a wonderful support to shape. You mustn't be afraid of it or too sparing. It only asks to be worked vigorously. There are all kinds of interesting postproduction processes, the only thing is not to lose your head. I love playing with material. I don't necessarily pursue over-definition. I like things to remain organic, to possess the slight imperfection that brings the image to life in my view. »

Eponine Momenceau

Director of photography
Born in 1985, Eponine Momenceau was the cinematographer of *Dheepan* (2015), directed by Jacques Audiard, winner of the Palme d'Or at Cannes that year.

« The choice of lens is still more crucial in the digital era. The sensor of a digital camera may capture too much detail. Digital technology tends to be cruel to actors and actresses, to see blemishes of the skin that the eye misses. I love its sharpness but not its cruelty. What I like in digital is what is most similar to film. I prefer soft lenses without too much contrast. At the same time, I don't like filters and excessive romanticism. David Cronenberg and I chose digital for *Cosmopolis*, which came out in 2011, my first digital film.
It's true that I felt closer to film cameras. In shooting, I really had the impression that the camera was part of my body. Today I have to check the image at regular intervals on the monitor. There isn't the same sensitivity as with film. I would say that digital technology has distanced me from the camera. And I often use remote control. The image on the monitor lets the director of photography see exactly what will appear on the screen. The director sees it too. There is no longer any possibility of doubt. In the era of film, it was the cinematographer that saw things first. The rushes were visible at the earliest the day after shooting on emerging from the lab. Some might say that this is a loss of power for the cinematographer. That doesn't bother me. When I see some of my old films again, I'm shocked by their loss of quality. The film of a movie is a perishable article subject to scratching and change discoloration. And the copies are seldom made from the original negative. For all these reasons, the quality of a film copy is often far removed from the cinematographer's work. The choice of one film or another was, of course, part of the identity of an image. The sensors of digital cameras don't offer this artistic dimension. What remains of the choice of lens, of focal length, of framing and lighting to create the identity of an image. To tell the truth, I think I'm definitively converted to digital. I no longer see any point, either esthetic or economic, in using film. »

Peter Suschitzky ASC

In a career of over fifty years, Peter Suschitzky has contributed with his talent and artistic sensitivity to more than forty major productions, an extraordinary filmography that has earned him numerous international awards.

« I am a native digital cinematographer. To shoot on film is a challenge for me. I think the greatest challenge of digitisation is the lack of convergence of processes. Previously, for it was very simple to do colour correction in post-production. One person could handle the entire process. Now, the equipment by every manufacturer uses different formats. We need to consider the entire workflow process during our planning phase—what format is used for monitoring, what LUT to apply, what format to store the footage, and what hard disk capacity to prepare; during editing, what format do we export in, what platform do we use for colour correction? In such a process, if communication within the team is not seamless, it will affect the quality of the finished work. »

Cécile Zhang

Part of the interview given to Jebsen Industrial (Angénieux distributor in China).
Cécile Zhang was the first recipient of Angénieux's award to encourage young talent in cinematography, presented during the Pierre Angénieux ExcelLens in Cinematography ceremony at the Cannes Film Festival in 2018 .
Born in 1994, Cécile Zhang is a graduate of the Beijing Film Academy.

3D Cinema

The advent of digital technology rekindled interest in 3D stereoscopic cinema. In response to this demand, Angénieux introduced the Optimo DP in a pack of two lenses with very similar optical performance in 2009. Perfectly suited to filming in 3D, which involves superimposing two shots, these zooms can be rapidly attached to 3D rigs, thus reducing the time required for calibration before each shot and proving very popular for stereoscopic productions.

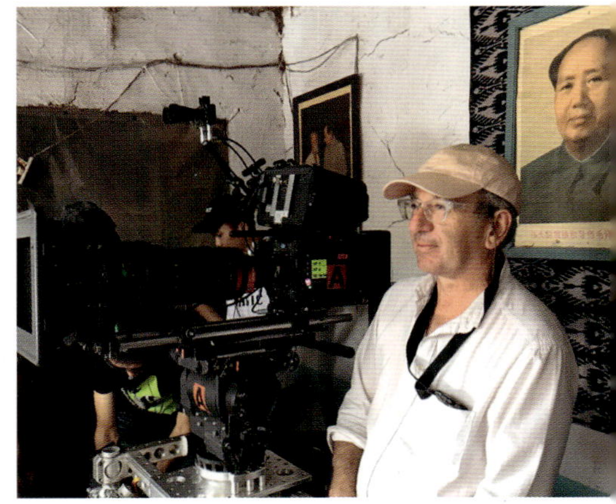

The use of an Optimo 44-440 A2S to film *Magic 7* in Chins (photo provided by Laurent Dailland). Laurent Dailland AFC is a French filmmaker particularly known for his work on *Place Vendôme* (1998), *Le Goût des Autres* (2000), *Asterix et Obélix: Mission Cléopâtre* (2002), *Welcome* (2009), *Man to Man* (2005), *Le Concert* (2009) *Papa ou Maman* (2015), *The History of Love* (2016), and *Jusqu'ici tout va bien* (2019).

Anamorphic Cinema

In the all-digital era, Angénieux is also concerned for the esthetic future of the anamorphic image. The widescreen format, for which the standard today is an image with the aspect ratio of 2.39:1 (width: height), fits in naturally with the format of the large panoramic sensors of the new digital cameras. The use of optical compression for improved resolution, which served in the era of film to form a widescreen image, could no longer have any reason to exist. It is, however, precisely this optical compression that contributes to the peculiar esthetic of these images. In order to ensure that filmmakers are still provided with tools capable of preserving the subtleties of a panoramic image in this era of ever-larger screens and improved projection quality (4K or above), the Angénieux teams set about developing anamorphic zooms based on the spherical Optimo 15-40 and 28-76, recipients of the Academy's scientific and technical award. The Optimo 56-152 A2S thus appeared in 2013, followed by the wide-angle Optimo 30-72 A2S in 2014.
The Optimo 56-152 A2S and Optimo 30-72 A2S are still the two smallest anamorphic zooms with 2x compression ever produced and are used today to capture new anamorphic images with shoulder-held cameras, Steadicam, drones, and other pieces of equipment where the weight of a system is a critical factor. Angénieux has since brought out two 10x anamorphic zooms, the Optimo 44-440 A2S in 2016 and the Optimo 42-420 A2S in 2018.

Increase in Sensor Size and Multiplicity of Formats

Until 2013 and the introduction of the RED Epic MX 5K camera equipped with a sensor with a diagonal of 31.4mm, the Angénieux cinema zooms traditionally manufactured for 35mm film (30mm diagonal) covered the formats of the sensors of the first generations of digital cameras. Their levels of precision were more than sufficient. It is the evolution and large-scale use of digital technology that led Angénieux to rethink its ranges of lenses, revise its formats of image coverage, and increase the performance of its zooms still further in step with the increasing resolution of digital sensors.
In 2012, at the NAB in Las Vegas, Angénieux supplemented its Optimo 24-290 and 17-80 with the Optimo 28-340 and 19.5-94, offering an image coverage diagonal of 31.4mm, compatible with the latest generations of digital cameras.
The number of cameras available on the market rocketed with the advent of digital technology. In the absence of an industrial standard for camera sensors, each manufacturer—and often each model of the same make—used sensors of different size. This situation created a major problem for directors of

The filmmaker Vladislav Opelyants with an Optimo 30-72 A2S (photo by Irina Saeda. Production Kinostudia Slovo Ltd).

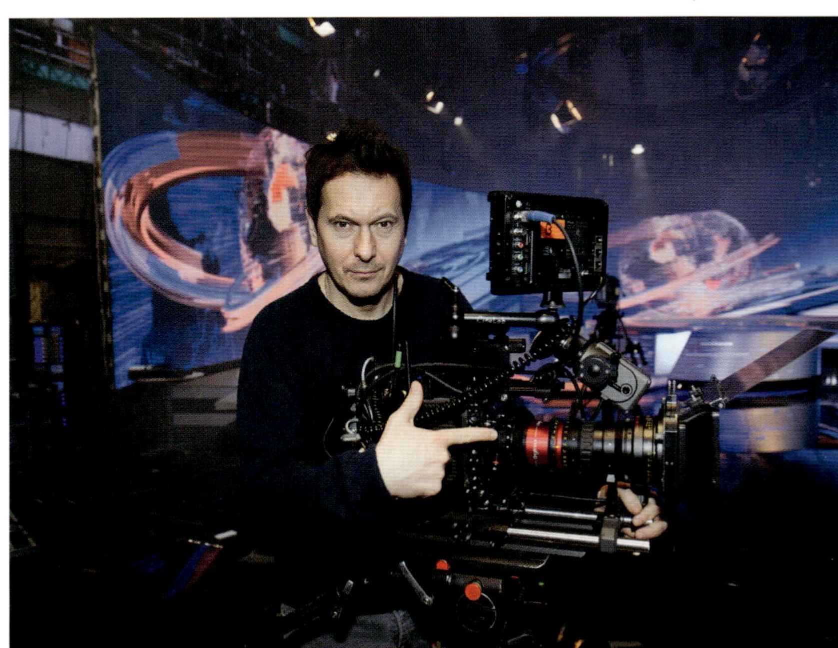

photography and equipment rental companies, as a lens could be mounted on a camera but would not necessarily cover its sensor entirely. Producers of top-of-the-range cameras like Arri, Sony, and Red followed the 35mm and S35mm with still larger sensors like Full Format, VistaVision, and so on. The obsolescence of equipment became a recurrent problem.
Movie equipment rental companies now had to earmark a regular slice of their budget for the purchase of latest-generation cameras. Faced with this complex market situation, and drawing on the experience acquired during the 1990s with its 11.5x zoom, configurable both for the S16 film format and for digital HD formats, Angénieux decided to offer its customers modular lenses adaptable both from a spherical to an anamorphic format and from S35 to formats of larger size. The new Angénieux products were therefore equipped with innovative IRO (Interchangeable Rear Optics) technology, enabling rental companies to expand their range of lenses at an attractive level of investment. The application of the concept first to the Optimo 44-440 A2S (T4.5) and then to the 42-420 A2S (T4.5) permitted their conversion to the spherical zoom (Optimo Style 25-250 T3.5). The latest Angénieux lenses owe their success to IRO technology.

New Zooms for the 35mm and Full-frame Market

New cameras with large-sized sensors were available at very affordable prices as from the middle of the 2010s, thus permitting numerous independent filmmakers to adopt these tools. Angénieux launched the Type EZ series in 2016 to meet the demands of this new market. The 35mm 30-90 EZ1 and 15-40 EZ2 are compact, lightweight lenses ideal for shooting with shoulder-mounted or moving cameras and available in the S35 version at a price in line with the budget of these new customers. Thanks to IRO technology, these two zooms have a Full Frame/Vista Vision option enabling the operator to mount them on a large range of cameras while optimizing field of focus and aperture for each format. A partnership was formed with the US company Band Pro Film & Digital Inc. and the Chinese Jebsen Industrial Technology Co. Ltd to finance and distribute this series of lenses.
The same period saw the development of a new 12x zoom, the Optimo Ultra 12x, a product of particular strategic importance designed to replace the Optimo 24-290 as the new standard for the market. This new lens also incorporates IRO technology and can adapt to three different formats of camera sensor. Thanks to its modular design, it can be configured in three different versions: S35 (24-290mm Ø31.1mm), U35 (26-320mm Ø34.6mm) and FF/VV (36-435mm Ø46.3mm), by simply changing the rear optics and replacing the focusing, zoom and aperture rings. The optical performance of the new Optimo Ultra 12x with Full Frame and focal lengths from 36mm to 435mm is far superior to that of an Optimo Ultra 26-320 equipped with a 1.4x multiplier.
In September 2018, just a few months after its appearance on the market, the Optimo Ultra 12x received the year's highly prized Cinec Award in Munich.

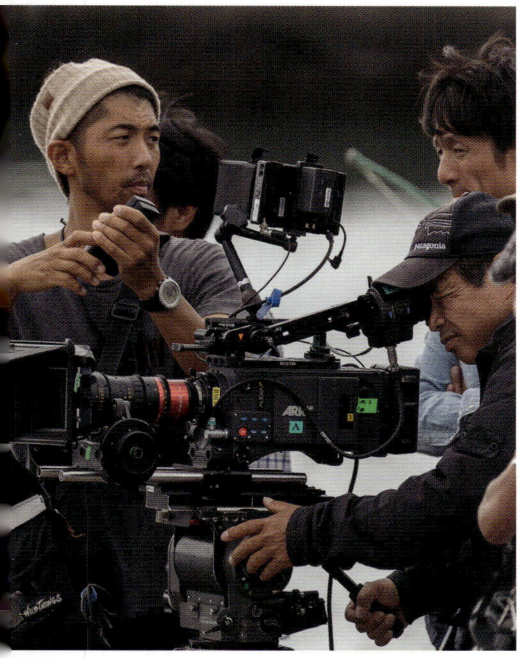

Optimo 30- 72A2S and cinematographer Norimichi Kasamatsu JSC for *IKari* (2016) directed by Sang- Il Lee (photo by Yasuhiko Mikami).

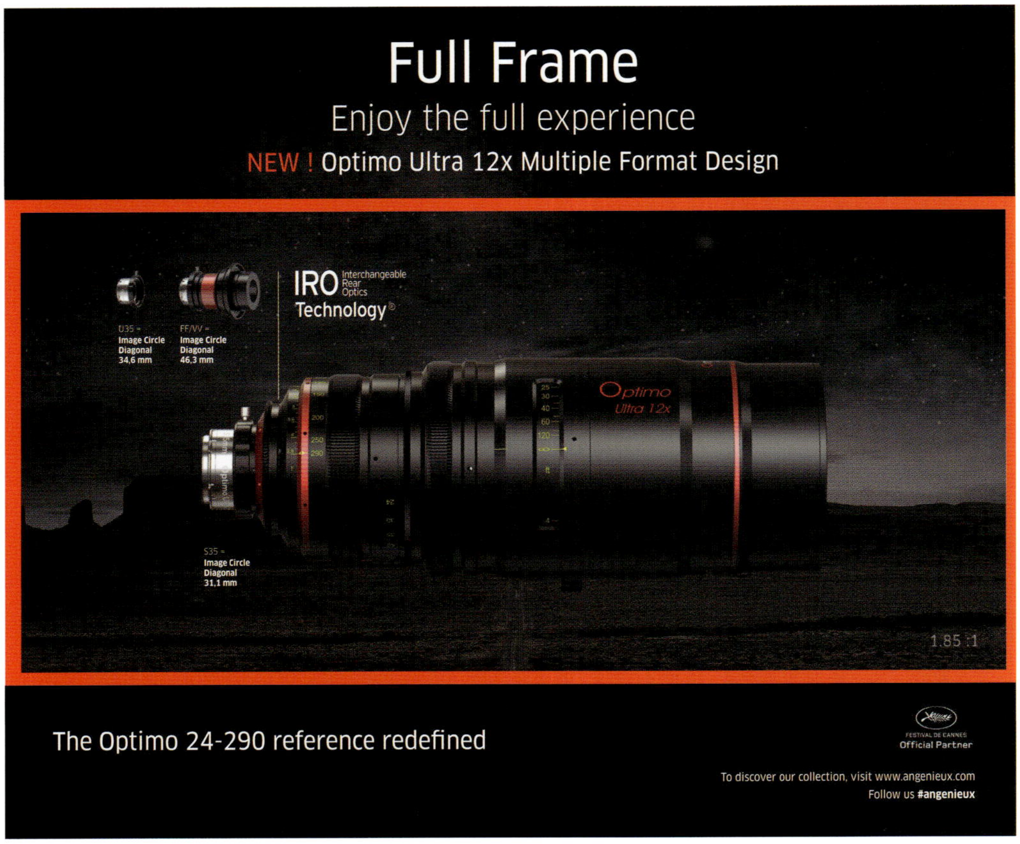

Advertisement for the launch of the Optimo Ultra 12x at IBC 2017 trade fair.

« Enabling customers to configure their lenses optimized for projects and/or cameras is truly a unique feature that is only available from Angénieux. The IRO technology is the result of the amazing talent of the Angénieux's design team, and our ability to design and manufacture large diameter and powerful spherical and aspherical lens elements, in order to increase the optical power of the front element, to reduce the sensitivity of the rear elements, which become easily interchangeable. Building optics with the IRO systems was a real challenge. The main target is to have a simple system allowing the user to swap the rear optics with only minor re-alignments on the lens. The interface between the zoom and the rear section is very critical on zoom lenses. Specific tolerances have been applied on the rear optics mechanical interface to achieve good image performances after swapping the rear group. Optical groups are factory aligned with a dedicated lens front and zoom sections in order to minimize spherical aberration and unevenness of the image. The lenses having IRO technology also need to have an easy access to the zoom, focus and iris rings allowing a quick swap of these rings without optical calibration. The IRO technology represents a future-roof solution in a continually moving world. »

Christophe Remontet

Technical & Engineering Director and Cinema Product Line Manager

The Angénieux Look

Some Cinematographers call it creamy, warm, organic... some talk about a round and delicate roll-off, with flares... some others would say it is very cinematic, warm and with character.

What is the so-called Angénieux look?
When Kees Van Oostrum, President of the ASC (American Society of Cinematographers), the biggest association of cinematographers in the world, defines the look of Angénieux lenses, in an interview with Jon Fauer from *Film & Digital Times*, he says:
"I think it's the way they manage human faces. When you photograph a face with an Angénieux lens, it seems to be gentle. It is very 'kind' even without putting diffusion in front. It's not only the definition. It's also the way that it treats the skin tones—the tonality of the skin—that have an influence over me."
And he adds: "Of course, we use sharp and ultra-sharp lenses for certain scenes, but if I had to live on an island alone with just one lens, I would take an Angénieux right now."

This quote stands for all. Such a result comes from more than eight decades of amazing know-how and expertise. Building a good lens is complex, but can be done by some optical manufacturers who have the knowledge of making optical elements and putting them together. Building a lens that makes a pleasant image which can be used in the cinema productions is another step. The lens needs to please first the cinematographer who is selecting optics very carefully in order to create his/her art.
Angénieux's philosophy is to give freedom to the cinematographers, to enable them to do almost anything they want. The idea is to give the very best, making outstanding lenses that they can degrade if needed, or modify with filters or nets. You cannot think technical only, it is important to think artistic too. This approach goes back to Pierre Angénieux's old times.
With specific parameters such as focal length, aperture, size, weight or reliability, the look of the image is one of the most important parameters for the DP when he/she selects a lens. Especially in the new digital age, where the image made through the digital sensor and the signal processing is very standard, the choice of the optics based on its look is becoming more and more important.
Years after years, Angénieux's optical engineers have been working on their design in order to build products which provide an emotion on the image, a unique look for telling a story. They design a collection of optics which are all very homogeneous, making not only a perfect image with good contrast and minimal chromatic aberrations, but also optics having a distinctive look, not too clean, not too sharp, with a good balance between resolution and contrast. Optimo lenses have also a nice and smooth transition between the area in focus and the area out of focus, providing a pleasant organic look. The glass and the coatings are selected carefully in order to generate a warm image giving an appealing rendition of the skin of the actors and actresses. The depths of field are also smooth and well balanced from the front to the rear of the scene.
Angénieux's optical designers are like the winemakers who are crafting a grand cru by blending different type of grapes or grapes providing from different field. They are designing a new product following years of experience making lenses for cinema and exchanging ideas with users in order to give them a product they will be pleased to use. That is the so-called State of the Art.

The Testimony of Cinematographers

« The choice of lens is crucial for me [...] The need for zooms soon became apparent on *Saisons*, as we knew that numerous takes would be impossible when filming the animals. The producers wanted to break away from the standard practice of long shots in animal photography. The idea was to work as close as possible to the animals. Everything was done in the film to ensure that no harm came to the animals, from blue tits to bears and hedgehogs to musk oxen, and to minimize any disturbance of their normal behavior. I wanted zooms capable of giving a rounded effect to offset the excessively high definition of digital photography. I had made Belle et Sébastien on film with Angénieux zooms and thought they would be perfect to obtain the look I wanted for *Saisons*. »

Éric Guichard

More than a documentary on animals, *Saisons* (2016) is an animal fiction film made by Jacques Perrin and Jacques Cluzaud.
Éric Guichard is vice-president of the AFC.

« Ever since I was a student of Cinematography in the 80's at the Film and Television Institute of India in Pune, the Angénieux Zooms have been trusted companions. In the later years the Optimo 24-290 became an integral part of my kit. Of late I have been using short zooms too, for tight corners and for a more fluid style of filming...Cinematography continues to be an exploration and Angenieux a worthy fellow traveller. »

Anil Mehta ISC

Among many others, cinematographer Anil Mehta's recent works are Jab Tak Hai Jaan (2012), Beyond the Clouds (2017), Secret Superstar (2017), Sui Dhaaga: Made in India (2018).

« On my three films (*Melancholia, Nymphomaniac & The House That Jack Built*) with Lars Von Trier the Optimo 28-76mm has been the true workhorse. The look, which in my opinion is beautiful for faces, combined with the all round focal lengths and the small size, makes this lens perfect for the kind of way Lars and I like to work, giving the actors as much freedom as possible. »

Manuel Claro

Cinematographer Manuel Claro lives in Copenhagen. *Melancholia* and *The House that Jack Built* were selected in Official Competition at Cannes Film Festival 2011 and 2018.

« I'm a big fan of Optimo. For me they are really convenient and time saving lenses in terms of operation. I have been using them extensively in action scenes and a lot of emotional scenes especially when I need to zoom in gently to draw attention towards something that is important in the scene.
I'm also a big fan of the 28-340mm Optimo zoom now.
I used it extensively in the Amazon prime series recently. »

Aseem Mishra WICA

Cinematographer Aseem Mishra WICA also known for his work on *Once Upon a Time in Mumbai* (2010), *Band Baaja Baaraat* (2010), *Ladies vs Ricky Ball* (2011), *Phantom* (2015), *Bajrangi Bhaijaan* (2015) and *Tubelight* (2017).

« Their short zooms are revolutionary. If you have to make a low-budget film, you can take one camera, and you can take those two zooms, and you can shoot the entire film with two short zooms. You can hand-hold, you can use them as variable primes.
They are practical. »

Dariusz Wolski

Extract from Jon's Fauer *FD Times* full article.
The Polish cinematographer Dariusz Wolski has got an impressive filmography: *Crimson Tide* (1995), *Dark City* (1998), the *Pirates of the Caribbean* series, *Sweeney Todd: The Demon Barber of Fleet Street* (2007), *Exodus: Gods and Kings* (2014), *The Martian* (2015) and *All the Money in the World* (2017).

« I used the same package of compact Angénieux zooms (15-40mm, 28-76mm, 45-120mm) on both movies, *Happy Death Day* and *Get Out*, but on *Get Out* we also had the 12x Optimo. We did have limited time, so using zooms does definitely save a few minutes a day; In the case of *Get Out*, entirely shot with Optimo zooms, it was also a directorial choice to prefer zooms over prime lenses. On *Happy Death Day* I was so happy with using the Optimos that I wanted to do it again! »

Toby Oliver

Get Out (2017) directed by Jordan Peele got four nominations for 2018 Academy Awards. Toby Oliver is also known for his work on *The Dirt*, *Insidious: The Last Key* and *Happy Death Day 2 U*, all with Angénieux zooms.

« The 3 short Angénieux zooms 15 to 40, 28 to 76, and 45 to 120 are revolutionizing the way we shoot television. With their fast aperture, light weight and close focusing they are the perfect tools for the organic handheld digital age. »

Rodney Charters ASC CSC

Rodney Charters is the cinematographer of numerous episodes of TV series like *24*, *Shameless*, *Criminal Minds*, *Dynasty* and, more recently, *Magnum P.I.*

« Overall technical requirements for live event productions vary greatly for this kind of work. Typically we'll be shooting with anywhere between 6 to 24 cameras, depending on the scale & ambition of the show. Combining a variety of grip solutions, from traditional cranes & dollies, to specialist robotic & aerial systems. One constant factor is always the need for versatile, fast, high quality zoom lenses. Filming any live event with a paying audience, the director and cinematographer do not have the freedom to put cameras wherever they want, or move them at will during the performance. So every camera position needs to offer a variety of shot sizes, often in low light.
In recent years the introduction of digital cinema cameras with S35mm sensors has been changing the culture of how live events are filmed. For some years HD TV cameras (with 2/3" sensors) were very dominant, as the relatively high cost of stock & processing pushed S16mm or 35mm film out of this market. The new digital revolution is allowing cinematographers to once again capture live events in a more cinematic way. Cameras like Arri Alexa, Sony F55/F65 & RED Weapon are way better equipped to handle the extremes of theatrical lighting than TV cameras, as they offer much greater latitude of exposure, and can also tolerate extreme colour saturation (very common in the way live music is lit).
The clarity & beauty of the images produced by superior glass, combined with shallower depth-of-field from using larger sensors, gives cinematographers the opportunity to capture the scale and drama of live events in a more 'filmic way'.
During the concert of Roger Waters, we had the great opportunity to try out the newest Optimo Ultra 12x.
I felt confident enough in the new lens to allocate it one of the most key positions in our plot, a panther dolly in the pit, which offered the most heroic angles for main vocal and lead guitar coverage of Roger and the band. I also put my most trusted crew on that camera - veteran operator John Simmons, my main focus puller for the last two decades Oleg Poupko, and one of my favourite grips of all time, Johnny Donne. Putting this beautiful lens in their hands gave us some great results, they absolutley killed it and their footage will be on screen much of the time in the finished film. »

Brett Turnbull

Brett Turnbull has an incredible experience in filmmaking: music videos, drama, documentaries but also multicamera live events, mainly concerts: Beyoncé, The Rolling Stones, Madonna, Coldplay, U2, Bruno Mars, Taylor Swift, Paul McCartney, Iggy Pop, The Who, Rihanna, Ed Sheeran, Justin Timberlake, Adele, Peter Gabriel, David Gilmour, Radiohead, Nick Cave, The White Stripes, One Direction, Queen, Pet Shop Boys, Kings of Leon, Bjork, Noel Gallagher, Shakira, Kylie Minogue, ELO, Tom Petty. Very recently he photographed Roger Waters' concert in Amsterdam at the Ziggodome during the *Us and Them* tour.

« Zoom lenses have a huge benefit in the air for many reasons.
The ability to start very close to a subject and do a large pull away with the helicopter as we zoom out is one way we take advantage of a zoom. The other obvious advantage is not having to land to change lenses. Changing lenses is time consuming as we have to rebalance the system and get to a suitable landing site. My favorite Zoom is the 24-290 Optimo Ultra it has the 3 things that I am sure all aerial DP's look for in a zoom. Quality, range and speed. We are looking forward to the new 24-290 Optimo that we will be taking delivery of soon. The other zoom that I use for aerials is the 19.5-94 Optimo. This zoom is like a variable prime lens. It has nice wide angle with good coverage for the larger formats like Arri's open gate and Red's 8K. The 2.6 T Stop makes it perfect for night work.
Recently I used the 25-250 Optimo style lens for *The Irishman* a new Scorsese feature, *Succession* a TV show and I used the 19.5-94 just this past Monday to shoot a cruise ship commercial for Royal Caribbean. All 3 of these projects were completely different but all utilized an Angénieux zoom.
I think aerials have become a standard to establish location and for transitions to move the story from one place to another as seen in all of the *Jason Bourne* movies. Sometimes they can add some real drama like in a Bollywood film we did last year called *A Gentleman*. We used the 44-440 Angénieux Anamorphic lens to approach downtown Miami low on the water and then fly up to a single building in push in to reveal a man swinging from a rope on the top of the building. Dramatic reveals like this are really only possible with a long zoom.
Most of the projects are requesting 4K. We have several different camera's we offer in conjunction with the Optimo zooms to achieve 4K and even up to 8K. However a higher K number does not seem to be as important as a high quality lens in my experience but I am sure 'the higher K race will continue'. »

Pat Longman

Pat Longman - Active Camera Systems Inc. (Miami)
Pat Longman is member of the SOC - Society of Camera Operators. He made aerial shots of a huge list of movies. Among other recent ones: TV series *Ballers* (2015), *Daredevil* (2016), *The Strain* (2017), *Succession* (2018), *Jason Bourne* (2016), *Padman* (2018), *Starbright* (2018).

« The Optimos are my first choice in approaching the story I'm telling because they are such a loyal translator of my vision. For me, they are literally the most important pieces of equipment, along with the camera. What I see with my eyes is perfectly translated by the lens and, because of its truthfulness, either the film gets printed with that or if it's digital, the media registers very, very accurately what I'm intending to do. »

Guillermo Navarro ASC AMC

Guillermo Navarro won in 2006 the Academy Award for Best Cinematography for the film *Pan's Labyrinth*. He photographied an incredible number of other major movies, among them: *Star Trek:Discovery, Night at the Museum: Secret of the Tomb, Pacific Rim, Twilight Sagas, I am Number four, The Resident.*

« We shot *The 33* movie in the above-ground scenes in Chile and underground in mines in Colombia. For this film, we used Angénieux lenses. The lenses were Angénieux Optimos. We had 24-290, 17-80, 28-76 and 15-40. We didn't use any filters underground.
With wide zooms, everybody said, "Oh my God, you're crazy, you should use primes for depth of field and the lack of light." But the director wanted to show the immediacy of the process, which you could only get with the freedom of doing one take.
Another reason was for the look. Most of the lighting was done with available light, we had the miners with their flashlights, and we had a few little lights hanging. There's a point with 33 miners with flashlights pointing at camera or at each other. Zooms have a more gentle flare. Some primes have a very pronounced flare, because of the sharpness of the optics. We had very soft halos, a romantic kind of flare. And that was given by the zooms. »

Checco Varese ASC

ASC - Extract from Jon Fauer's *FD Times* full article.
The 33 (2015) directed by Patricia Riggen is based on a real-life event, when a gold and copper mine collapses in Chile, it traps 33 miners underground for 69 days.
Checco Varese recently shot three seasons of Tom Clancy's *Jack Ryan* TV seires as well as feature *Replicas* by Jeffrey Nachmanoff.

« The lenses provide a very beautiful image quality, which is very lively and not too contrasty. They are also very well balanced and of great mechanical and optical quality. The Optimo Style 16-40 and the 30-76 have the exact same dimensions and weight, so lens changes are extremely quick and easy to do. The lenses are very light and quite short compared to other zooms of the same range. So that definitely helps for long verité days with lots of handheld shooting. »

Tom Bergmann

Tom Bergmann mainly shoots documentaries, of which *Traffic Stop* by Kate Davis and David Heilbroner, *Abacus - Too Small for Jail* by Steve James, both documentaries were nominated for Oscars 2018, *Life, Animated* by Roger Ross Williams and Julie Goldman was also nominated in 2016 for an Academy Award. *Promised Land* (re-called *The King*) by Eugene Jarecki co-photographied by Tom Bergmann, Christopher Frierson and Etienne Sauret premiered at the 2017 Cannes Film Festival

« In the early days selecting equipment was a function of mostly weight. It wasn't possible to climb the walls without pairing down to the absolute essentials. Now I go for a combination of the size and weight but also the cinematic quality the cameras and lenses will produce. It's become a new athletic endeavor to carry a little extra weight in order to bring back higher end images from these remote high altitude places.
For lenses, I love both prime and zoom lenses and whenever possible I use higher end PL mount cinema quality glass. Often the situations I'm in dictate zoom lenses such as the Angénieux lightweight zoom lenses so I don't miss the action and still have the quality of a prime.
I've seen the Optimo Style 16-40 put up with sub-zero temps in the high mountains as well as wet and nasty jungle conditions. Recently we were shooting a National Geographic feature story in the jungles of Nepal about a culture of people who hunt honey on steep cliff sides. The lens got wet and covered in smoke, honey and attacked by gaint bees while dangling 300ft in the air on the cliff face. It held up to a month of this kind of abuse. I'm known as a gear killer but have never been able to break an Angénieux lens. »

Renan Ozturk

Cinematographer and director Renan Ozturk is a German-born, American rock climber, and visual artist, best known for climbing the Shark's Fin route on his second attempt to Meru Peak in the Himalaya in 2011 - co-founder and director of Camp4 Collective.

« Being an extremely visual director,
I was dreaming of adding a set of glass that would check all the boxes. I wanted a warm cinematic look, but I needed agility and speed. When I saw the specs initially I got butterflies in my stomach. It was like meeting the girl for the first time knowing that she may be the one. I went with my gut and a few months later we were married. These lenses have changed the game. They are the hardest workhorses we have, but their look and feel is so subtle and pleasing. There is always a best lens for the job, but these are the best for so many jobs. We fly them on the drone, we put them on the Movi, we shoot in low light, in bright light. It doesn't matter. Their workmanship and characteristics have been better than we ever expected. »

Aaron Brown

Director and founder of Onion Creek Productions, a new breed of production company based in Austin (Texas), about the Angénieux Type EZ lens.

« The intention of the short *Drift* was to create a visually fascinating film where drone-shots play a part and add value in the story but at the same time are not the main character.
That got us thinking on what we could do with a drone that has not yet been done before. That's when we came up with aerial retro zooms, anamorphic retrozooms. *Drift* won an incredible number of awards worldwide. Flying the anamorphic Angénieux zooms was a big part of *Drift*'s success. »

Frank Dewaele

Frank Dewaele - Kelly's Heroes
Drift is a short film Kelly's Heroes - a Belgian collective - shot as a demo project to try out some of those new possibilities in aerial filming with a drone.

« With the aspect we wanted for the film (*El Cielo del Centauro*), our first requirement was to have anamorphic lenses. Then, in analysing the storyboard Hugo Santiago had put down on paper, where he described all the shots in the film precisely, we realized we needed a zoom. In addition to simplifying the production with its versatility and cost, we needed to combine the physical camera movements with optical movements of the zoom to get a wide variation of shots within a take. And because of that variation we were looking for, the zoom's minimum focusing distance of 0.63 m was indispensable. »

Gustavo Biazzi

Gustavo Biazzi is also known for his work on *Castro* (2009), *The Student* or *Récit d'une jeunesse révoltée* (2011), *Los Dueños / The Owners* (2013), *Paulina* (2015) and *Los Vagos* (2017) as director.

« Angénieux new Anamorphic zoom lenses are lighter, they are compact, they can be used for shooting from hands. It is very convenient in use for the zoom effect for action scenes, on suspensions, on cranes, without changing the optics you can make a visual accent. You can create video effects that were not available before, for example, dolly zoom, when you ride the cart forward, and zoom out on the zoom. When the focal length changes, the depth of field changes, this gives a beautiful optical effect. In *#Selfi* film, the Angénieux optics revealed excellent plasticity, depth, crystal transparency, compatibility with fixed focus lenses. Some episodes I shot both on zoom and prime. For other creative solutions, it was stylistically better to use zoom. For all its sharpness, where it's necessary to be sharp, the optics of Angénieux are absolutely cinematic. »

Vladislav Opelyants

Cinematographer Vladislav Opelyants is native from Russia. His latest movie *Leto* (2018) was presented in Official Competition at Cannes Film Festival. He has just finished *Prityazhenie 2* (release planned for 2019). He is also known for *#Selfi*(2018), *Hostages* (2017), *(M)uchenik* (2016).

*A dolly is a device running on wheels or tracks upon which the camera is mounted in order to take an uninterrupted tracking shot.

« I have been shooting with my Type EZ-1 since I received it in June 2017 and really love it. It is an Angénieux lens by all means but it would be unfair to expect Optimo quality given the price difference. The colour, warmth, pleasing skin tones are there, but missing is that creaminess, round and delicate roll-off, and flares that have made Optimos famous in the industry. The Type EZ lens are beautiful lenses and in my opinion have a preferred look over other lightweight cine zooms in the market. (...) I am extremely pleased with my investment and look forward to many years with these lenses. I think these lenses are perfectly suited to people outgrowing or upgrading from their still lenses and moving up in their career or also perhaps investing on glass that is future proofed for our current rapidly changing camera technology world. »

José Alkon

José Alkon is a young Peruvian born and Sydney based cinematographer, one the first purchasers of an Angénieux Type EZ lens.

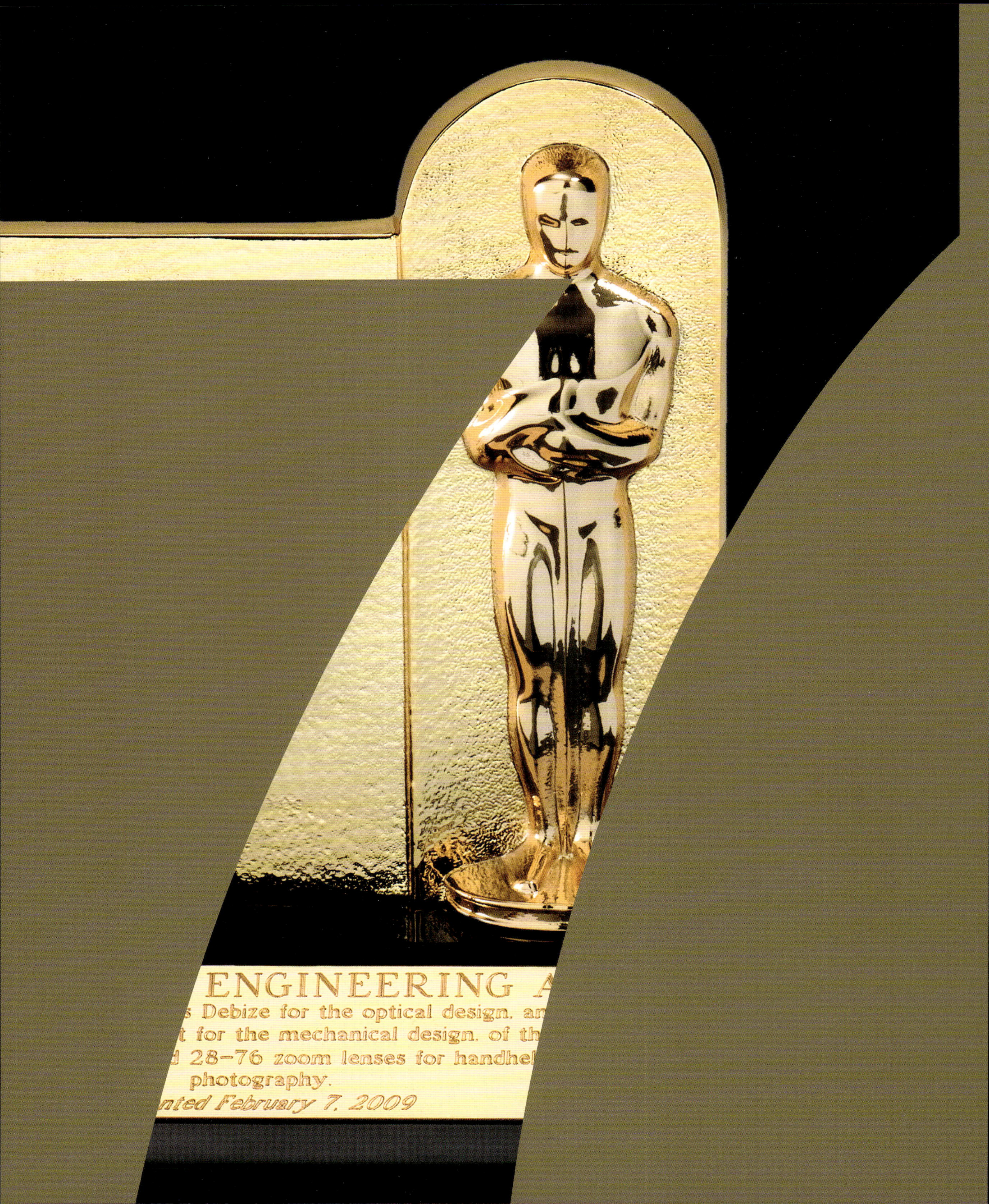
ENGINEERING
Debize for the optical design.
for the mechanical design. of
28-76 zoom lenses for
photography.
February 7. 2009

HIGHLIGHTS
TRIBUTES AND AWARDS

The **esthetic and technological demands** of cinematographers have ensured the success of Angénieux lenses for over 80 years. Their exceptional quality is regularly hailed in the greatest events in world cinema. Be it a full-length feature **film, a TV series, an advertisement, a documentary or a video clip**, the image translates a drive for narrative creation. Cinematographers are the **essential link** in this **creative process**. It was to pay tribute to these professionals of image-making, **without whom cinema would not exist**, that the **Pierre Angénieux ExcelLens in Cinematography** award was born at the Cannes Film Festival in 2013.

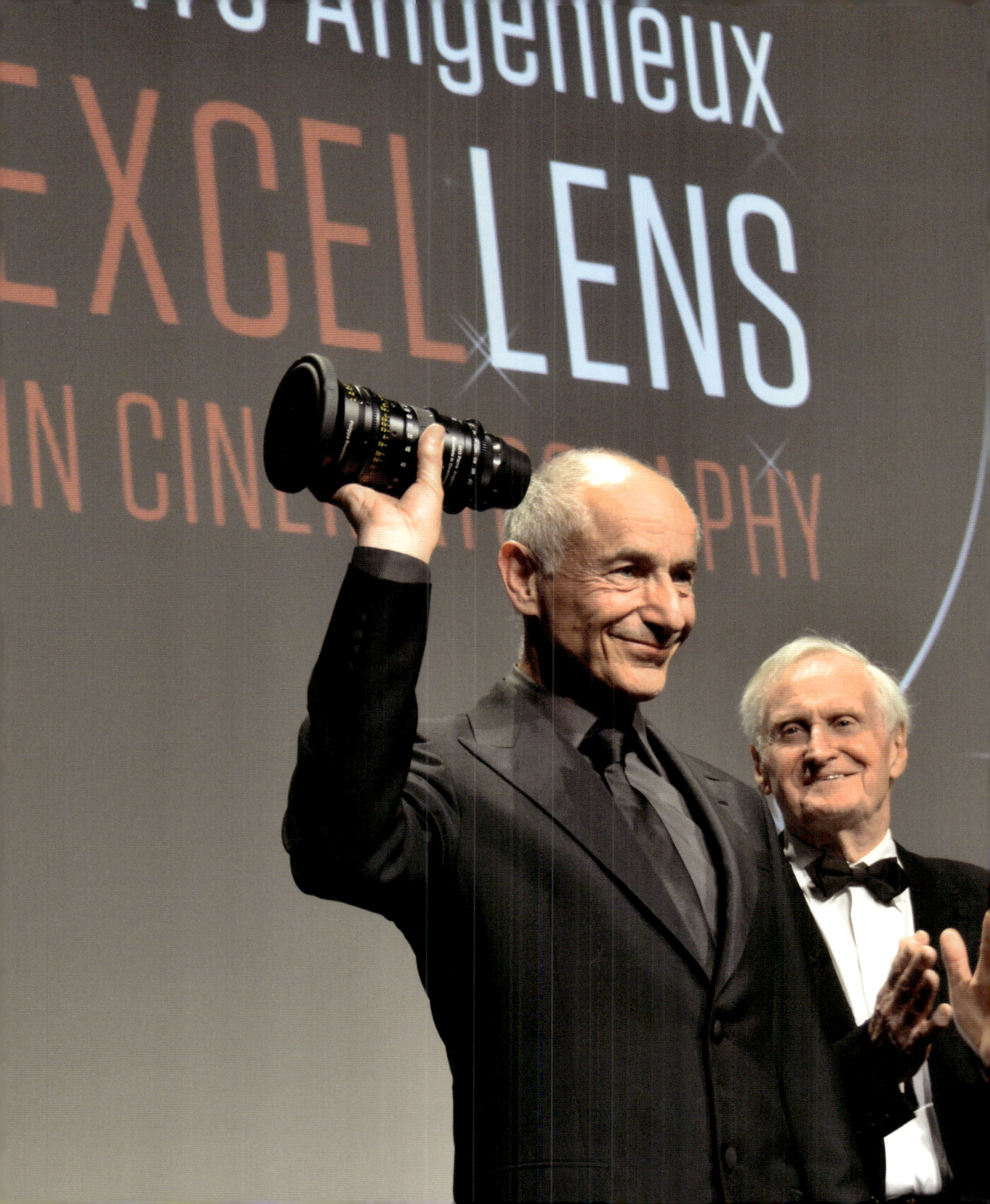
EXCELLENS

Cinematographers unanimously Praise Angénieux Optimo Zooms

Launched in 2001, thanks to its exceptional performance, the Optimo 24-290 immediately won favor in the world of filmaking, becoming the lens of choice on many of the biggest films in Hollywood. Its incomparable quality soon made it the benchmark lens of world cinema.
In September 2002, just one year after its introduction, the Optimo 24-290 received a **CINEC Award** in Munich.
The Optimo series has expanded over the years with new zooms. The Optimo 17-80, Optimo 15-40, Optimo 28-76, Optimo 45-120 offer a variety of new possibilities to cinematographers and have become synomymous with excellence, quality, and reliability.

CINEC Award

2002

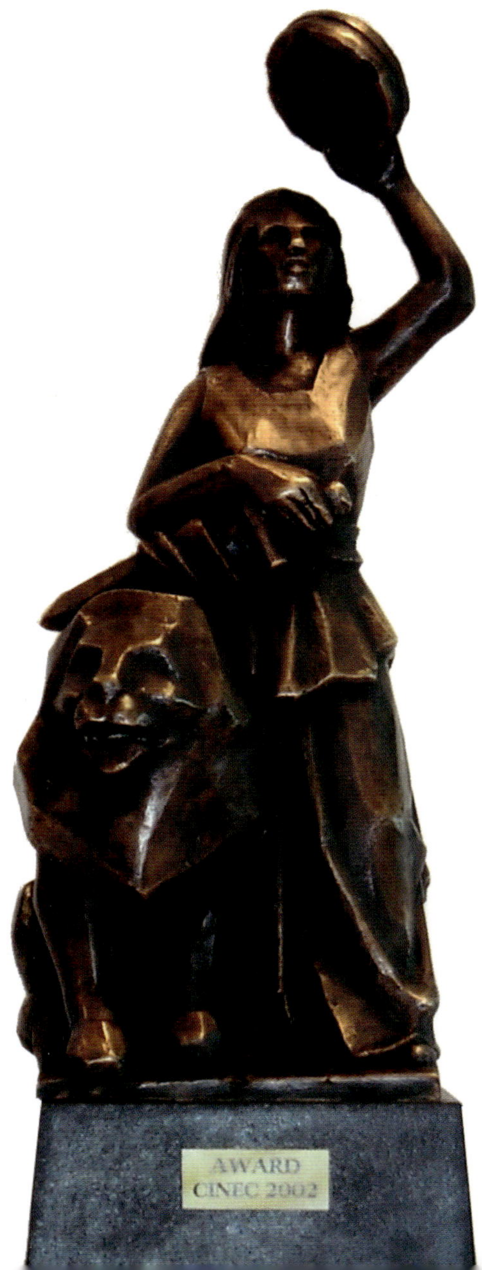

In Hollywood in 2009, the Academy of Motion Picture Arts and Sciences hailed the exceptional performance of the Optimo 28-76 and 15-40 lenses, and awarded a **Scientific and Engineering Award** to the engineers responsible for their development: Jacques Debize, Bruno Coumert, Christophe Reboulet, and Dominique Chervin. These lightweight, compact zooms were immediately adopted and particularly appreciated for filming with hand-held cameras, Steadicam, and now also drones.

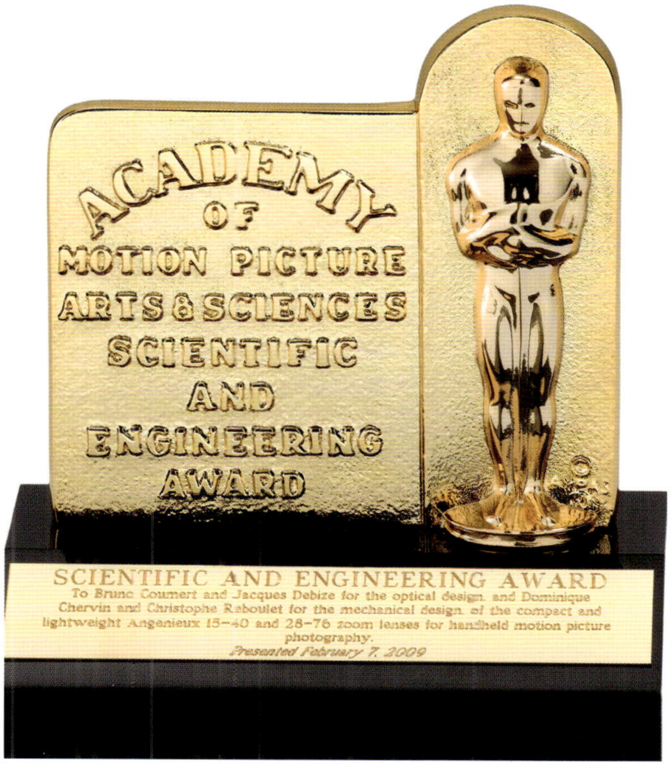

Hollywood, February 7, 2009. Dominique Chervin and Bruno Coumert, two of the four award-winning engineers, accompanied by Philippe Parain, President of Thales Angénieux, are presented with their Scientific and Engineering award by the actress Jessica Biel (Angénieux archives, photo by Jean-Marc Bouchut).

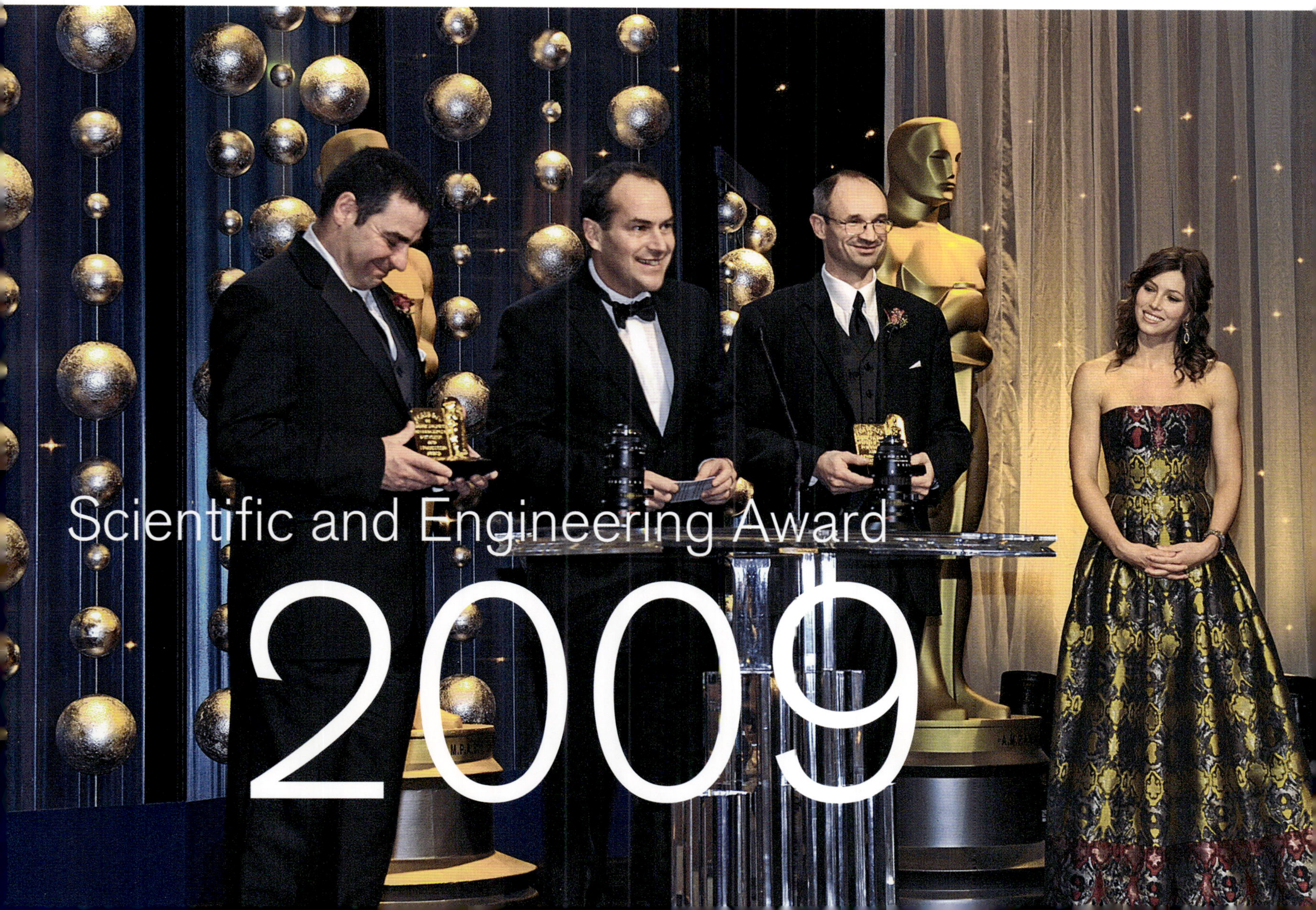

Scientific and Engineering Award 2009

CINEC Award
SOC Technical Achievement Award
2012

Rodney Charters (cinematographer of the TV series *24* and, since 2012, of *Shameless, Dallas, The Last Ship, Criminal Minds, The Fosters, The Shannara Chronicles,* and *Dynasty*) with Pierre Andurand, president of Thales Angénieux, and Dominique Rouchon-Picariello, then Sales Director (photo SOC).

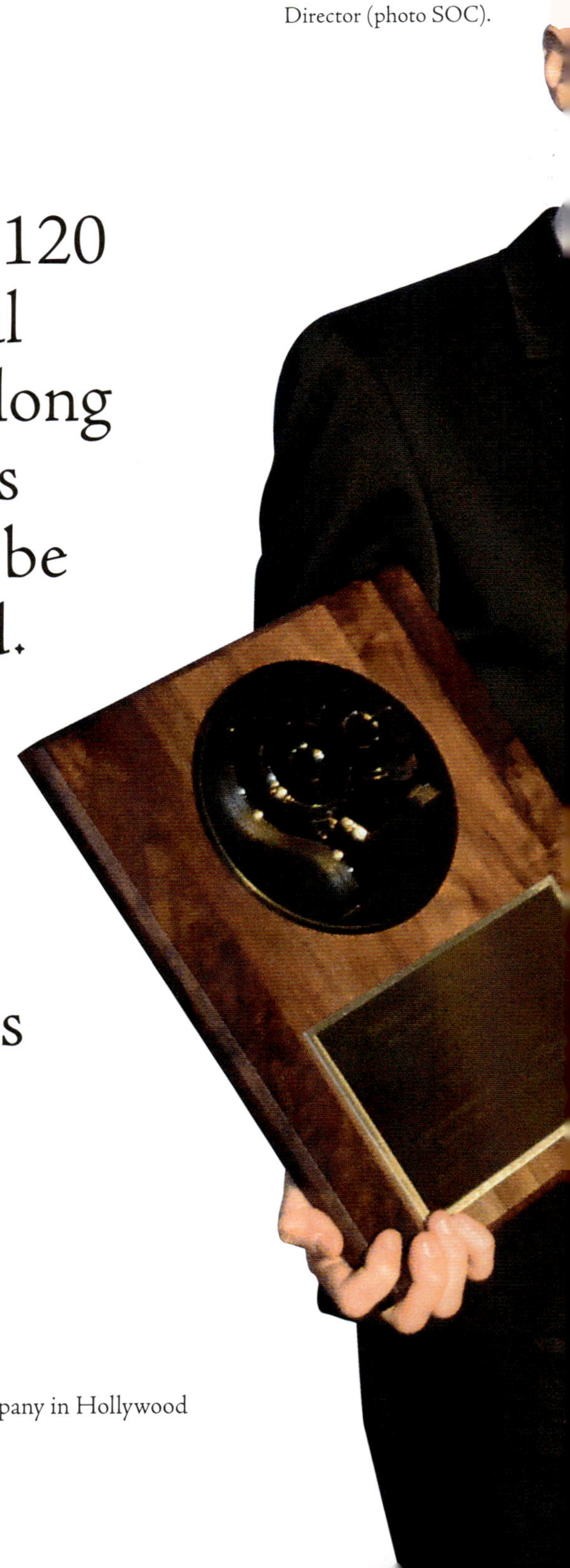

The Society of Camera Operators (SOC) hailed the appearance of the Optimo 45-120 in Hollywood in 2012 and awarded Angénieux with a **Technical Achievement Award** for the unrivaled qualities of the Optimo 15-40, Optimo 28-76 and Optimo 45-120 portable zooms for use on hand-held cameras due to their compact size and weight. The award was presented by cinematographer Rodney Charters ASC CSC NZSC on February 19, 2012.

Angénieux received a second **CINEC Award** in September of the same year for the exceptional performance of the Optimo 45-120 from the hands of Denny Clairmont, founder of the illustrious Clairmont Camera rental company in Hollywood.

ASC *American Society of Cinematographers*
CSC *Canadian Society of Cinematographers*
NZSC *New Zealand Society of Cinematographers*

« The Optimo 45-120 is amazing. The focal range of this lens is long enough for close-ups and light enough to be used on a hand-held. Cinematographers can get their shots without being in the actor's space and this makes it more comfortable for the actors. »

Denny Clairmont

Founder of the illustrious Clairmont Camera rental company in Hollywood

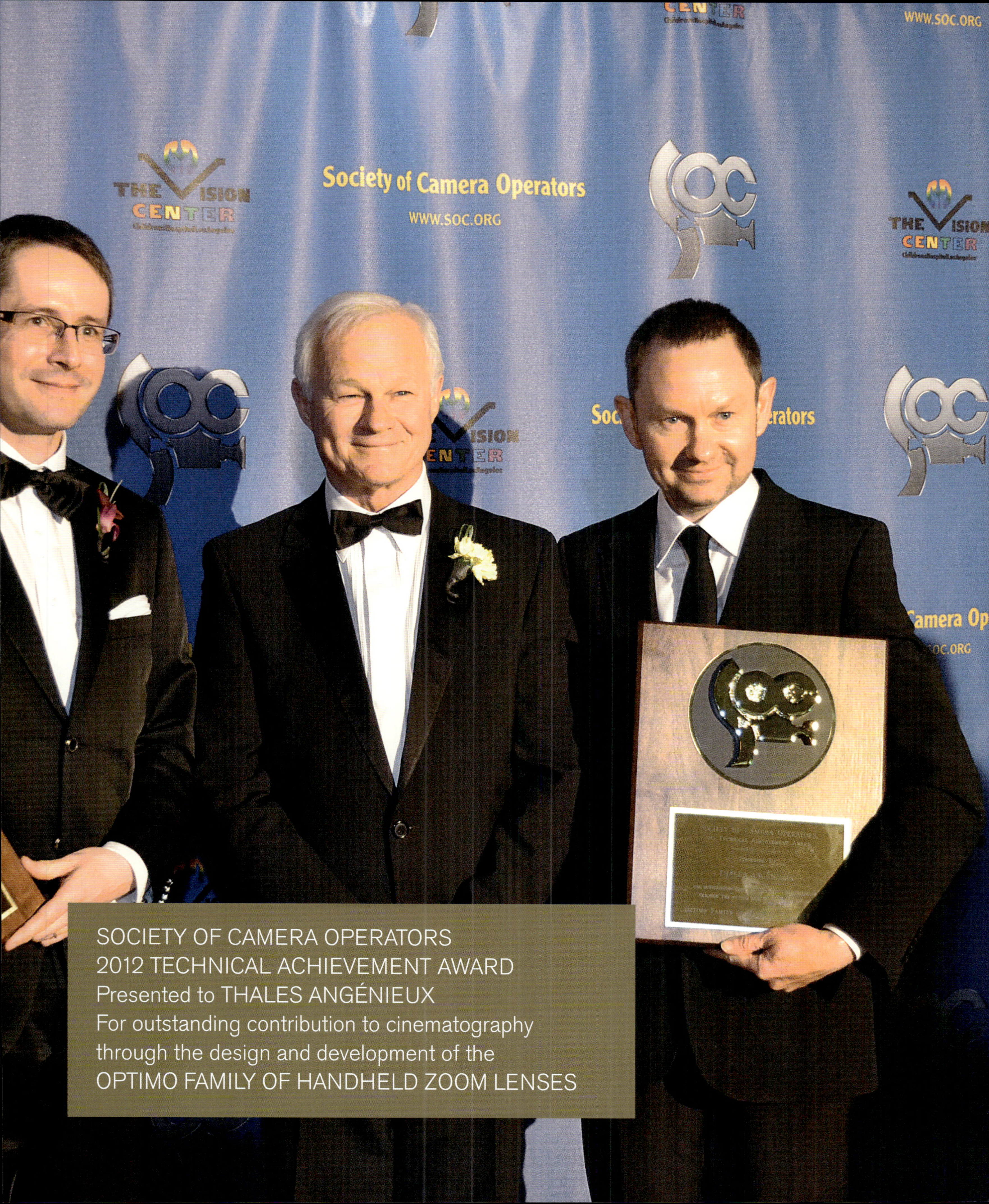

SOCIETY OF CAMERA OPERATORS
2012 TECHNICAL ACHIEVEMENT AWARD
Presented to THALES ANGÉNIEUX
For outstanding contribution to cinematography
through the design and development of the
OPTIMO FAMILY OF HANDHELD ZOOM LENSES

Angénieux received the **Bert Easey Technical Award** in London at the British Society of Cinematographers (BSC) annual Operators Night on November 28, 2014. This is the highest technical distinction awarded to a person or a company in the United Kingdom for contributions to the film industry. In presenting the award, John de Borman, then president of the BSC, praised Angénieux lenses for their lasting excellence in filmmaking all the way from the 1960s to the digital era.

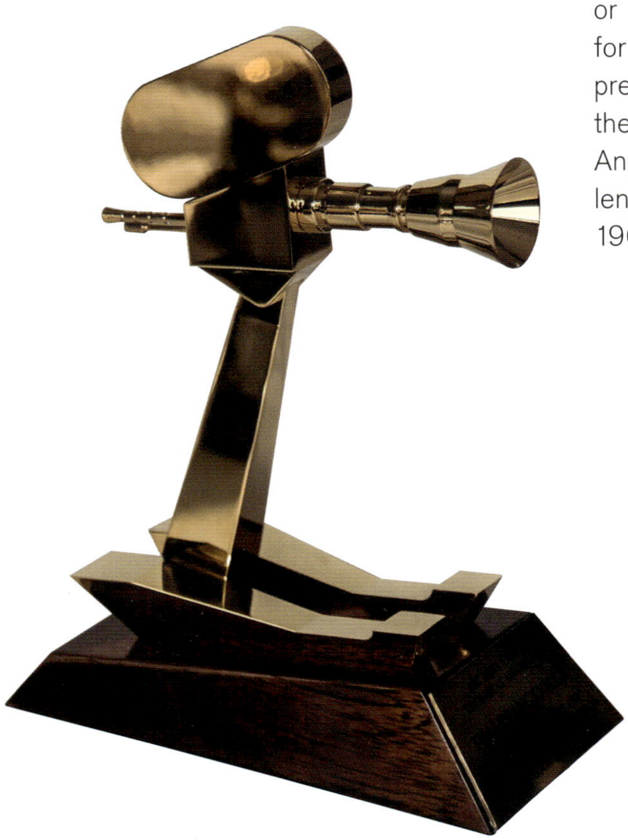

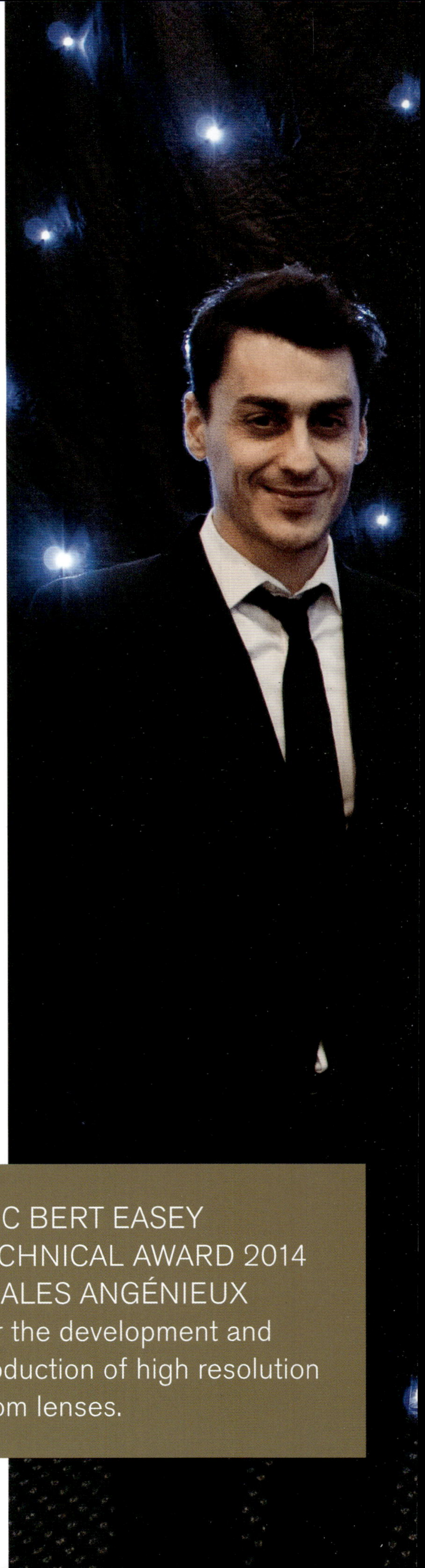

From left to right: Davy Terzian, head of sales for the United Kingdom, Dominique Rouchon, Pierre Andurand, and John de Borman (photo Claire Pie).

« We are delighted to present this award to Angénieux for producing zoom lenses which have become a major asset in the world of cinematography. »

John de Borman

President of BSC - British Society of Cinematographers (2011–2014).

BSC BERT EASEY TECHNICAL AWARD 2014
THALES ANGÉNIEUX
For the development and production of high resolution zoom lenses.

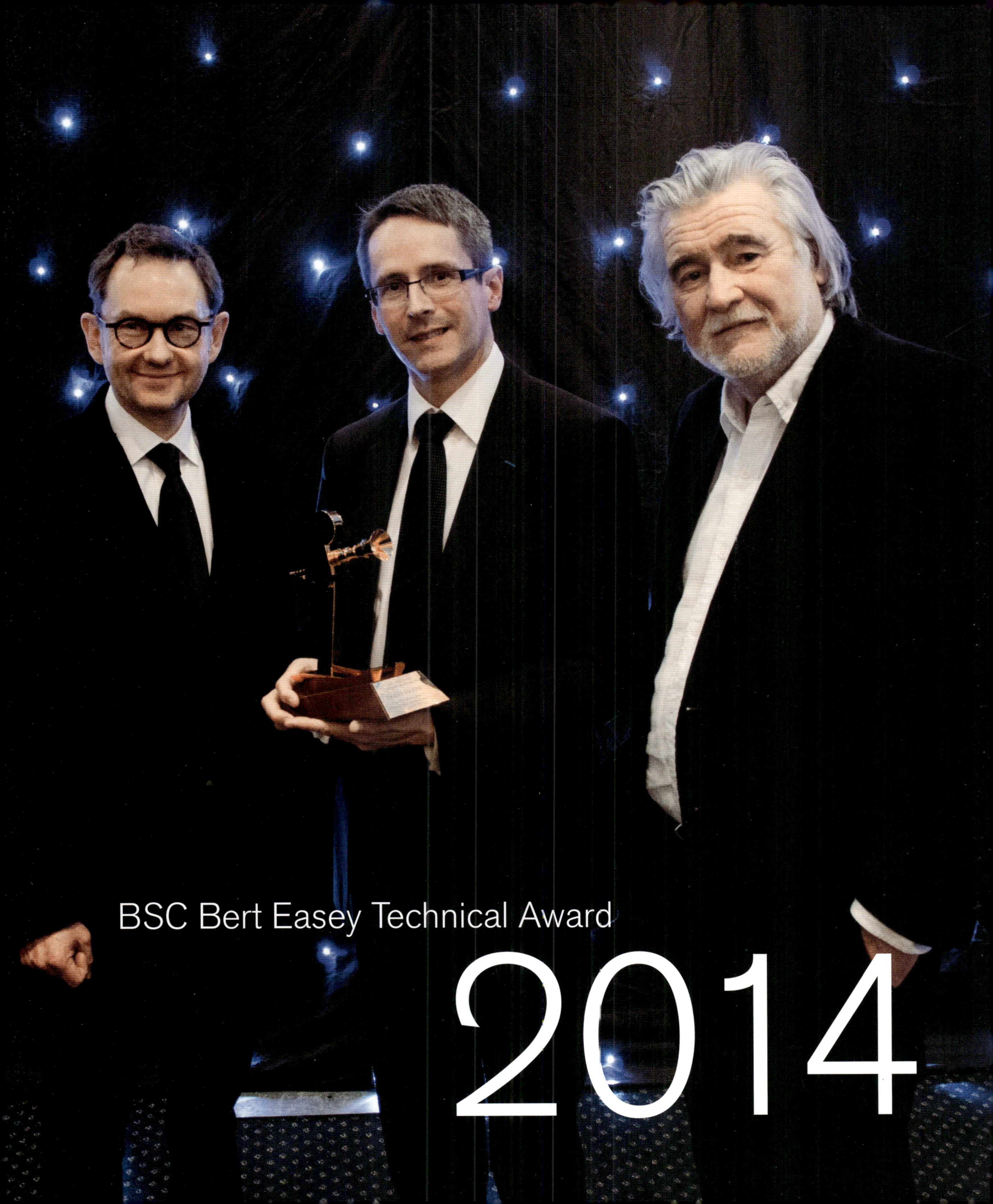
BSC Bert Easey Technical Award
2014

In 2002, just off the Angénieux production lines, the Optimo 24-290 received a Cinec Award and immediately won favor with film-makers. In 2017, Angénieux presented the Optimo Ultra 12x, a zoom designed to replace the 24-290. Precisely sixteen years after the CINEC Award of 2002 (in the lens category), Angénieux won the **CINEC Award 2018** for the Optimo Ultra 12x. The CinecAwards are for innovative products breaking new ground within the film industry. The prize rewarded all of the Angénieux cinema teams for their hard work and efforts in addressing this technological challenge.

The Optimo Ultra 12x represents an entirely new generation of zooms that Angenieux provides to the film industry, designed to meet the needs of constantly evolving market. This lens is a high-ratio modular zoom that retains the company's distinctive look and offers the standards of mechanical performance, handling, and durability so highly appreciated by Optimo users.

IRO – Interchangeable Rear Optics technology enables this particularly innovative zoom lens to cover a range of different camera sensor formats.

« A radical departure from the industry standard Optimo 24-290 zoom lens, the new multi-format Optimo Ultra 12x integrates state of the art optical design and manufacturing as well as truly advanced mechanics, while retaining the intrinsic qualities of the tool, that has contributed to shape the work of a generation of filmmakers, for the new formats of the future. »

Danys Bruyère

DGA Technologies-group TSF, member of the jury of the Cinec Award 2018

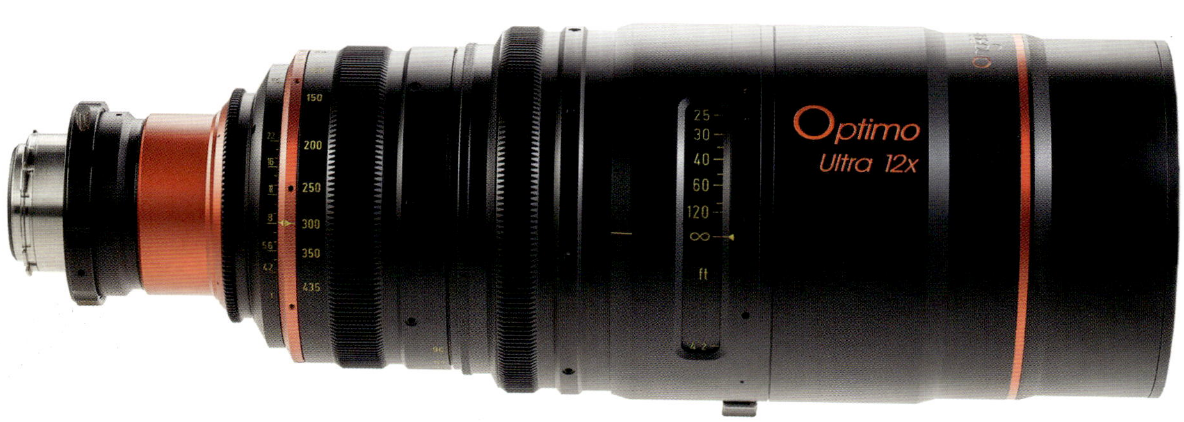

Optimo Ultra 12x (photo by Sylvain Madelon).

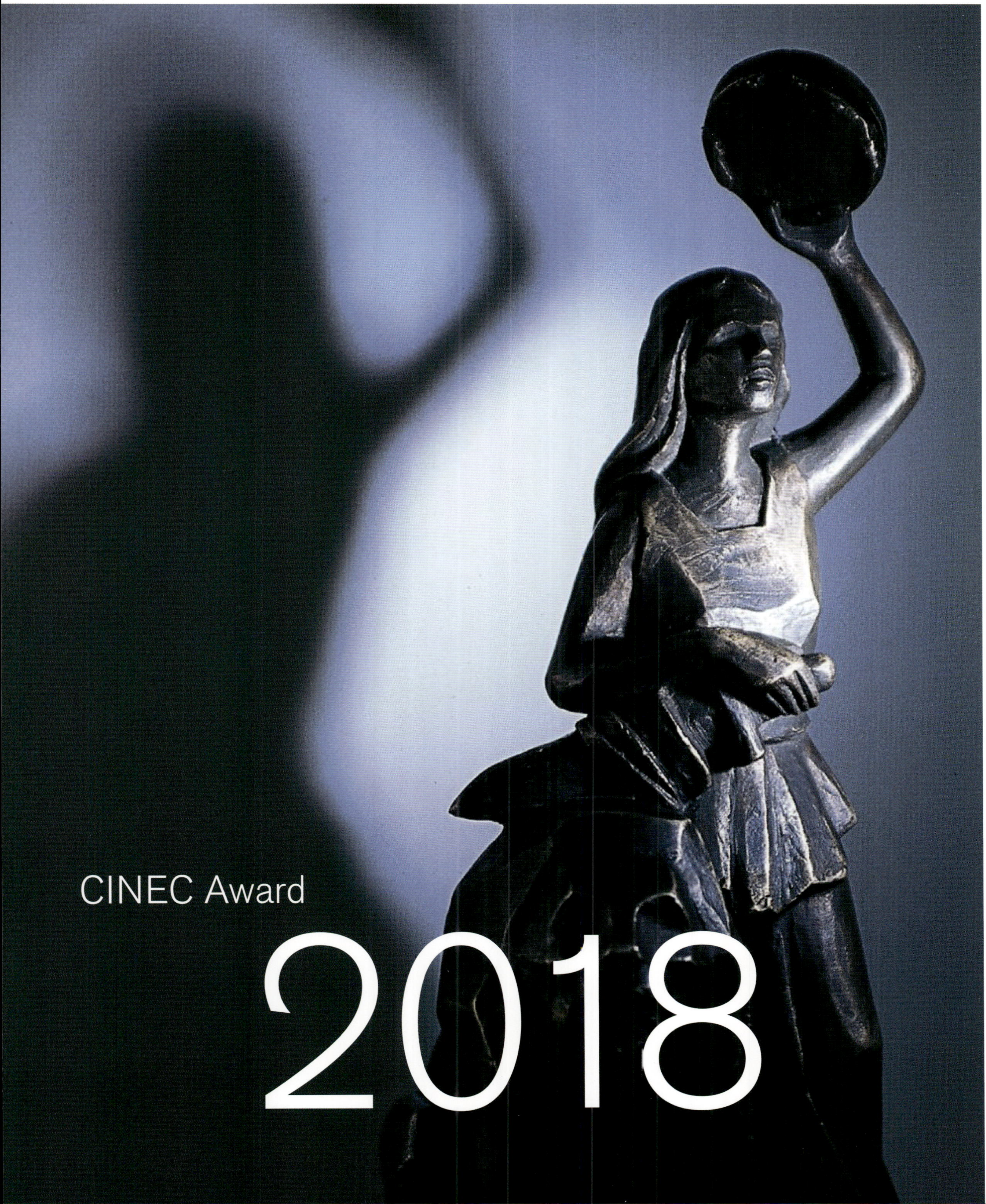
CINEC Award
2018

«ExcelLens in Cinematography» at the Cannes Film Festival

Angénieux Pays Tribute to Cinematographers

In 2013, for the first time in its history, Angénieux became an **Official Partner of the Cannes Film Festival**. On this occasion, the company instituted the Pierre Angénieux ExcelLens in Cinematography tribute to highlight the work of directors of photography. During an exceptional gala at the Palais des Festivals, in the presence of outstanding personalities and talents of the world of cinema, the winning cinematographer is presented with an Angénieux zoom specially engraved in the Saint-Héand factory with his or her name. Beginning with Philippe Rousselot AFC ASC in 2013, Vilmos Zsigmond HSC ASC in 2014, Roger Deakins BSC ASC in 2015, distinguished recipients are Peter Suschitzky ASC in 2016, Christopher Doyle HKSC in 2017, and Ed Lachman ASC in 2018.
In 2018, Angénieux decided for the first time to highlight a young talent of cinematography during the tribute presentation to Ed Lachman. Cécile Zhang, a Chinese cinematographer just 24 years of age, who had the great privilege of receiving the Angénieux special encouragement from the hands of Severine Serrano, Angénieux Director of Sales and Marketing. The only woman in her year at the Beijing Film Academy, Zhang graduated with honors and is now a member of the International Collective of Female Cinematographers (ICFC). In honoring the work of two cinematographers, one already enjoying unanimous international acclaim and the other at the start of her career, Angénieux also expresses its gratitude to the great figures that have contributed to the company's renown and to its interest in the new generations of professional cinematographers.

Cécile Zhang with the actor Xu Zheng and Severine Serrano (photo by Pauline Maillet).

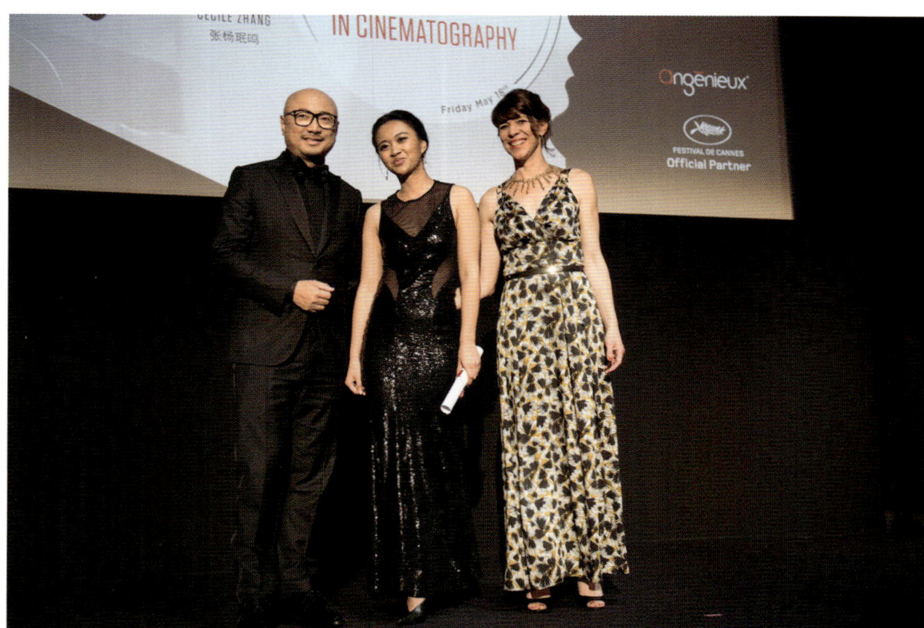

AFC *Association Française des directeurs de la photographie Cinématographique*
HSC *Hungarian Society of Cinematographers*
HKSC *Hong-Kong Society of Cinematographers*

« And I wish to say again that Angénieux has had a beautiful idea, and that this event has become a rendezvous to gather and honor one of the great cinematographers.
Little by little, this palmares is sketching another map of world cinema, another 'Map of the Stars' – the great stars of cinema lighting.
The Cannes Festival is therefore very pleased to partner this award. »

Thierry Frémaux

Delegate General of the Cannes Film Festival, at the 4th edition of the Pierre Angénieux ExcelLens in Cinematography award in 2016

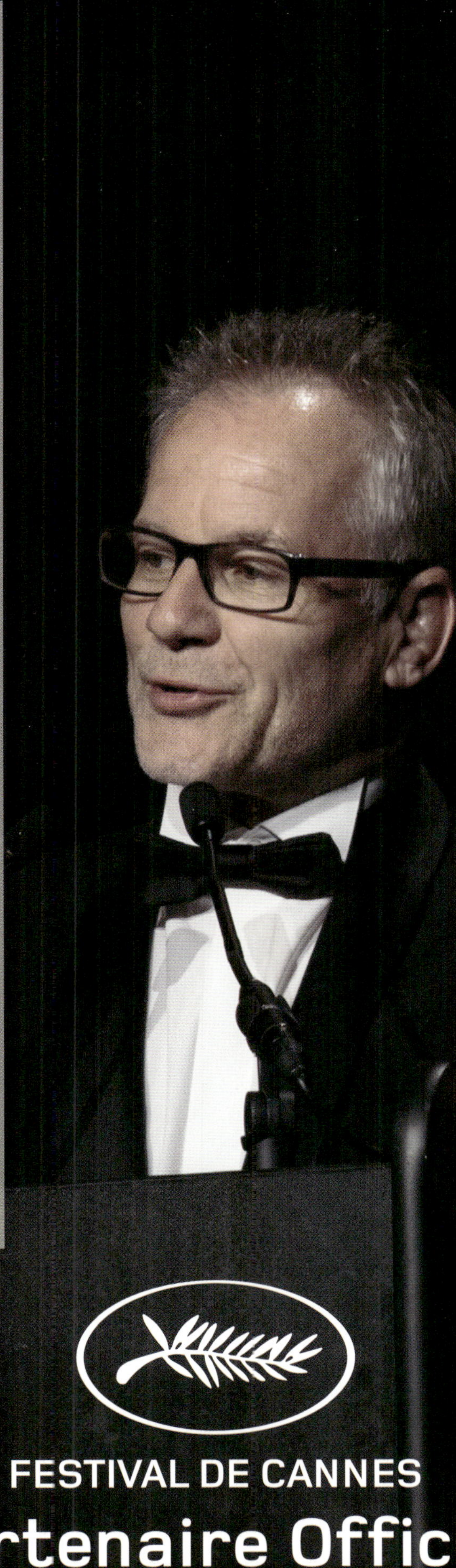

A look back in images on the editions of the Pierre Angénieux ExcelLens in Cinematography award at Cannes from 2013 to 2018.

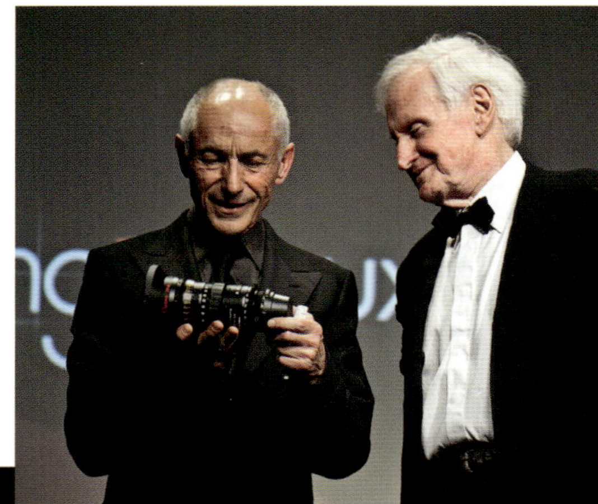

Cannes 2013: tribute to Philippe Rousselot in the presence of Carmen Chaplin, Kristin Scott Thomas, Victoria Abril, Uma Thurman, Jean-Marc Barr, and the director John Boorman (photos by Jon Fauer, left, and Dominique Charriau, below)..

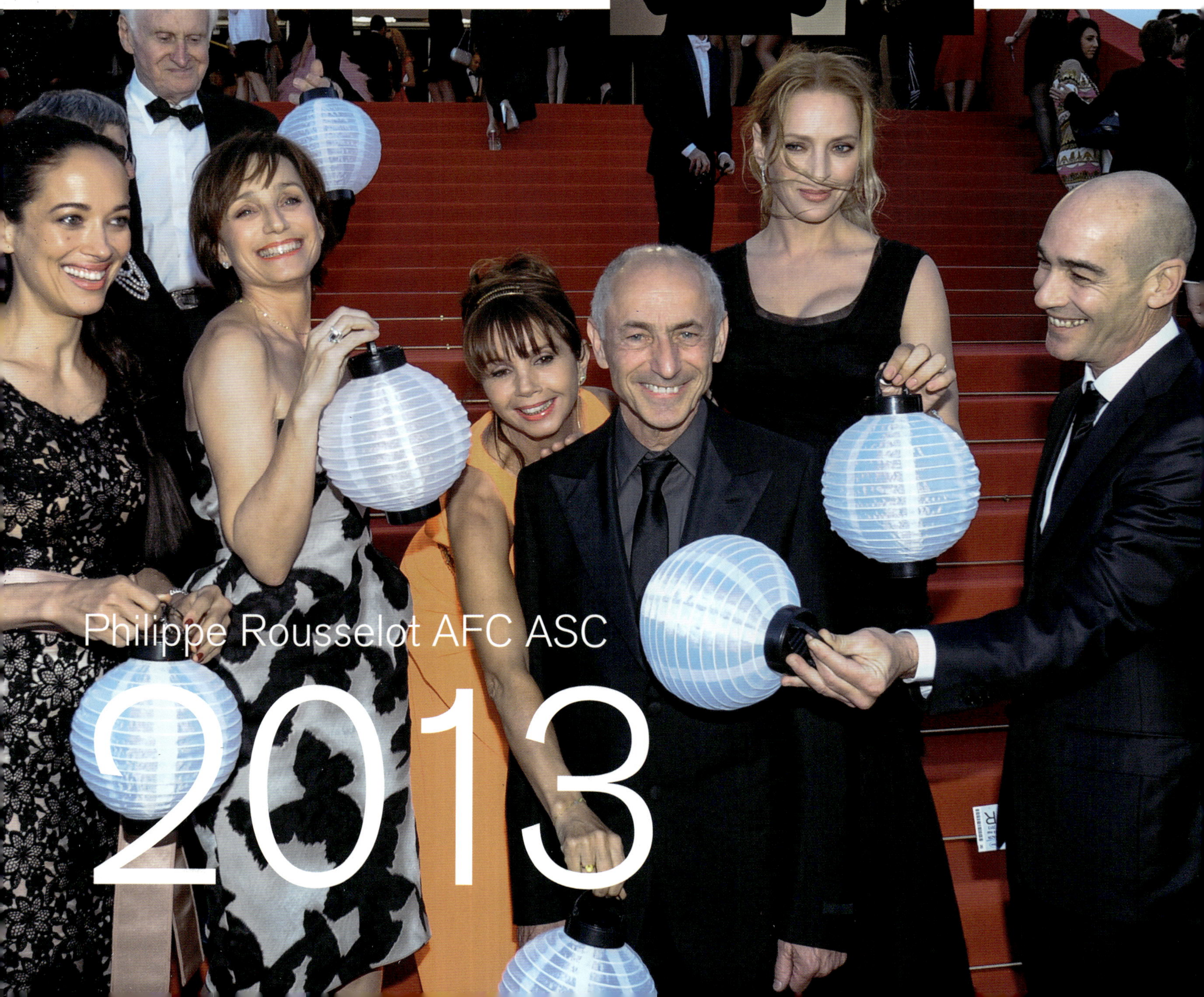

Philippe Rousselot AFC ASC

2013

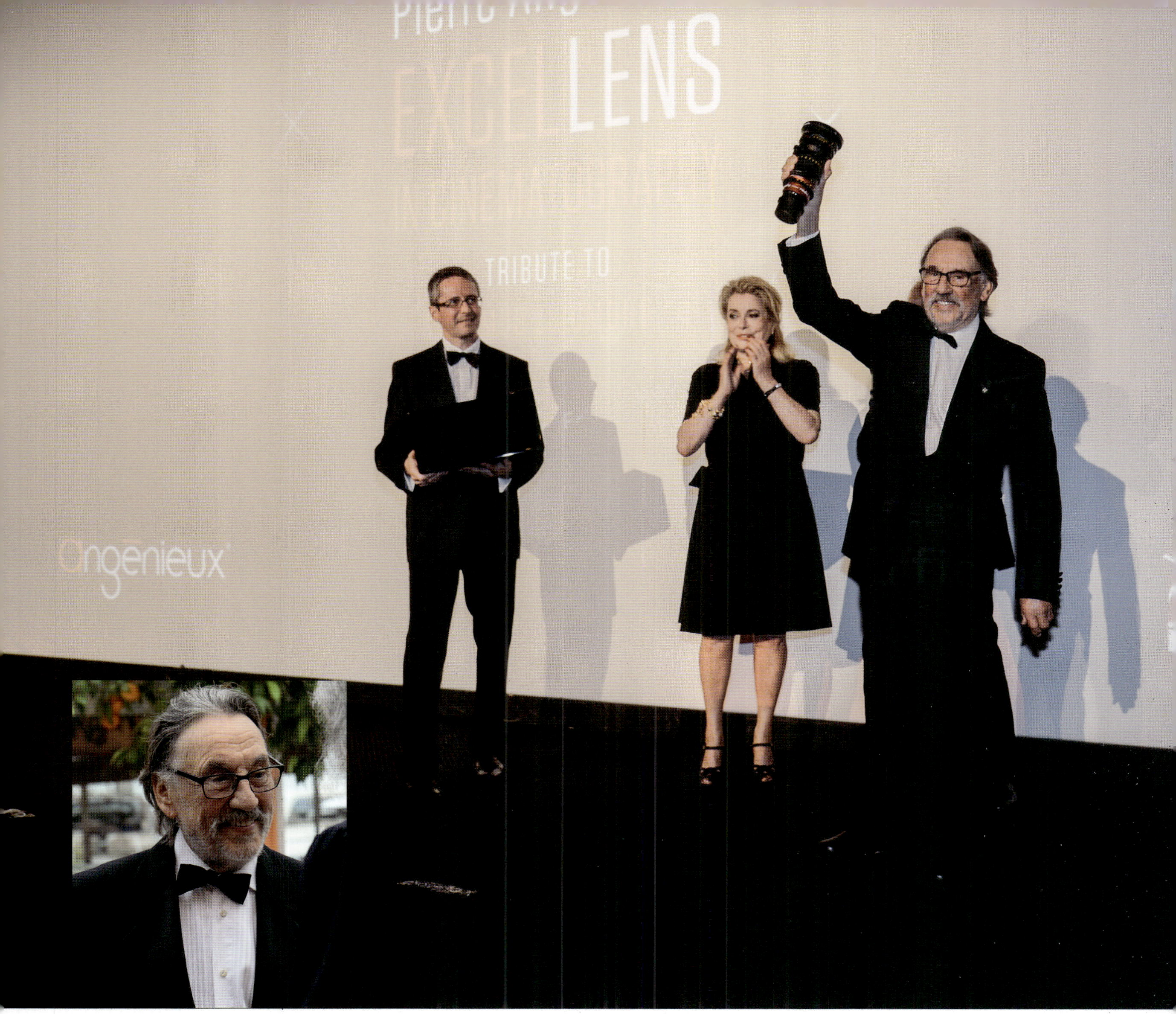

Vilmos Zsigmond HSC ASC

Cannes 2014: tribute to Vilmos Zsigmond in the presence of Catherine Deneuve and the directors Jerry Schatzberg and John Boorman (photo by Pauline Maillet).

2014

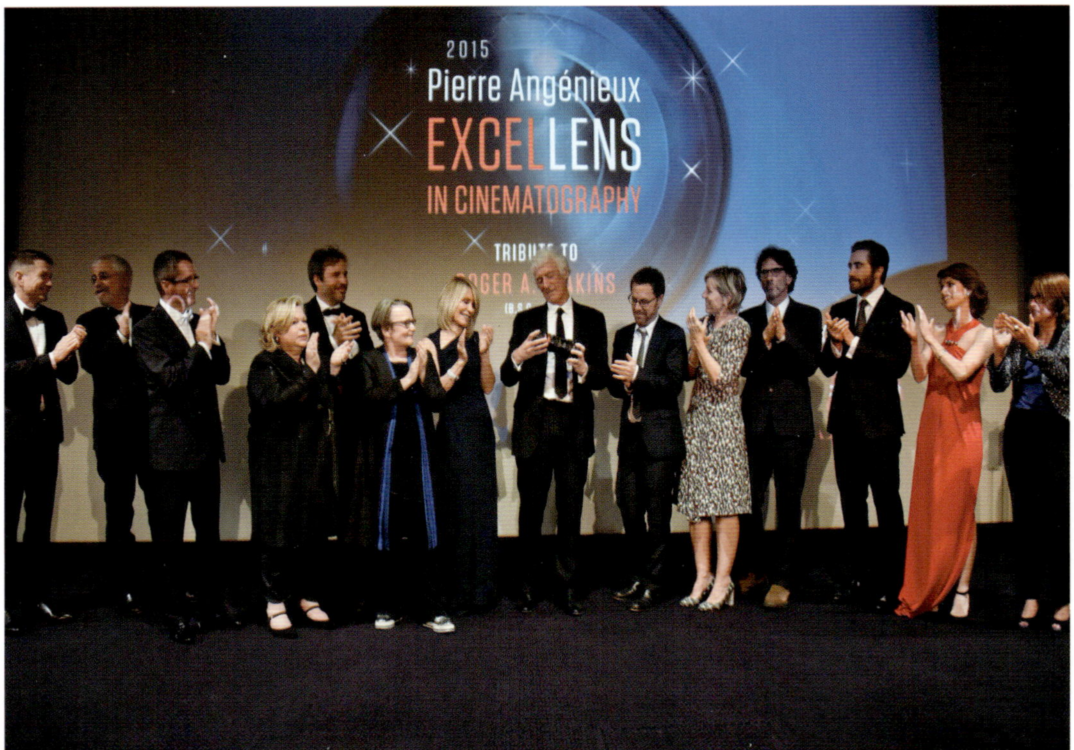

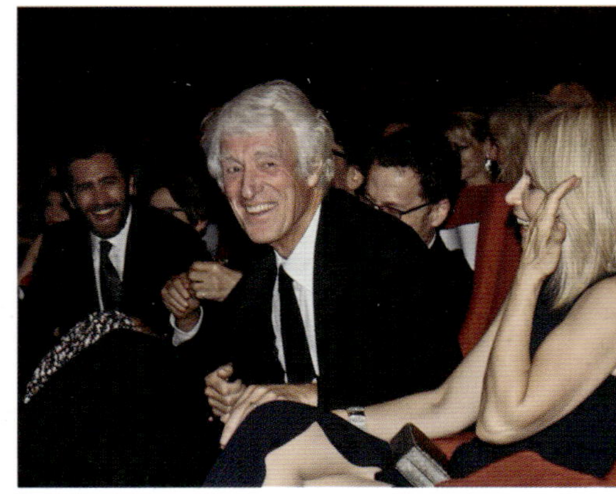

Cannes 2015: tribute to Roger Deakins in the presence of Frances McDormand, Irène Jacob, Jake Gyllenhall and of the directors Agnieszka Holland, Joel and Ethan Coen, and Denis Villeneuve (photos by Pauline Maillet and Dominique Charriau).

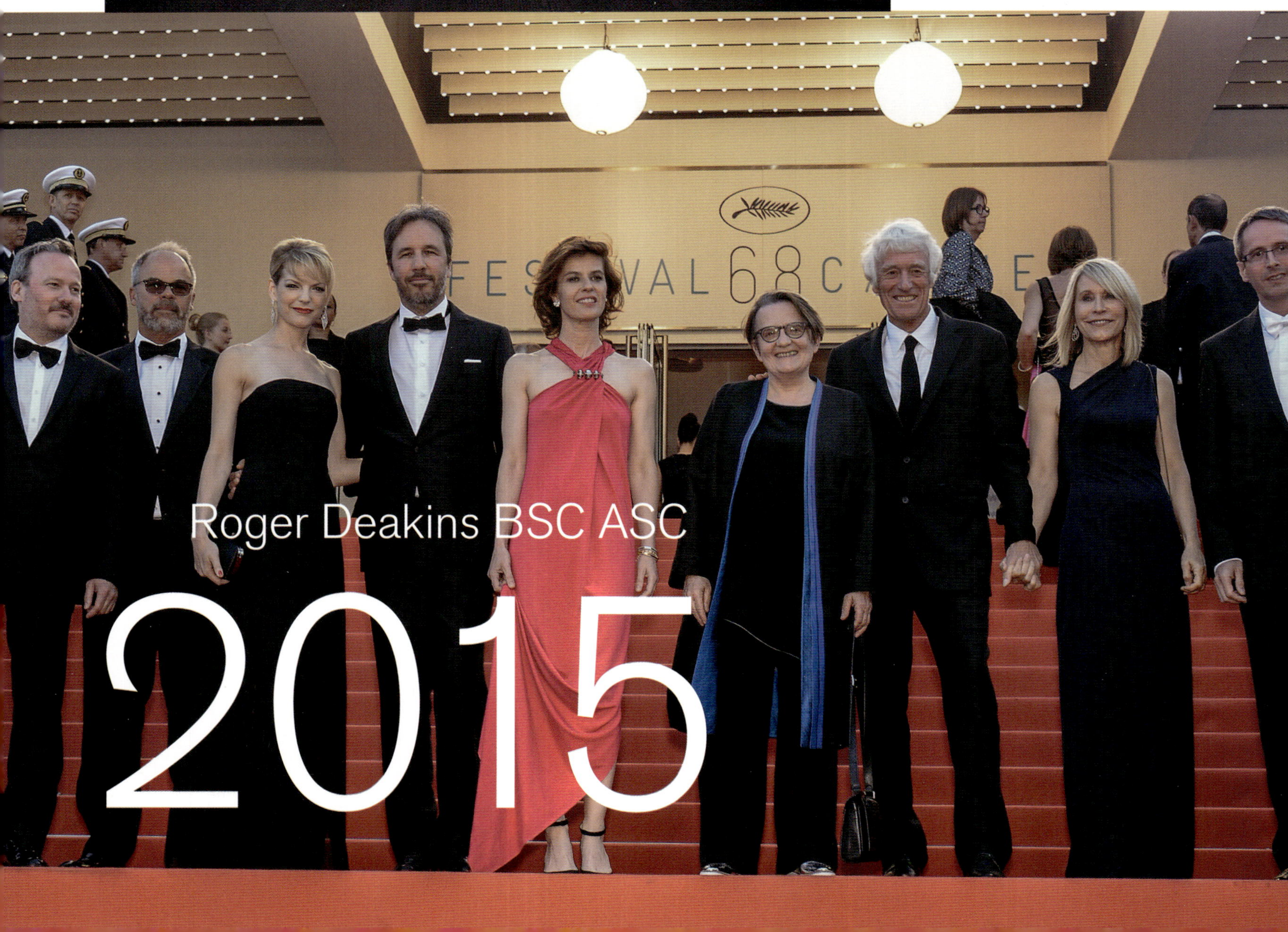

Roger Deakins BSC ASC

2015

Cannes 2016: tribute to Peter Suschitzky in the presence of Juliette Binoche, Alba Rohrwacher, Viggo Mortensen, and the director Matteo Garrone (photo by Dominique Charriau and Pauline Maillet).

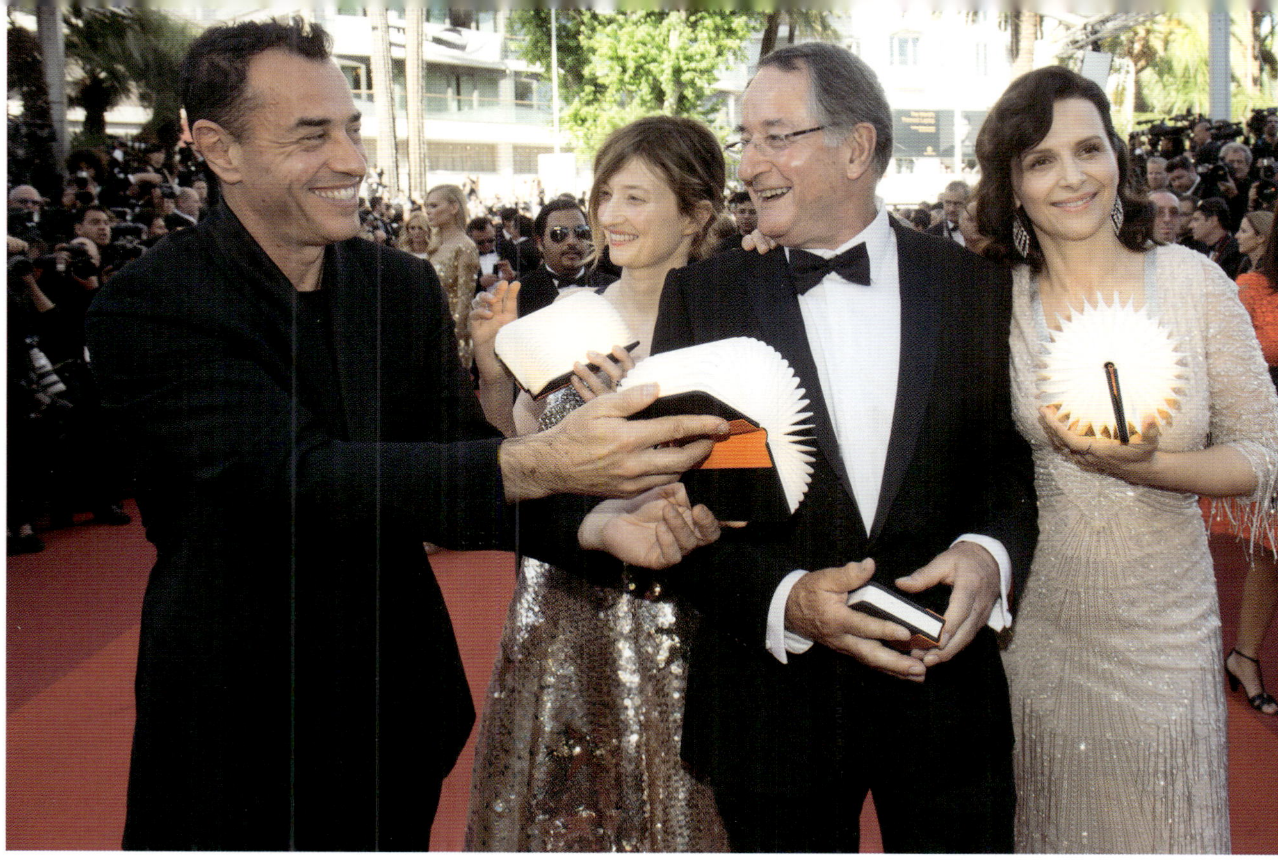

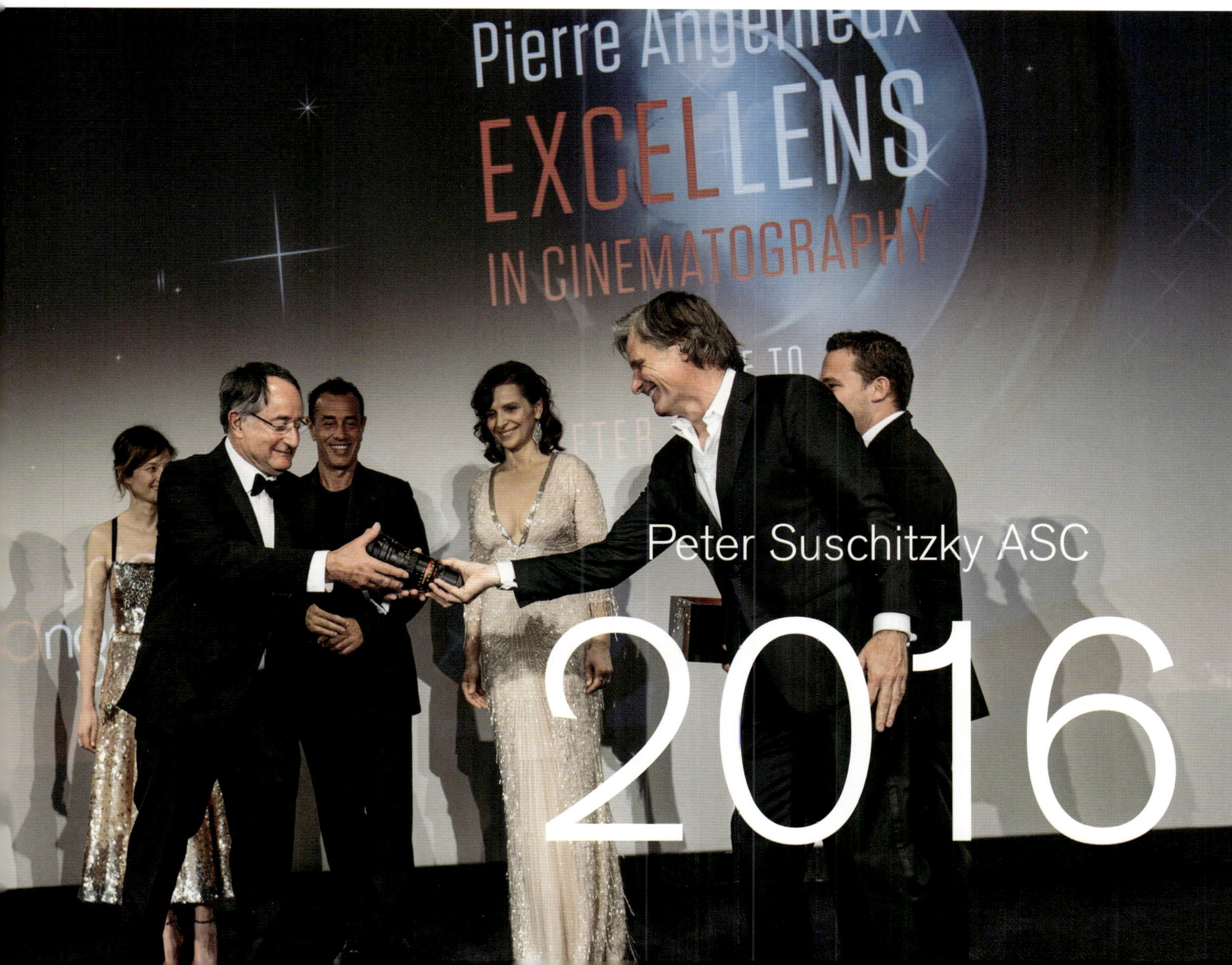

Peter Suschitzky ASC 2016

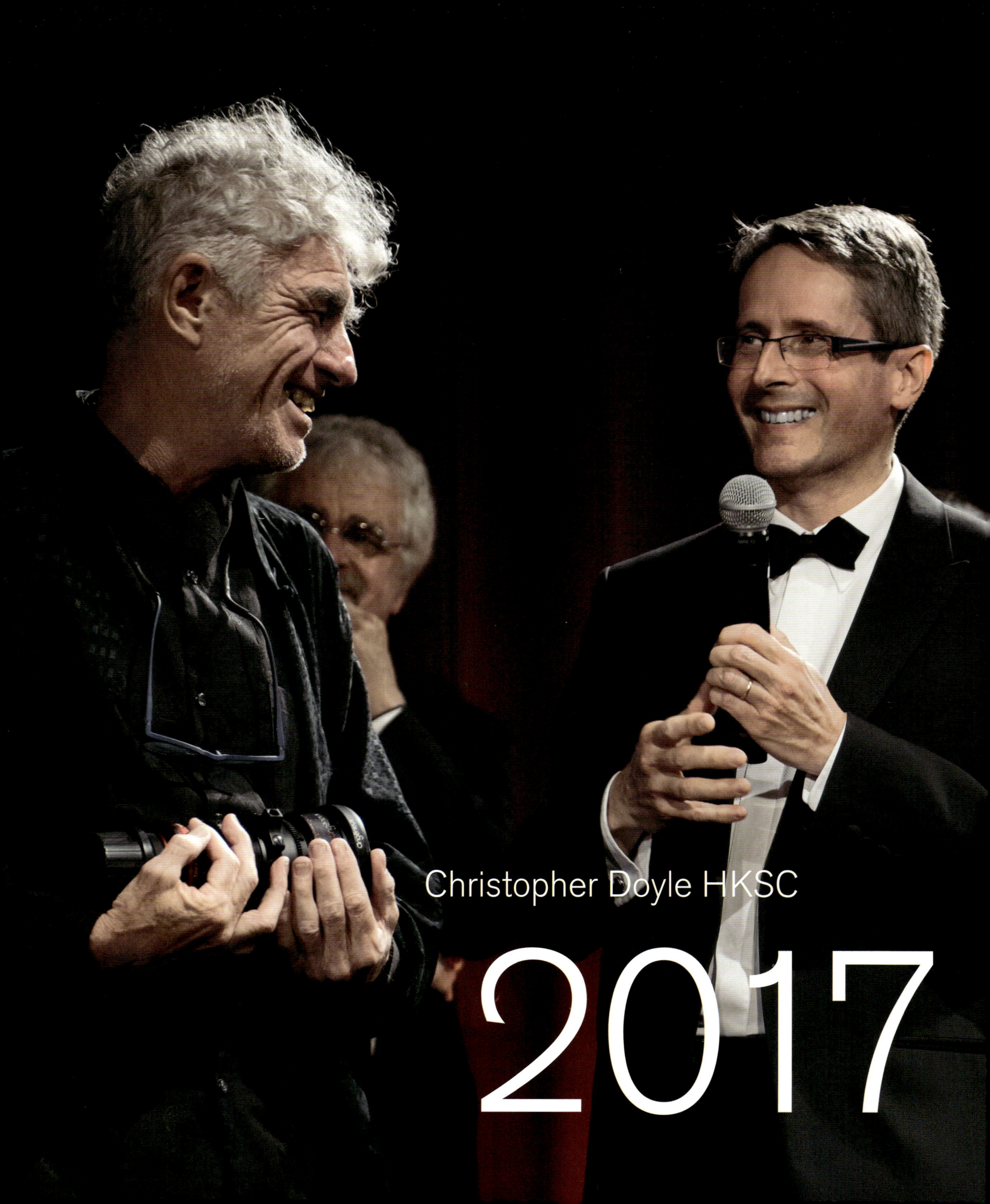

Christopher Doyle HKSC

2017

« Most awards are some kind of symbol: a horse for example or a little golden man. This award means so much more than just an honour for my work. This award has been created by craftsmen and women with precision and care. It's the bridge between their craft and my endeavour for art that I will be using every day on every film to come with pride and respect. With this award, Angénieux has passed on to me their excellence, now it's my responsibility to be excellent in return. »

Christopher Doyle

on receiving the Angénieux ExcelLens award in Cannes in 2017

opposite
Christopher Doyle and Pierre Andurand.

right
Cannes 2017: tribute to Christopher Doyle in the presence of Juliette Binoche and the directors Jenny Suen, Olivier Assayas, and Frédéric Auburtin (photos by Pauline Maillet).

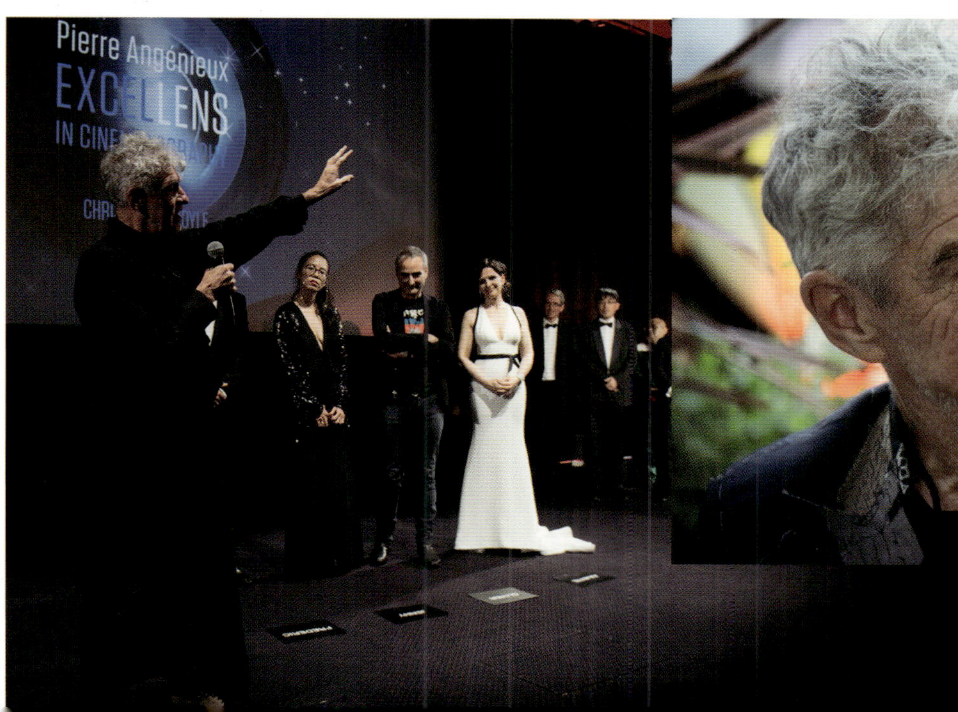

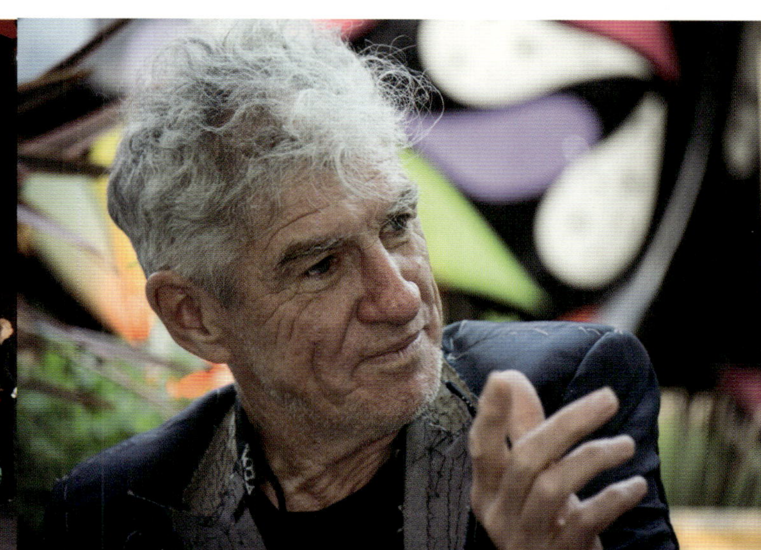

Presentation of the Angénieux Tribute to Ed Lachman ASC on May 18, 2018

Ed Lachman was accompanied at the ceremony by the director Todd Haynes, his friend, with whom he has already made four films: *Far From Paradise* (2002), *I'm Not There* (2007), *Carol* (2015) and *Wonderstruck* (2017).
In the presence of Thierry Frémaux, Director of the Cannes Film Festival, and Raphaël Keller, CNC Director for innovation, video and technical industries, Ed Lachman was presented by Emmanuel Sprauel, Angénieux President, with the Angénieux trophy, an Optimo 28-76 specially engraved with his name.
The sixth edition of the Pierre Angénieux ExcelLens in Cinematography award included a particularly moving event with the showing of a whole series of video testimonials from personalities from the artistic world, film directors, and actors who were unable to be present in Cannes but still wished to congratulate their friend Ed Lachman: Christopher Doyle, Julianne Moore, Mark Wiener, John Malkovitch, Volker Schlöndorff, Peter Sellars, Ulrich Seidl, Michelle Williams, Werner Herzog, Willem Dafoe, Sofia Coppola, Greta Gerwig, Steven Soderbergh, Meryl Streep, and Wim Wenders.
Another splendid surprise awaited Lachman at the end of this series of tributes, namely a video homage from Thomas Bangalter of Daft Punk, for whom Lachman directed *The Collaborators* series of episodes for the album *Random Access Memories*.

« As President of the company, I can tell you for certain that everyone at Angénieux is equally determined to provide you with the tools needed to capture your vision. We all endeavor, every day, to make the best lenses for your eyes but also for your hands, so that you can simply forget all about them because this is what you most need. »

Emmanuel Sprauel

President of the Angénieux brand

The final message addressed to Ed Lachman by Thomas Bangalter in the video specially made in his honour and shown during the ceremony.

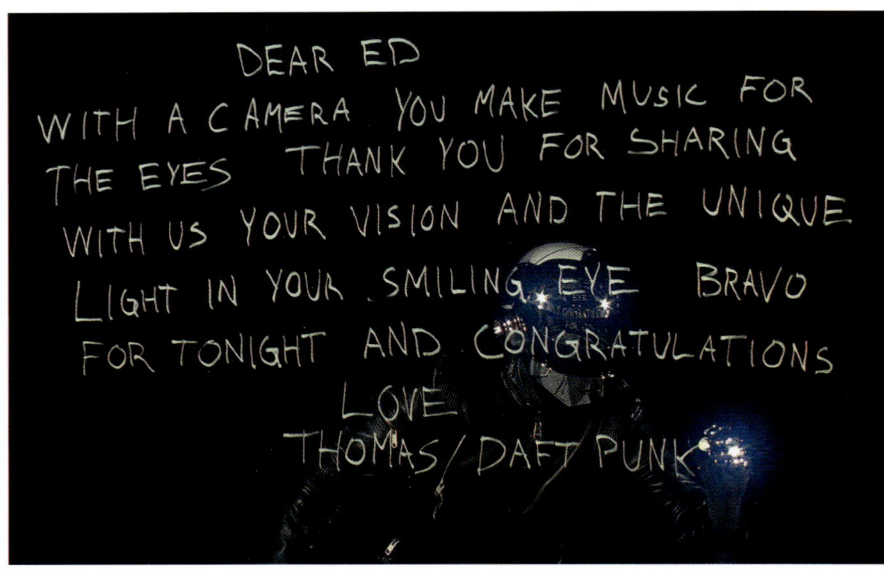

Edward Lachman ASC

2018

On the red carpet, from left to right, Cécile Zhang, Emmanuel Sprauel, Dominique Rouchon, Ed Lachman, Todd Haynes, Severine Serrano and the actress Elodie Bouchez (photos by Pauline Maillet).

below
Ed Lachman and Emmanuel Sprauel.

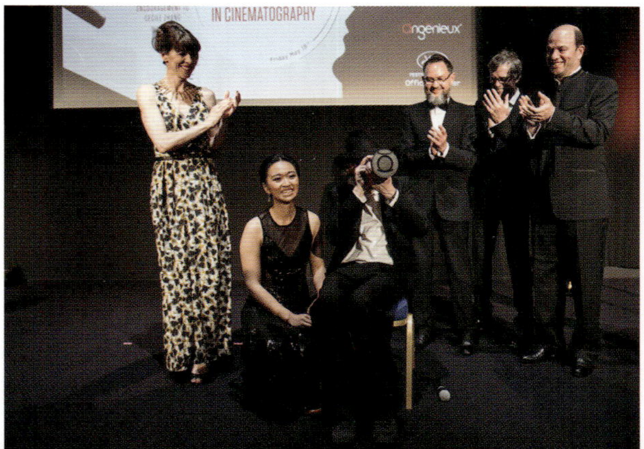

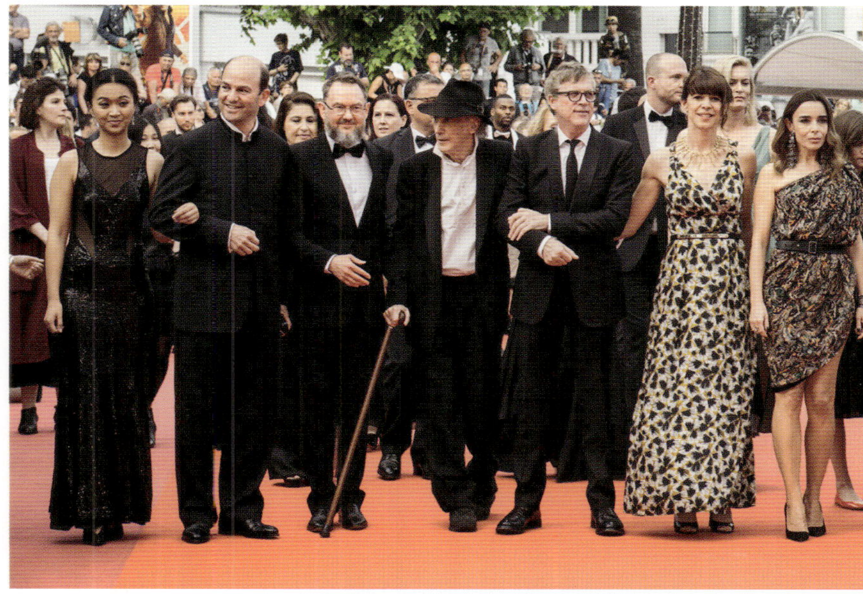

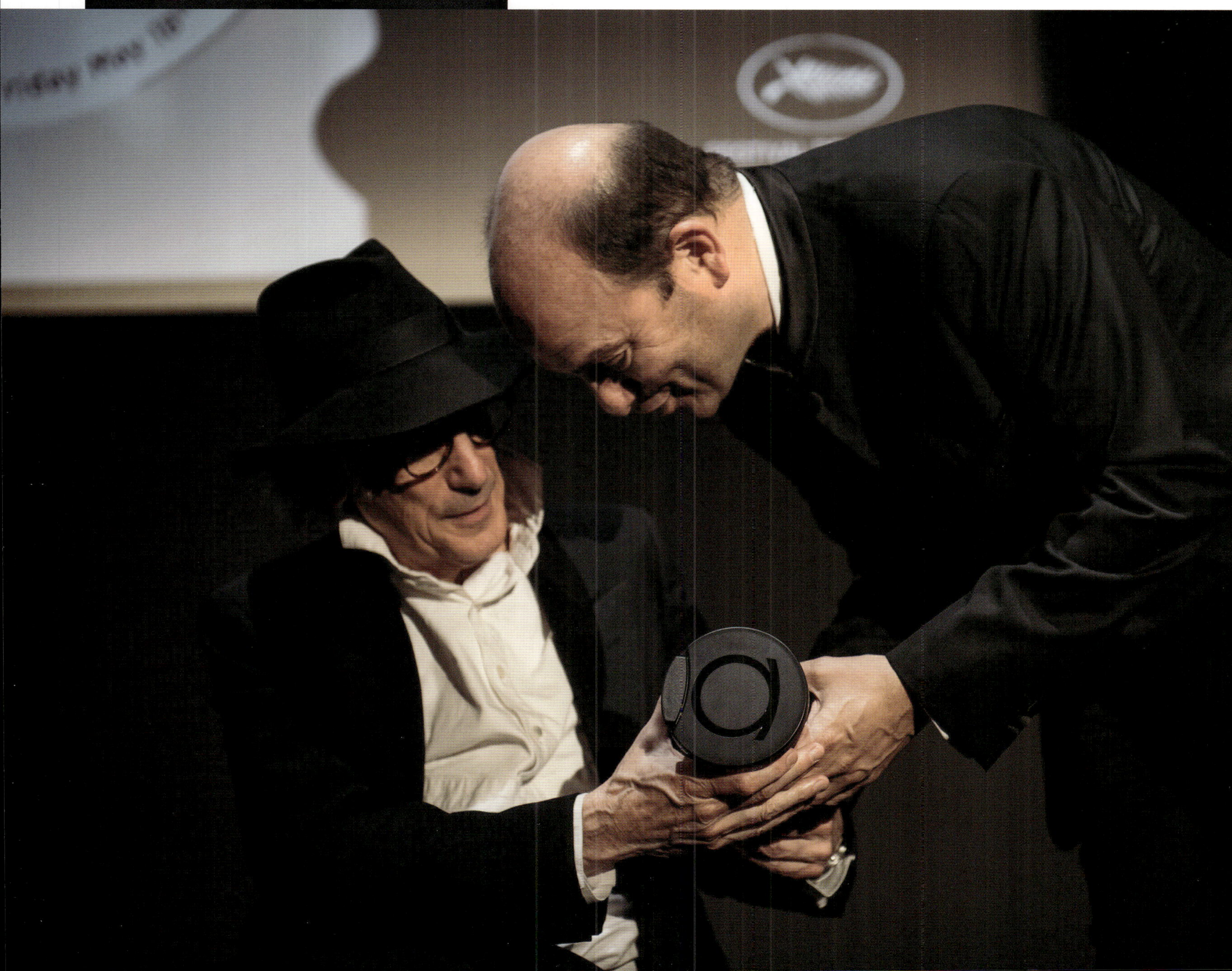

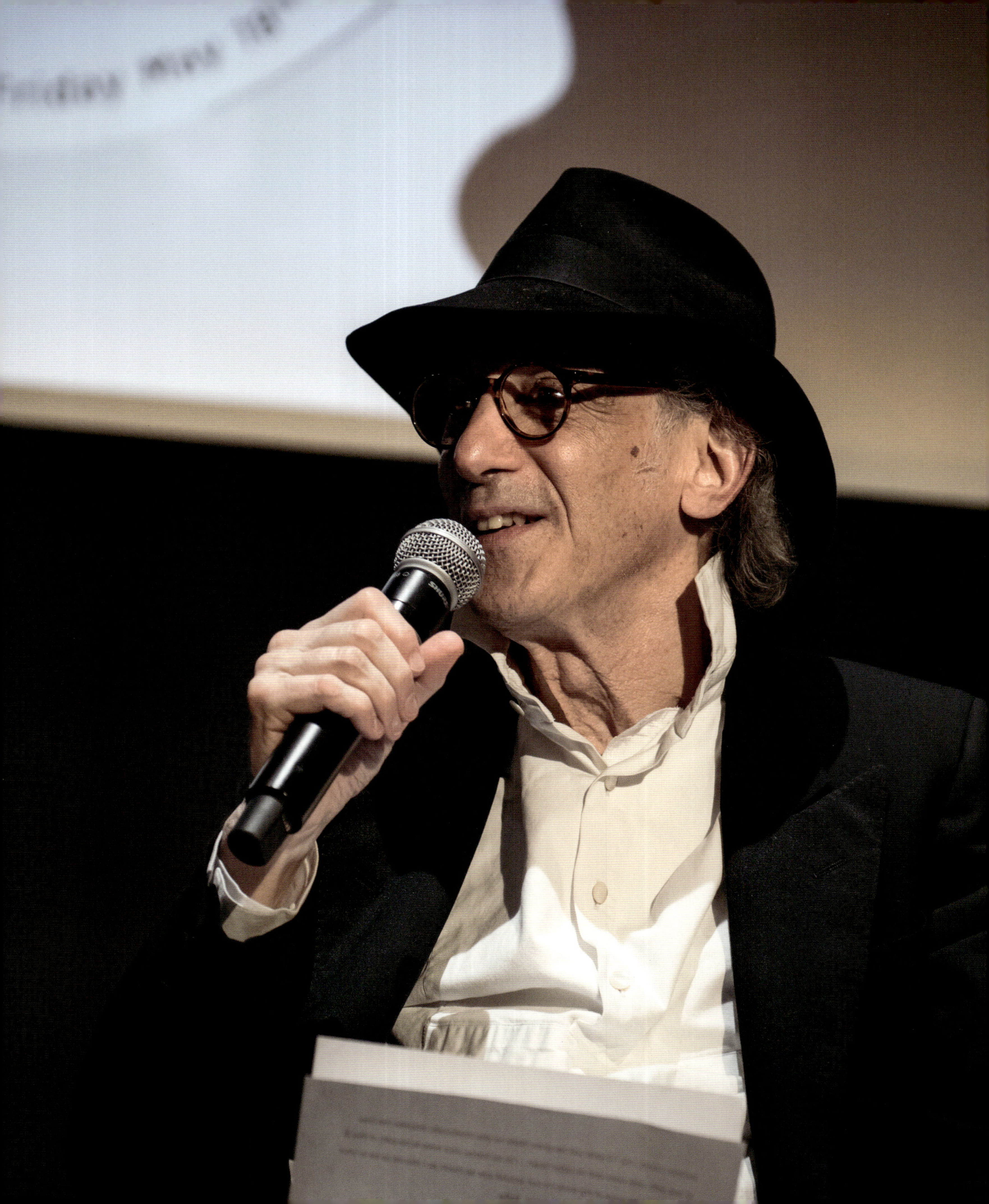

« For me the challenge of cinematography is revealing or trying to evoke the poetic or psychological authenticity of the image in the story you are telling ...
Images aren't only around pleasant pictorial aesthetics but also can be the projection of the emotions for the characters in the story.
Images are about experiencing something with your camera, it is the immediacy of that experience inside yourself that you are able to translate, and you can share in a documentary or narrative form. »

Ed Lachman

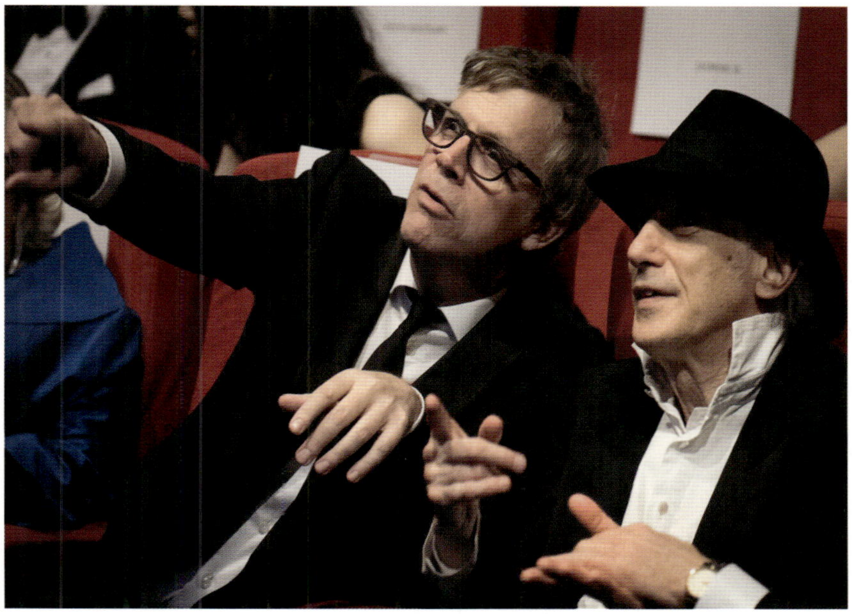

Ed Lachman and director Todd Haynes, with whom he has already made four films films (photos by Pauline Maillet).

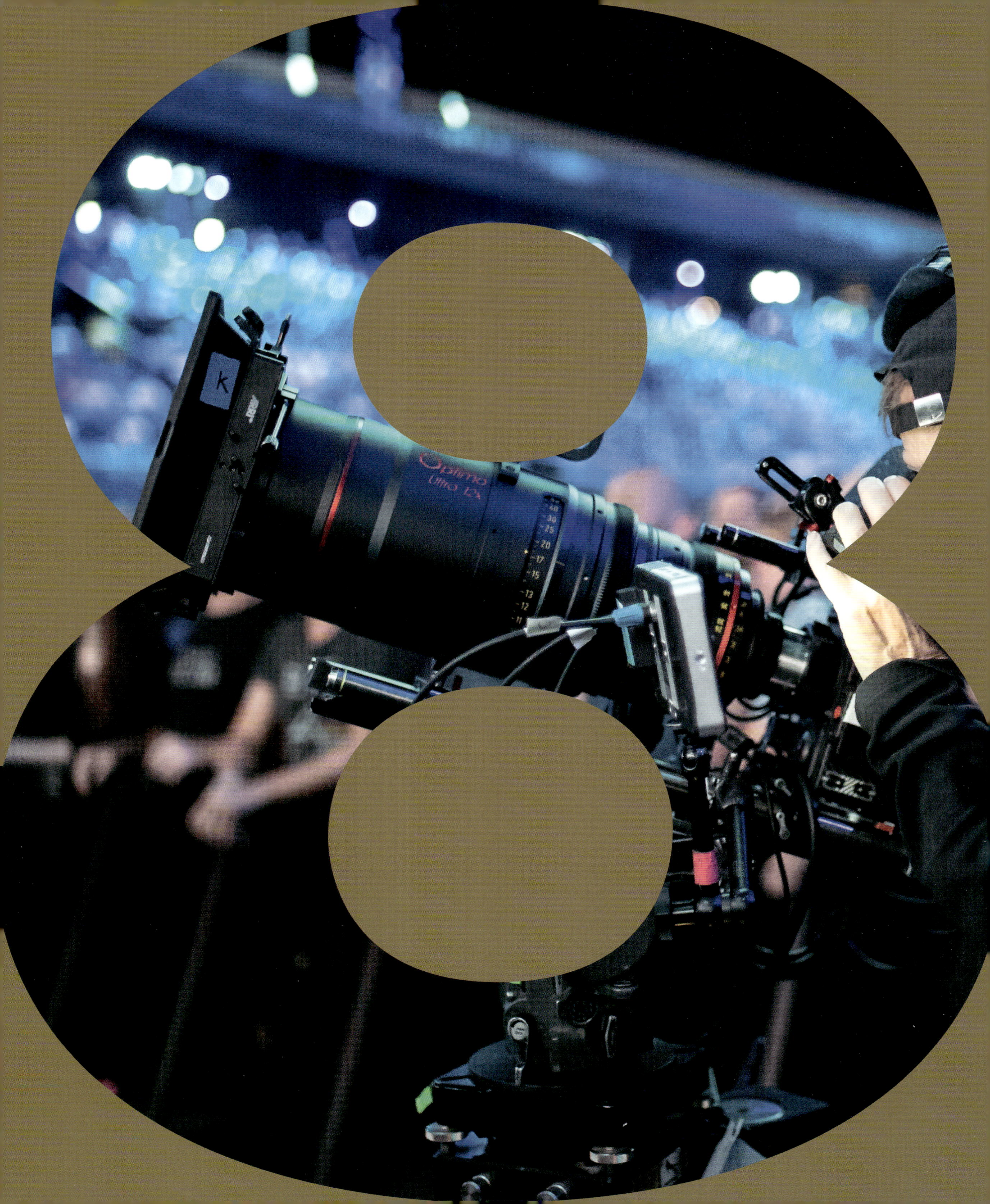
Optimo
Ultra 12x

TOMORROW …
CHALLENGES AND INNOVATIONS

The incredible legacy bequeathed by Pierre Angénieux is rooted in an exceptional **spirit of inventiveness**, which the teams in Saint-Héand endeavor to keep constantly alive. Innovation lies at the heart of their everyday activities in a highly competitive market.

This innovation is primarily **technological** in terms of solutions as well as practices and approaches undergoing continuous **digital transformation**. Application of the pertinent solutions developed in the most appropriate fields is essential to survival in the digital era.

In this all-digital world, however, nothing can replace the **direct contact** with customers, the present or future users of Angénieux products. This is crucial to the reciprocal transmission of knowledge.

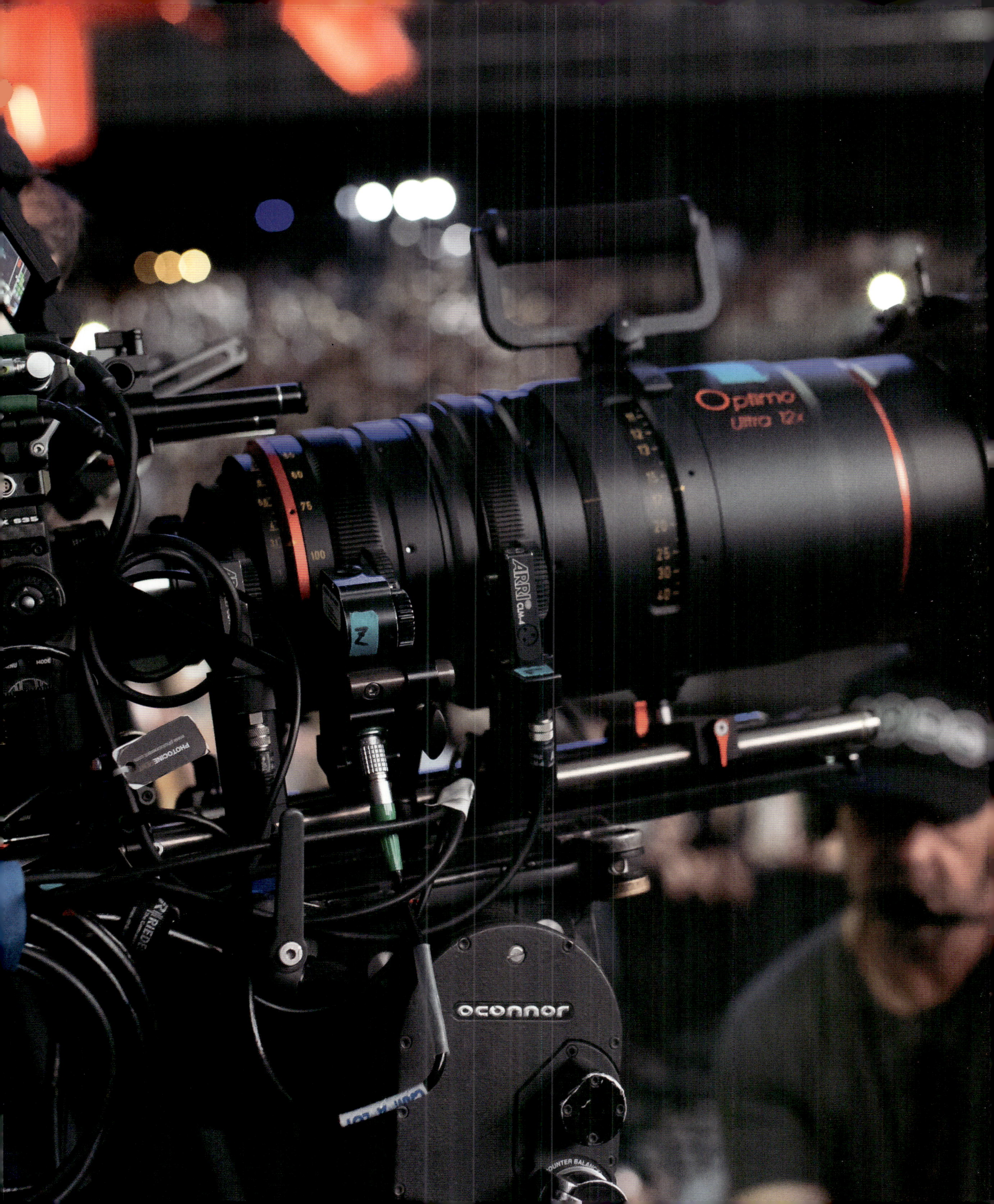
Optimo
Ultra 12x
oconnor

Towards what Kind of Cinema Image?

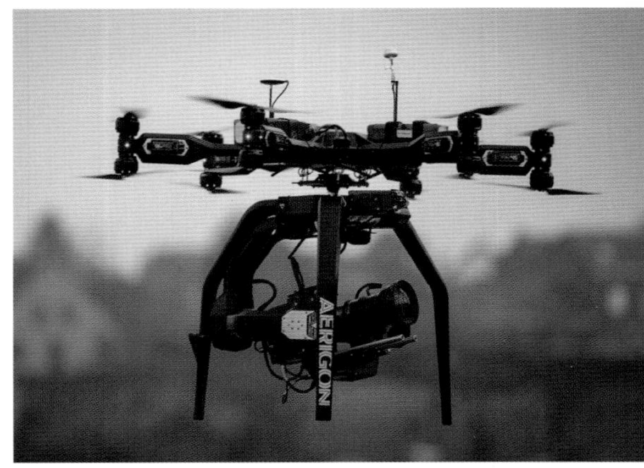

Optimo 30-72 A2S on Arri Mini camera and Intuitive Aerial's Aerigon drone. Production "Drift" by Kelly's Heroes. Photo Anthony Boon.

More than any other art, cinema is a matter of technology. Since its very beginnings, the evolution of the techniques applied has contributed to the emergence of new forms of visual writing. Cinema is indisputably digital today and evolving at top speed, just like our own lives.
New technologies affect all our lives a little more each day. Our telephones foresee needs and facilitate choices. While filmed images once appeared solely in cinemas, screens are now omnipresent, from the smallest to the largest.
The techniques of image capture are not impervious to this evolution. As a firm working in the image industry, Angénieux has no alternative. It must identify the future trends as soon as possible and invent innovative solutions at the service of creativity and the requirements expressed.
Like our smartphones, digital cameras are undergoing constant renewal in terms of ever-improving ergonomic qualities and performance, and endowed with variable and configurable capacities as regards both resolution, especially the size of the sensor, and the format for recording the image in digital memory. New systems are also appearing in addition to the camera, such as drones that can carry cinematographic equipment and make previously unimaginable images possible by placing new viewpoints at the service of new narratives. Once sharply distinct in technological terms, the universes of television and the cinema have now merged. While TV sets offer access to countless movies, films produced for television, like documentaries, series, video clips, and recordings of concerts and other cultural events, borrow the codes of cinema and use cinematographic equipment.
The arrival of Internet giants like Apple, Netflix, Amazon, YouTube, Facebook, and many others still continues to disrupt the economic models of cinema and television. By creating a direct link online between content producers and subscribers, these "game changers" influence the worldwide production of images, the types of content, the way of filming them, and our modes of image consumption.
The filmed image is omnipresent, now accessible on our telephones, tablets, computers, television sets, and game consoles. We must project ourselves into a world where the multimedia diffusion of images will be crucial to the profitability of a work for a producer. Today the tendency is towards Full Format allowing largest scales of projection.
In addition to these upheavals, a whole movement is gathering momentum. Questions of diversity are no longer ducked. Innovation and creativity can only be boosted by a greater multiplicity of approaches, viewpoints, and ideas. In 2018, Rachel Morrison ASC was the first woman director of photography to be nominated for the Oscar for best photography.

Closer to Young Filmmakers

In order to meet the challenges of the future successfully, Angénieux attaches priority to close contact with the cinema community as represented by filmmakers, directors of photography, and their professional associations. It is also to this end that the firm is committed to working with film schools and supporting them in their mission to teach the various techniques of image capture, a precious heritage that must not be lost in the era all-digital technology and apparent easiness of filmmaking.
While some filmmakers are self-taught, others learn their trade in film schools, which are the first to defend the standards of image production in all its technical, esthetics, and artistic dimensions. How do these schools see their mission of teaching cinema today?

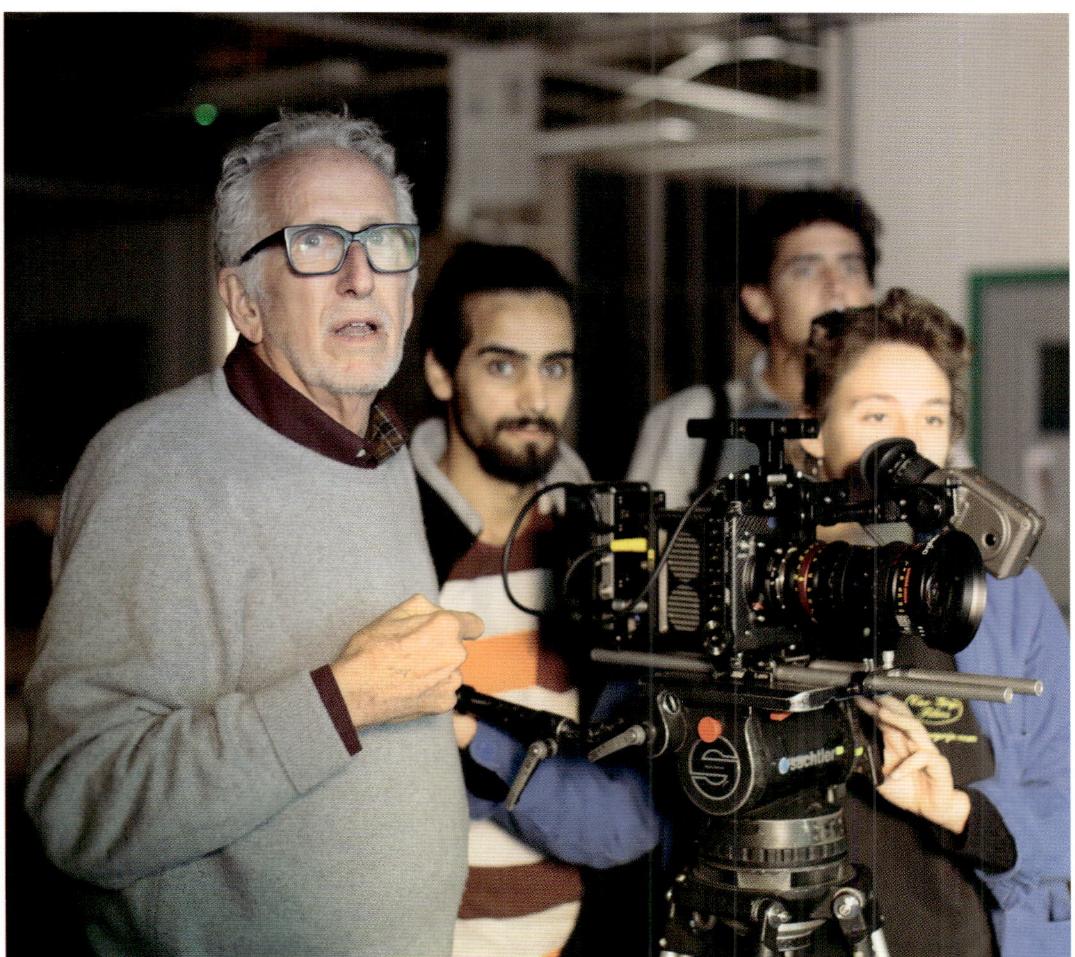

American cinematographer Tom Stern best known for his work with Clint Eastwood and a group of Cinéfabrique students in Lyon (France), November 2018.

How to Teach Filmmaking Today?

École Nationale Supérieure (ENS)
Louis-Lumière - France
Vincent Lowy - director
Pascal Lagriffoul, cinematographer

Like all the great film schools, the ENS Louis-Lumière is at a turning point. We have gone from shooting on film to digital movie making. We are faced with the profusion/confusion of screens on the set and in every phase of image production through digitally-controlled devices of every kind. We do not believe that using only digital cameras is necessarily proof of up-to-date training in line with new practices. We are absolutely determined to preserve the culture of photochemical processes as a guarantee of rigor and a certain form of alchemy that it is our task to transmit. The ENS Louis-Lumière was founded in 1926 and we were thus the first school of camera operators in the world. This entails its own obligations. It can be said in this sense that the teaching of the profession at the ENS Louis-Lumière is also a matter of heritage. It is, however, precisely this continuity in the field of technological expertise that must enable us to approach these new technologies boldly and with full understanding. It must be clearly understood that the digital revolution is comparatively recent and that the technicians now active are still digesting its various aspects. We must ensure that our students are capable of high performance in the digital environment, and this will be directly proportional to their mastery of the fundamentals already taught in the era of film (sensitometry, colorimetry, photometry, instrumental and applied optics, etc.). This is what prevents standardization and precisely what makes the difference between our training and that of the other schools. The watchword is innovation, both in technological and artistic terms, and as regards educational practice. We believe that in order to achieve these various goals, we will have to intensify our rhythm of production while allowing students to devote more time to practical application. It will be necessary in the long term to double and even triple the number of films shot in the school precisely in order to allow our students to become truly familiar with these new tools and the associated practices. At the same time, we are fully aware of the constant technological acceleration: the digital techniques themselves are constantly evolving and the teaching of a standard digital support (such as RAW) is not yet crucial to the teaching of a standard workflow. What does this mean? Film gave this coherence by definition. There was in any case a support, a physical reality of the image produced by the cinematographer, even though it could be ruined by inadequate printing. The impact of the revolution of the digital image extends far beyond cinema. We refer here to the widespread diffusion of the image but also to its status, its "truth", to the programmed weakening of everything bound up with an idea or understanding of the image ... The probable evolution of technologies, which will make digital special effects (as regards sets, make-up, figuration, and characters) easy to obtain and integrate into the basic workflow, will alter the profile of our professions and must be addressed in terms of teaching. This affects all the cinema professionals, including actors, many of whom are now working in completely different conditions. The unprecedented profusion of filmed images requires us to raise the standards set in the production of cinema images, otherwise it will become impossible to distinguish between them ... The primary challenge for the ENS Louis-Lumière today is to problematize these new digital practices and guide our students through this watershed. The other very important change is not technological but regards the equality of women and men in the professions of filming and filmmaking in general, which are, as we know, still largely seen as male professions (cinematographers and assistants but also chief electricians and grips). Our students must learn about lighting, framing, and calibration. We must teach the history and physics of images as well as the history and physiology of vision, and above all what will always continue to determine the quality of an image: the ability to look and see. Though this may appear obvious, it is necessary to beware of confusion between looking at a screen and looking at "reality", to beware of the temptation to look "only" at screens. It is necessary to go on teaching the artistic fundamentals of cinema and image, the

history of the image, of painting, of photography ...
We would also like to strengthen teaching about the theory of the image, especially in relation to the new technologies, which we see as rich in promise but also dangers. Extended, virtual or augmented realities: everything that is immersive raises questions of a philosophical nature. What remains the human dimension? Will the machine not come to rule the world in the era of artificial intelligence? Is it already too late? We want to place the cinematographer at the heart of cinema because we are convinced that this is still an activity that produces unique rather than standardized objects.
While we do not underestimate the importance of the D.I.T.* (digital imaging technician), we think that the relationship with cameras and lenses will remain always intimate and passionate in movie making. A cinematographer stated at Camérimage in 2017 that the set is now constantly invaded by people commenting and interfering in the realm of the director of photography, and that the lens it the only part of the cinematographer's magic still to survive. This is true, but if the digital part of the work is not controlled and decided upon by the director in the closest possible collaboration with the cinematographer, there will be little hope for visual creativity in the long run. This is why the ENS Louis-Lumière bears particular responsibility with regard to these evolutions.

* The digital imaging technician or D.I.T. is a new filmmaking professional who works in close cooperation with the cinematographer to handle all the problems of digital workflow on the set so as to ensure rushes in accordance with the set aims.

Fémis - France
Nathalie Coste-Cerdan - director

The Fémis had the good fortune to benefit from an ambitious plan for the digitization of its entire production system, initiated in 2011 and continued until 2018. Successively regarding the stages of editing, shooting, and post-production, this made a radical transformation of teaching both possible and necessary. Technical basis, processes, and workflow were reconsidered for all the professions and every course. Partnerships were forged with industry so as not to be left behind by technological evolution. The school must make still further progress in transmission of the new organization of work introduced by digital technology. Internships with firms are an opportunity to achieve this. It is on the "artistic act" that teaching focuses at the Fémis, the mastery of digital technologies being seen only as a means, not an end. The use of film, super 16 or 35mm, is still an option for students. It is necessary to start with the fundamentals of cinema, which rest largely on the analysis of films, in order to stimulate curiosity, as well as the other arts and the history of cinema, thus enabling all the students to develop their own style. They must also, learn to work in a team and to communicate their artistic intentions through a highly innovative form of teaching, widely imitated throughout the world, based on practical exercises in which a student occupies each position in the movie-making chain. For the professions of filming and lighting, we agree with Ed Lachman on the need to seek something with the immediacy of inner experience with the camera, and the importance of capturing and sharing this. Like all the teaching in the school, this involves theory and then pratical application. In the first year, the study of geometrical and applied optics makes it possible to control the basics. Students of film directing, image and editing then collaborate to understand the importance of the choice of framing and hence of lenses, above all through a workshop on the use of natural light, designed for second-year students of image, which given them that opportunity to compare different cameras, lenses, and formats on location for a week.

A unique and crucial experience for their future as directors of photography. Priority is given to teaching about visual effects, not least in accordance with the wishes of students of all the professions. Technologies evolve very quickly, faster in any case than the financial capacity of schools, which makes it necessary to develop key partnerships with the major actors in this field. It is also necessary in teaching to foster links between cinema and TV series, cinema and the visual arts, so as to offer students greater flexibility in an increasingly complex professional world.

CinéFabrique - France
Claude Mouriéras - filmmaker and director

As a filmmaker, I have always liked to work with zooms. Willy Lubtchansky and I loved this way of gliding towards a face, approaching it slowly, as imperceptibly as a caress. For me, one of the most moving zoom-ins is onto the face of Monika at the end of the Bergman film. Her look into the camera becomes inevitable, unstoppable, and it is the zoom that gives it this power. The zoom has always played a key part in the structuring of a scene. It is what provides this rhythm, this addition of heart. Today at the CinéFabrique, I believe that this generation of aspiring filmmakers seeks a more visible and deep approach with the zoom. They want abrupt changes in focus that make statements and cuts. They use the zoom like an effect of editing, radical transition from one shot to another. The use of the zoom, like any other technical means, is right when it accords with and reinforces a director's decisions. As for writing with a tendency towards less discursive narratives, they seize on the zoom with joyful inventiveness and radicality in search of greater freedom of action. In more general terms, however, digital technology has altered practices as regards not only production but also teaching. Students at the CinéFabrique seek much greater lightness and versatility in screenplays and operative practice, and want their films to reflect this freedom of creation. In the third year, for example, with have end-of-course films that involve major resources of funding and production but also the "filming at all costs" sequence, which makes it possible for each student to invent new processes of filmmaking and production. I believe that it is easy for a school to give a theoretical and technical grounding. What is not so easy but so much more precious is to give students the self-confidence to dare. To dare at the risk of making mistakes. A school serves this purpose, giving students the courage to make mistakes so that they can learn from their errors.

Beijing Film Academy (BFA) - China
Mu De Yuan - former director of Beijing Film Academy (BFA) An Image Department, former director of BFA Academic Affairs Office, winner of Best Cinematography in 18th and 23th Chinese Golden Rooster Award, Chairman of Chinese society of cinematographers

The Cinematography department of Beijing Film Academy has a research laboratory used specifically for the development of film production, film technology and film education. We have long predicted the transition from film to digital.
In the early days of the development of digital technology, we saw some pictures taken using digital film cameras, knowingly predicting that the prosperity of digital technology would come within next 15 years. So as educators, we decided to recruit some new and proficient professors to strengthen our team. So we recruited two engineers. Within the department, we also jointly established a new major "video engineering" with Cinerent Group in the same year. When we started studying digital film cameras, we noticed that the entire workflow was very different from that of the old ones. And it would require an engineer to manage all electronic equipment. The aim of this option was to train video engineers to the digital era. Right after establishing this option, China openly embraced digital technology with the film age being frequently replaced. I believe that the work that has been done by the department at the Academy to fuse film and digital education together was pretty far-sighted even on a world scale. So is the speed of its learning on digital technology. The teaching system of the Image department is based on the four means of cinematography modeling—lighting, motion, optics and composition. These four tools are excellent "housekeeping skills" of a cinematographer. So we use them as basic courses, each of which includes theory, course development and technical practices. This teaching system is a cornerstone of the Academy that was in development for more than 60 years. It is difficult to say what the direction of reform will be leading us in terms of China's film education, because everything in China is based on the State system. Since the reform and opening up, the system has not fundamentally changed. Problems remain in the Chinese education system, where we stress the importance of production, teaching and research. If we only prioritize production in the first place; education would lose its core mission as talents-cultivation. China pays much attention to inheritance. History and humanities need to be passed on to the next generation. So does education. Therefore, we should, on the one hand, strengthen and maintain a team of experienced professors as some kind of "messengers" to assist the growth of generation after generation. On the other hand, speaking from the meaning of education, we should first see talents-cultivation as its core mission. Then productivity of the talents comes after. And I personally think that this is the direction educators should lead the way.

Statements recorded for Angénieux by Cécile Zhang, a former pupil of Mu De Yuan

The Testimony of Cinematographers

What do they think?
Some of the brightest or most promising cinematographers speak about the future of cinema.

« I am very optimistic about the future generations. Something unquestionably needs to be reinvented. The younger generation seems to want to master the tools of its time, which is a good thing. There are so many things to experiment with and tools to compose with. It makes your head spin and stimulates you at the same time. It is just necessary to find the right tools, ones you feel at home and familiar with, to find your musical instrument. I often compare cinema to music because it is there that everything started for me. I studied the harp, musical theory, and singing for sixteen years at the conservatory, I played in an orchestra, in chamber ensembles, and the dynamic is similar to that of making a film. It is music you play together. You interpret different works of different composers. There is a conductor who has his or her own sensitivity, a repertoire he or she loves. It all depends on finding your niche, your repertoire, the music and the instrument you love to play. »

Eponine Momenceau

« What counts in making a film is the screenplay. If the story is good and the characters are interesting, why not shoot a film with a smartphone? I find it hard, however, to imagine a whole film shot on a hand-held smartphone with no camera footage. And then, the primary function of a smartphone is above all to be a telephone, whereas there are wonderful tools to make films.
What worries me is rather the shorter time allowed for filming. The shooting of *Dead Ringers*, my first film with David Cronenberg, took twelve weeks. It took us only 29 days to shoot *Maps to the Stars*, my latest one with him. »

Peter Suschitzky ASC

Winner of the 4th Pierre Angénieux ExcelLens in Cinematography award. at the Cannes Film Festival in 2016

« For tomorrow ? ... I hope more women behind the camera: to give us a more balanced view, a view that has been fought for (and is therefore stronger). A vision that is resolves with images, not guns, with an embrace rather than a first fight. To young people willing to shoot movies, I would say: don't listen to people like me. Learn who you are. Why would you want to be someone else and actually you can't be? Learn from your mistakes. JUST DO IT. »

Chris Doyle HKSC

Winner of the 5th Pierre Angénieux ExcelLens in Cinematography award. at the Cannes Film Festival in 2017.

« In the era of all-digital technology, smartphones and multi-support distribution, I would tell young people to beware of the trivialization of their vision of cinema and communication, and to fight against digitally-induced standardization and formatting. Anyone can make a film with a smartphone. Not everyone can create an image for cinema. A cinema image is a construct, a combination of light and movement. It must have a style that expresses meaning, which is instead lost in the proliferation of meaningless images that invade the world with the shortness of commercials and accompanying acoustic pollution.
I believe in the cinema of tomorrow. The craftsmen have not lost the battle. The rigor of shooting on 35 mm film has not been lost and the quality of lenses is indeed higher than ever, as demonstrated by the Angénieux range above all. With the support of manufacturers who set such high standards, the technical artists known as cinematographers will always remain the guardians of the quality of the images they create both artistically and technically.
The constant stream of talented young directors of photography joining the AFC is the cornerstone of my confidence in the future. »

Pierre -William Glenn AFC

Director of photography and director
Co-founder of the Association Française des directeurs de photographie Cinématographique (AFC)
President of the Commission supérieure technique de l'image et du son (CST) from 2002 to 2018
Co-director of the Fémis Image department

« In the digital age, the number of people entering the industry is increasing but the number of people who are passionate about movies is actually decreasing. I hope to see more people who are passionate about movies and serious about filmmaking joining the industry. This will raise the standard of the industry. I also hope that movies can be popularised as an art form. I hope people will bring their children to cinema to give them a foundation in the arts, and not purely for entertainment purposes. For myself, I look forward to continue experimenting, to work with people who are passionate about movies, and to produce more great work. »

Cécile Zhang ICFC

Recipient of Angénieux's award to encourage young talents in cinematography in 2018.

« I don't have a crystal ball. There will certainly be a lot of changes, but the role of the director of photography is not only to produce an image but first of all to help the director tell a story as well as possible. This function is not going to disappear all at once.
I would tell young people who want to make films to do just that, even if it means selling their parents' car.
Joking apart, I would tell them to take photos, draw and paint, go to museums and cinemas, open their eyes and learn to see, to discover what they like and what they dislike. »

Philippe Rousselot AFC ASC

Winner of the 1st Pierre Angénieux ExcelLens in Cinematography award. at the Cannes Film Festival in 2013.

Angénieux: Embracing a World of Change and New Challenges

For Angénieux, in addition to the key markets of Europe and the United States, China and India offer huge developmental and growth potential.

« Even though France was the historical cradle of cinema with the Lumière brothers, Hollywood remains the key market in terms of production with a major influence on the European and emerging markets. Today, as a result of economic globalization and the entry of large countries like China and India into world cinema, technological demands have become identical in all markets. Areas like China and India today demand the most modern technical means for their productions. After launching, the new Angénieux products are unanimously employed and approved all over the world.
In these faraway lands, Angénieux is clearly identified as a guarantee of excellence, as attested by the foremost cinema equipment rental companies. These work in the closest contact with producers so as to prepare the lists of equipment necessary for shooting in good time and meet the needs of directors of photography in search of the latest tools to serve the purposes of their boundless creativity. »

Severine Serrano

Angénieux International Sales and Marketing Director

« Without Angénieux, we don't exist. As a brand Angénieux is the most dependable in India: trusted lenses which have been tested many times for every shoot and are still going strong. It (IRO) will take some time for the market to adjust to this technology as generally DP's want a single format lens without much complication of changeable adaptors or attachments. We have many 24-290mm and 15-40mm Optimo zooms. They are the most widely preferred lenses due to their high dependability and ease of use. We are the first to introduce digital cameras in the country. This shows we are well aware and ahead of time. In my opinion, web based - OTT (Over The Top services) platform projects will rule the market demand. »

Gowri Shankar

Prime Focus India General Manager

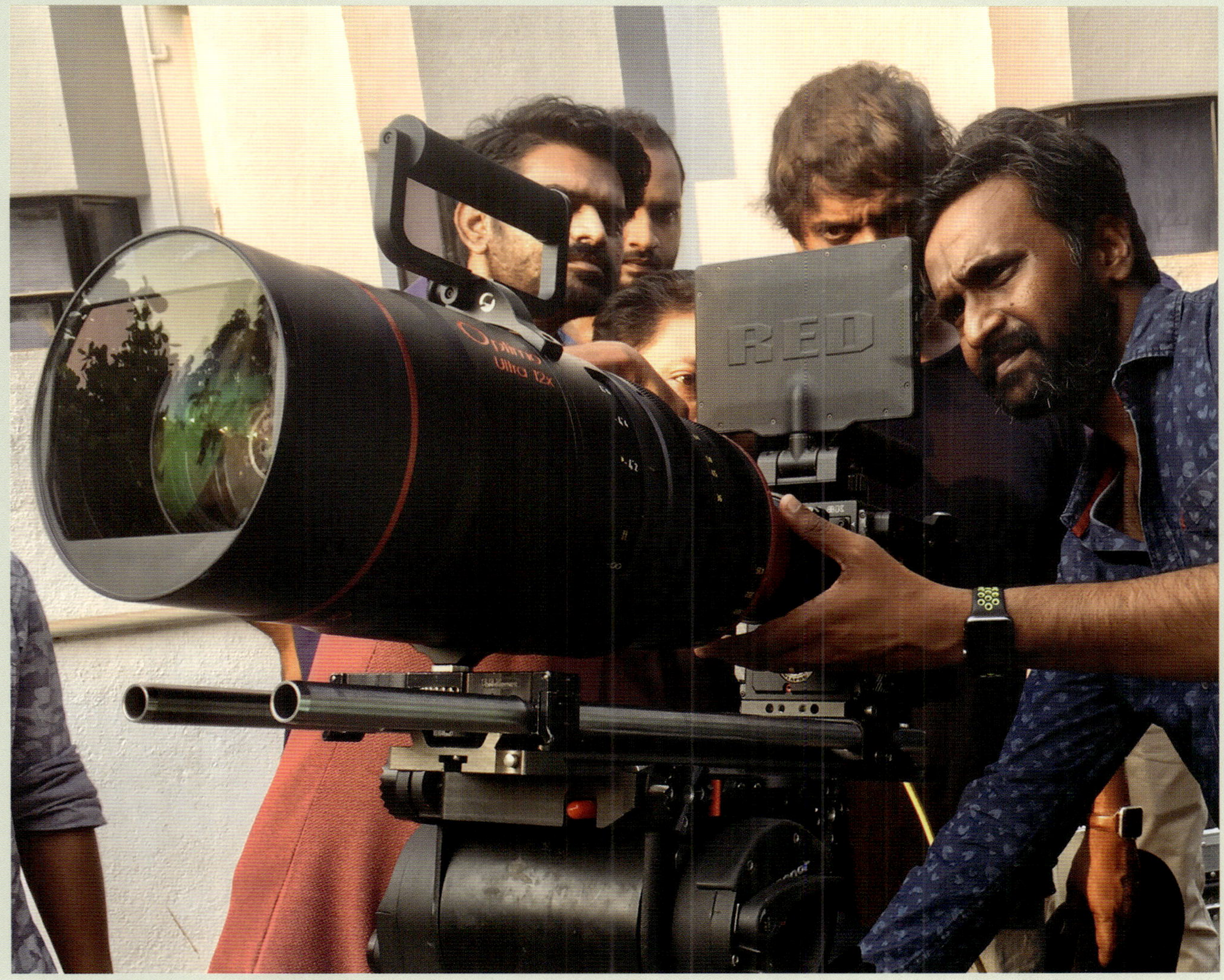

Cinematographer Manoj Paramahamsa and the new Optimo Ultra 12x. He is the first owner of a Red Monstro camera in India with an impressive filmography of high budget blockbusters, among them, *Vinnaithaandi Varuvaayaa "VTV"* (2010), *Race Gurram* (2014), *Dhruva Natchathiram* (2018), and *Amour* to come in 2020. Photo by Jean-Yves le Poulain.

« Their products are exceptional and capable of meeting the needs of the most demanding cinematographers and producers.
Angénieux lenses are very reliable and we can rent them with great confidence.
We know that these lenses are a good choice and one that meets the needs of directors of photography.
Thanks to Angénieux, we have become bolder in our procurement strategy. In terms of business development, these days, almost every production team would request for Angénieux zooms. We have encountered low-budget production teams that can only afford to rent one camera but still insist on the Optimo 24-290 or 28-340.
As regards equipment, I would say that the profession in China requires formalization and prioritization. The more we interact with cinematographers, the more we discover that they constantly want better equipment and more complete accessories. In the past, Chinese cinematographers and production teams did not have many requirements. But as they seek improvement and perfection, they have come to adopt a more holistic perspective when choosing their equipment. It is important that we are ready to meet their needs. We even adapt our products quite often to meet specific filming requirements.
They insist on authentic products. In China's equipment industry, there is no lack of imitations. Initially, people did not understand the difference but imitations are made cheaply and, more often than not, they create problems on set.
Production categories that are currently in the high-growth stage are High-end TV series. Also, because China is producing more well-made, Hollywood-style blockbusters, investments in movie production are steadily increasing. This leads to better movies.
This year, we participated in two shows, *Emperor's Conquest* and *The Legend and tThe Hero*. The equipment used by these two productions alone was equivalent to that of 20 to 30 regular productions. The team was very professional and very knowledgeable about equipment. Naturally, they had very high expectations. We were very happy to work with and serve them. I think this marks the start of a new turning point. Everyone says that Hollywood movies have an indelible place in people's hearts, but it probably won't take long for Chinese cinema to achieve this and more. Against this backdrop, a major driving force will be the industrialisation and scalability of the industry, in addition to formalisation. Personally, I would like to see people and culture movies that are shot and packaged in a more commercial manner. It is about taking a good story and telling it in a commercially appealing way. Movies that have succeeded with such a narrative would include *Wolf Warrior* and *Dying to Survive*. »

Yang He

Cinerent's General Manager

« To describe the Angénieux brand in one word, I would say 'quality'. As a cinematographer, when I am selecting zooms for a movie production, Angénieux is my first choice. Whether in terms of picture quality or stability, it leaves cinematographers with no complaints. Quality is also evidenced through the construction and technology of its lenses.

For a rental house that is just starting out, it is important to establish a foothold in the industry quickly. We must first analyse market needs and determine our development direction. This is my take — China's TV industry is at a turning point; TV production has to meet the needs of viewers and viewers have increasingly high expectations of picture quality. Every Angénieux lens is designed to serve the picture, which is why we chose to purchase more Angénieux zooms. This allows us to bring better quality equipment to the TV production sector, while creating more business opportunities for ourselves. With the collaboration of Jebsen team and the support of Angénieux, we were able to bring in the full set of Angénieux lenses in the shortest amount of time. This helped strengthen our value proposition and our brand image.

In China, web dramas are developing rapidly now. The risks associated with movies are becoming increasingly high; a mega-production may not necessarily achieve good box office results. But the advent of the Internet and web dramas have created new possibilities for people in this industry. Just the growing demand for short video clips, such as Taobao's video streaming site for sellers, has been able to keep many small rental houses afloat. Because they saw opportunities for growth and success, they entered the industry. As business volumes continue to grow, many of these companies may also grow in size and scale. »

Yihe Li

Flaming Mountains' General Manager

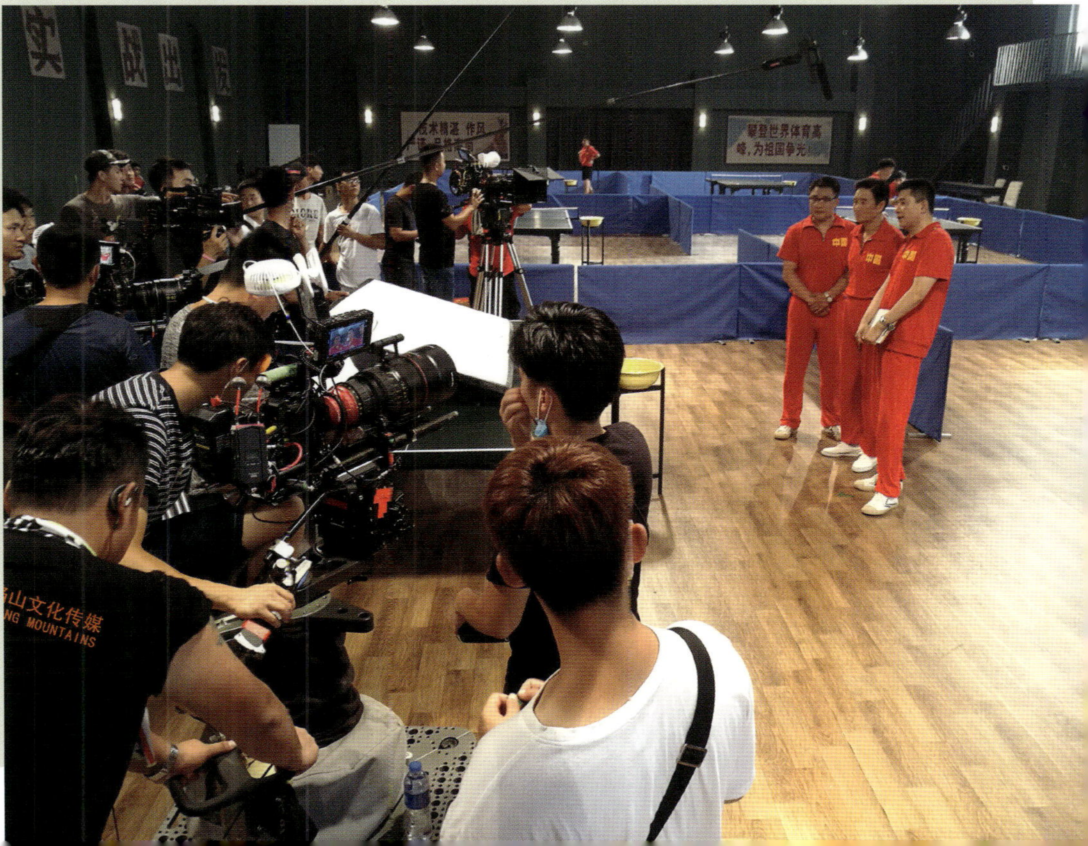

Optimo zoom during the shooting of the TV series *Go for Gold* directed by Zhang Qian, photographed by Yang Wei. Image provided by Flaming Mountains Co. Ltd.

« Angénieux is synonymous with premium cinematic lenses. When Scenefone first started out in 2008, we already purchased the Optimo zoom lenses. At the time, the Chinese cinematic industry was broadly divided into cinema and TV. The lenses available in the market were also highly limited. Angénieux lenses are a great product but inventory levels were low and rental costs were high. We saw this as an opportunity. In 2016, we purchased a large quantity of Angénieux lenses and invited cinematographers to trial the lenses. Today, 90% of local cinematographers would choose the Optimo zooms for their productions. Whether for TV, commercials, web productions or movies, the equipment and technology are now quite similar. Chinese cinematic equipment and technology are now on par with other countries'. Thanks to our domestic market, we are now in the golden age of Chinese film market. The success of even the best production ultimately depends on the public.

The prosperity of the market directly influences the development of the cinematic industry.
In the film age, Hollywood and Europe prospered. In the digital age, the volume and box office appeal of Chinese movies have far surpassed them. Propped up by the market, Chinese cinematic equipment and technology have been able to outshine many foreign rental houses. I believe that, in the future, Chinese cinema will become even more systematic, professional, and directed in its development. It is important that we continue to strengthen the 'soft power' of our technology.
I believe that Chinese cinema will continue to diversity. For example, the movie *Dying to Survive*, would not have been approved for screening two years ago. Art imitates life, and good movies need to be grounded in reality. That is why such movies are popular among consumers. »

Weifeng Sun

Scenefone's General Manager

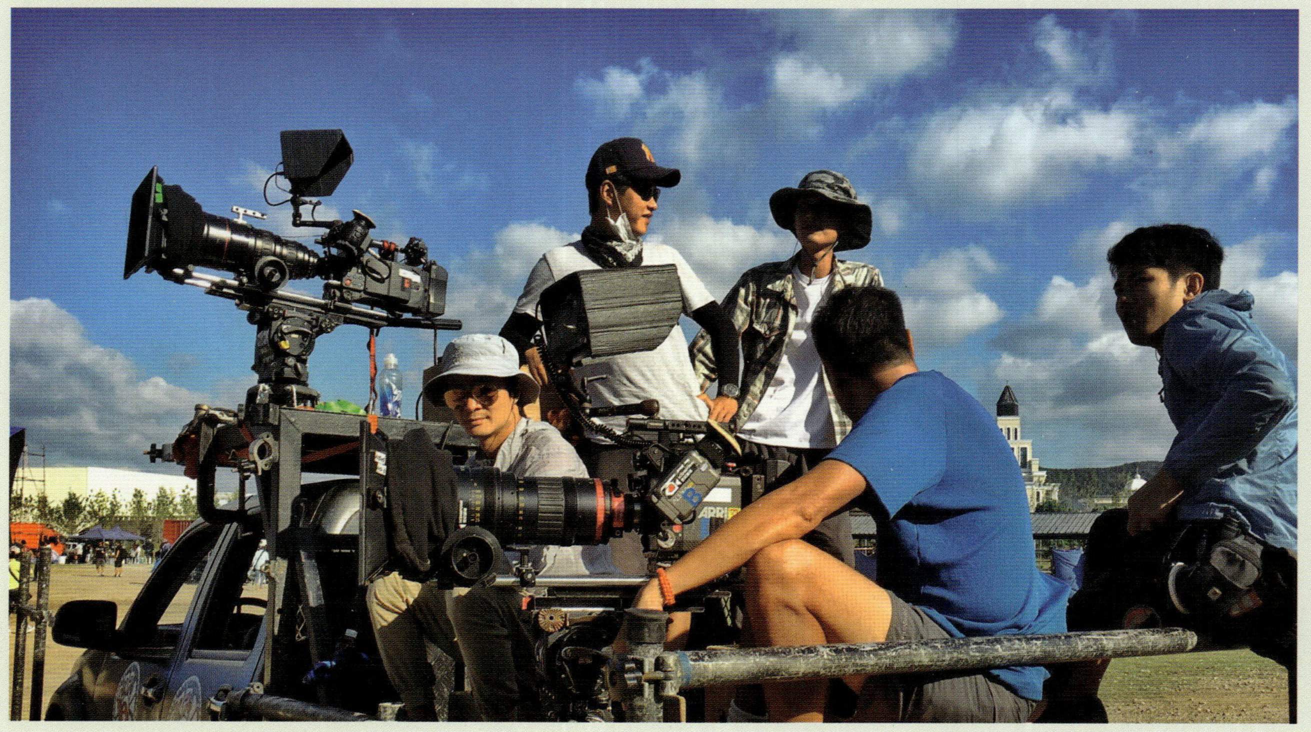

Optimo zooms during the shooting of *Dynasty Warriors* (2018), directed by Roy Chow Hin Yeung, photographed by Kenny Tse Chung To. Image provided by Scenefone Film Equipment Rental (Beijing) Co. Ltd.

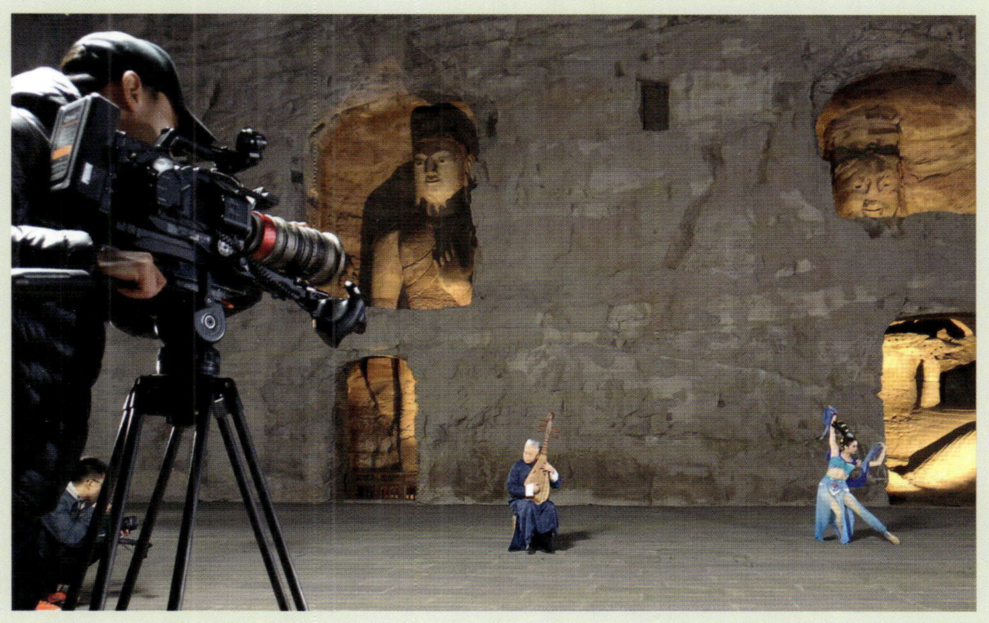

EZ zoom lens during the shooting of the show *Wonderful Night of the Museum*, directed by Fan Kai and photographed by Guan Peng. Image provided by Beijing Guang Da Ren He Film Equipment Co. Ltd.

« Everyone in this industry knows Angénieux but, for a long time, the brand was out of our reach. Then came the accessible Angénieux Type EZ. We were all quite surprised. Despite its affordable price tag, Type EZ does not compromise on quality. In fact, many cinematographers who have tried the lenses have nothing but praise for it. The strong brand name aside, what really matters to us is that Type EZ offers a practical focal length and at full frame too. Full frame lenses tend to have a longer lifespan, which is very valuable to rental houses.
Documentaries are a great format for Type EZ. The lenses are light and portable, and all you need are two zoom lenses. Easy! The image quality is also excellent — colours are good and images are rich and textured. The lenses are also a practical choice for TV shows and drama series.
For a rental house, the IRO technology helps us reduce costs and increase possibilities. A modular lens that can be used in two different ways is capable of meeting numerous production needs. From the user's viewpoint, these zooms make it possible to save time, which means saving money and labor.
The modular nature of the lenses, formats and mounts are the greatest advantages of this range. »

Cheng Wang

Guang Da Ren He Film Equipment's General Manager

The statements made by Yang He (Cinerent), Yihe Li (Flaming Mountains), Weifeng Sun (Scenefone), and Cheng Wang (Guang Da Ren He Film Equipment) are taken from interviews specially organized in China for this book by Jebsen Industrial, the distributor of Angénieux products in China.

New Technological Challenges

The general trend in filming equipment is toward greater lightness. One of the major challenges for a firm is to produce compact, lightweight lenses of high quality. For Angénieux, this means maintaining the level of excellence on which its reputation is grounded today. While Angénieux has undeniable advantages in this race, the watchword remains innovation.
Innovation is first and foremost the capacity to implement new solutions starting with the design of products. The use of aspheric, diffractive and freeform[1] lens elements makes it possible to design lenses of smaller size but makes calculation of the same terribly complex. The increased power of our computers makes it possible for designers to imagine systems that were still inconceivable just a few years ago. Today, assistance is provided by increasingly powerful tools of simulation, constantly optimized by our engineers, making it possible to ascertain the performance of these new elements digitally in specific conditions of utilization. It is possible, for example, to simulate and therefore adjust the reaction of a zoom exposed to extreme temperatures (from -20°C to +45°C) and different conditions of atmospheric pressure.
One of the major areas of innovation in which Angénieux is a forerunner, due to the company's historic and ground-breaking skills, is **wavefront coding***. This consists of taking into account all the elements involved in the design of a filming system: the lens, the camera and the digital processing in real time. This approach makes it possible to reduce the constraints and optimize the efficiency of a complete optical system.
Combination of all these new solutions will make it possible tomorrow to provide increasingly compact and lightweight products of ever-greater power to meet the needs of productions in which importance it attached to image quality.
Angénieux's mastery of the configuration of an image on a sensor, acquired over the years in response to the demands of film production, is recognized and regularly drawn upon in the Thales group in order to take part in the development of the optronic systems of the future for infantry and pilots, and especially communicating systems of augmented reality. The strength of Thales and its methodological and strategic support constitute a considerable asset.
Outside the sphere of Thales, Angénieux also takes part in research programs aimed at addressing the new challenges of the image. Angénieux is thus fully involved in the European REVE 5D project alongside other companies and research centers and in the development of devices for superimposing real and virtual images for both military and civilian applications. New connected goggles will enable infantry, for example, to visualize the position of strategic points in addition to their direct vision. By combining historical scenes with a real landscape, these goggles can also make visits to sites of cultural heritage more living and authentic. In the industrial sphere, personnel can be guided in the performance of tasks by helmets with visual display incorporated.
Angénieux is also involved, together with various industrial partners and universities, in the European research program EXIST aimed at improving the technology of CMOS sensors for 4K imaging. The broad range of different applications includes security and digital lifestyle as well as smart production, agriculture, health, and gas detection.

[1] A freeform lens element is produced from a spherical surface deformed with the aid of a polynomial XY or other equation. Divergence from the sphere is not rotational, as in the case of an aspheric lens element. These elements are used in HUD (Head-Up Display) systems for fighter pilots, for example, or in telescopes in which the principal mirror is off-axis.
They make it possible to correct the defects of off-center systems.
A diffractive lens element is produced from a spherical, aspheric or freeform surface through the addition of microstructures to modify the apparent index of the glass. They make chromatic corrections possible with no need for doublets and serve to lighten a system or to generate deviations of light rays without the use of a prism.

Perfect control over its lenses enables Angénieux to operate through wavefront coding*. Wavefront coding consists in creating a deliberately degraded image so as to correct or accentuate its parameters to better effect through digital processing.
This will make it possible to produce less expensive lenses capable of providing images perfectly in line with the aims pursued, artistic or otherwise.

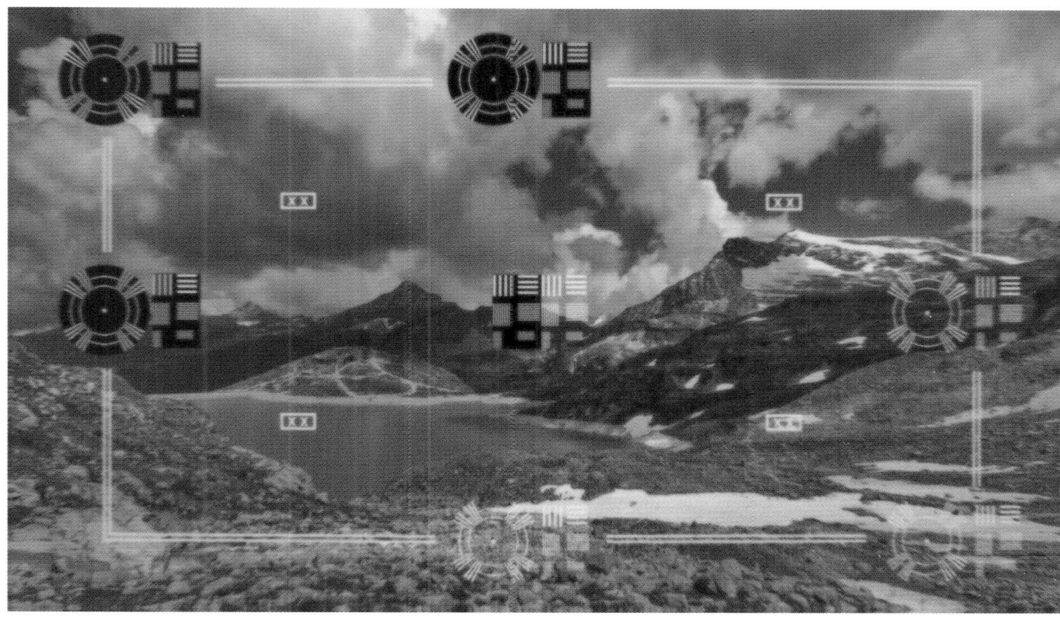

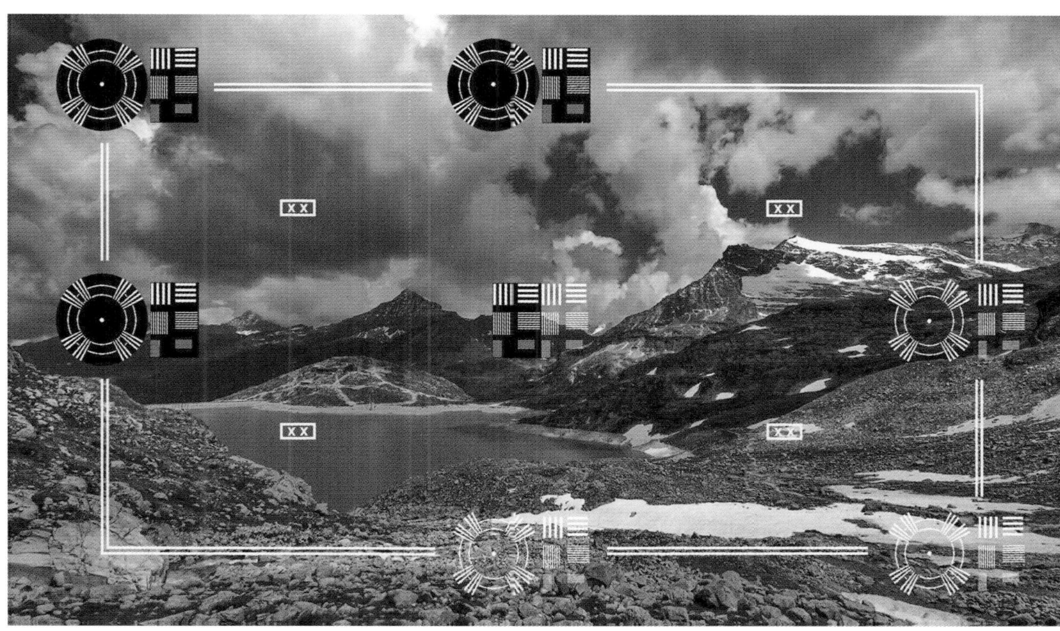

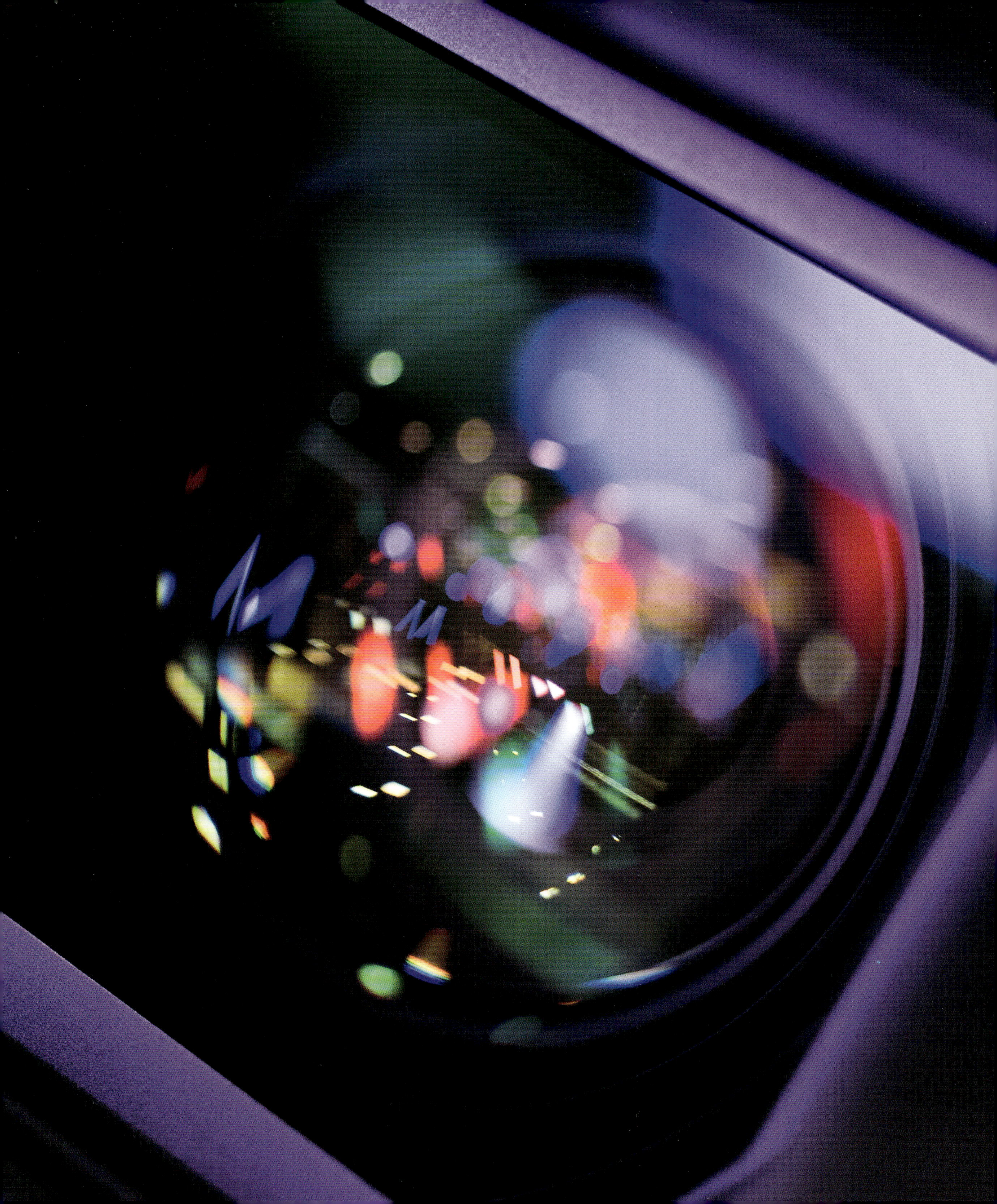

Tomorrow …

« Because its natural function is to stimulate our senses, memory, imagination, and emotions, cinema remains the place of all new requirements. And in this place, in a permanent race for innovation, at the forefront of know-how and technology, Pierre Angénieux remains a source of inspiration.
It is the very essence of our ambition that spurs us to go on supplying all our customers and all the users of our products unfailingly with the best technological solutions. These solutions must meet the requirements of their professions, also in economic terms, as well as the still more demanding requirements connected with the practice of their art. Because it is our passion and because we are faithful to the legacy of Pierre Angénieux.

We also know that nothing can replace the talent of directors of photography. Today, with the increasing importance of the digital realm and the impact of the associated transformations, they are the ones called upon to master and adapt these new technological tools, just as all their predecessors have already done in the history of cinema. They are women and men of art and they are also great technicians. By means of their talent, they can make the contribution of technology invisible, concealing all trace of effort so that only pleasure, involvement, and emotion remain on viewing the screen. »

Emmanuel Sprauel

President of the Angénieux brand

Thank you @adamvideoman (Adam Plowden – video production, based in Surrey UK) follower of @angenieuxlenses Instagram who published this beautiful picture of the Optimo 30-72 A2S.

Acknowledgements

We wish to express our gratitude to all those involved in making this work a reality.

Our thanks go first of all

to the curators of the exhibition *ZOOM! Angénieux : changez d'optiques* (Saint-Étienne, Musée d'Art et d'Industrie, March–November 2017):

Nadine Besse
chief curator and former director of the Musée d'Art et d'Industrie

Éric Perrin
assistant curator for industrial heritage

Alain Renaud
philosopher

whose writings and iconographic research constituted an invaluable resource for the work

to all the team of the Musée, especially Marie-Caroline Janand, *director Musée d'Art et d'Industrie* and Christelle Chandy Duranton, *administrator Musée d'Art et d'Industrie*

to Gaël Perdriau, *mayor of Saint-Étienne and president of Saint-Étienne Métropole*, and Marc Chassaubéné, *head of the Saint-Étienne department of culture*.

We also thank

the directors of the various cinematographic institutions who contributed to the work:

Thierry Frémaux, *delegate general of the Cannes Film Festival and director of the Institut Lumière*

Laurent Mannoni, *scientific director and curator for techniques of the Cinémathèque Française*

the presidents of associations of cinematographers

Pierre-William Glenn, *former president of the Association Française des directeurs de la photographie Cinématographique (AFC) and of the Commission supérieure et technique de l'image et du son (CST), joint-director of the Fémis Image department*

Sunny Joseph, *president of the Indian Society of Cinematographers (ISC)*

Kees Van Oostrum, *president of the American Society of Cinematographers (ASC)*

Gilles Porte, *president of AFC*

the managers of the Schools:

Nathalie Coste-Cerdan, *director of the Fémis, École Nationale Supérieure des Métiers de L'Image et du Son*

Vincent Lowy, *director of the École Nationale Supérieure Louis-Lumière*

Claude Mouriéras, *filmmaker and director of the CinéFabrique, École Nationale Supérieure de Cinéma et de Multimédia*

Mu De Yuan, *former director of the cinema department of the Beijing Film Academy (BFA), former director of the office for academic affairs of the Beijing Film Academy, president of the Chinese Society of Cinematographers*

the editorial contributors:

Jon Fauer, *director of photography and founder of the magazine Film and Digital Times*

Pascal Martin, *university professor at ENS Louis-Lumière and lecturer at La Fémis*

Priska Morrissey, *lecturer on film studies, Université Rennes 2, historian of professions and techniques in the film industry*

Marc Salomon, *historian of directors of photography*

THALES

Angénieux®

Ville de Saint-Étienne
L'expérience design

MUSÉE D'ART ET D'INDUSTRIE
SAINT-ÉTIENNE

all the members of the cinema community, directors of photography, partners, equipment rental houses and production companies, especially:

Christopher Doyle, Ed Lachman, Peter Suschitzky, Nancy Schreiber, Laurent Dailland, Cécile Zhang, Eponine Momenceau, Savita Singh, Govind Nihalani, Aseem Mishra, Anil Mehta, Manoj Paramahamsa, Slawomir Idziak, Vladislav Opelyants, Dariusz Wolski, Toby Oliver, Guillermo Navarro, Rodney Charters, Renan Ozturk, Manuel Claro, Checco Varese, Tom Bergmann, Gustavo Biazzi, Frank Dewaele, Luk Malu, Brett Turnbull, Shu Hirayama, José Alkon, Claire Mathon, Guillaume Schiffman, Eric Guichard, Jean-Marie Dreujou, Pascal Lagriffoul, Antoine Monod, Jérôme Dolbert, Denny Clairmont, Otto Nemenz, Fritz Heinzle, Brad Wilson, Russell Allen, Gowri Shankar, Pat Longman, Adam Plowden, Aaron Brown, Teresa Wong, Jack CK Yu, Anthony WH Chan, Nikolay Senin

and Agustina Llambi Campbell, Angelika Albrecht, Clare Spencer, Meredith Emmanuel, Ken Hui, Victoria Reese, Margriet Spikman, Mathilde Annaud, Xavier Castano, Tania Khali and Zhang Peng

the former Angénieux personnel whose writings, collections and memories provided invaluable assistance:

Bernard Angénieux and Jean Moret, *former directors*

André Masson, *former director*

Gérard Corbasson, *former director*

Bernard Philibert, Jacques Durand, Guy Gauthier, René Michalet, *former managers*

Jacques Debize, *optical engineer with the company for nearly forty years from 1969 to 2007, a precious scientific collaborator* who passed away recently, without seeing this work completed, and has a special place in our thoughts

the numerous Angénieux fans and collectors who placed their knowledge and/or collections at our disposal

in particular, the Masson family, André and Olivier Granet, Bernard Tichit, Patrice-Hervé Pont, Arnaud Saudax, Jean-Paul Bouchet, Sylvain Tomasini, the ACHDR (Association du Centre Historique de la Diffusion Radiophonique), the town of Saint-Héand (Jean-Marc Thelisson and Lucie Thomas), the Fondation Pathé Seydoux, Jacques Grandclaude, Georges Mourier, Jean-Loup Princelle, the École Supérieure d'Optique and the Notre-Dame de Valbenoîte school.

Finally, we thank the following for their support all through the creation of this book:

the senior management of the Groupe Thales and its department of communication,

the senior management and department of communication of the GBU, Thales LAS France,

the senior management of the BL OME and its marketing & strategy management, its sales management, and its communication management

the management of the Cinema sector of Thales LAS France SAS.

Among the teams of the plant in Saint-Héand, the following are thanked for their particular contribution to the success of this initiative:

Edith Bertrand (project leader and publishing supervisor), Jean-Marc Bouchut, Jean-Claude Guérin and Yasuhiko Mikami, Bernard Féraud, Jean-Yves Le Poulain and Dominique Rouchon, Grégoire Bastin, Thierry Bonfils, Alain Brunel, Emmanuel Buhon, Jacques Bouley, Jean Barthomeuf, Benoit Brismontier, Michel Breure, Bruno Coumert, Pierre Dugay, Christine Durand-Malavieille, Jean-Luc Espie, Paul Fuller, Frank Goh, Jérémie Krymalowski, Nick Hayes, Catherine Hun, Bernadette Jourjon, Laurent Montagne, Denise Raymond, Christophe Reboulet, Davy Terzian, Philippe Vignal, without forgetting the Club Photo Angénieux and in particular Pierre-Etienne Debard and Nicolas Grognet.

Finally, our thanks to all those not mentioned above who contributed in various ways to the success of this work.

We are also deeply grateful to all the people who have written pages in the history of Angénieux, who are doing so today, and who will continue to do so tomorrow.

Silvana Editoriale

Direction
Dario Cimorelli

Art Director
Giacomo Merli

Editorial Coordinator
Sergio Di Stefano

Copy Editor
Noa Strada

English translation from French
Paul Metcalfe and Meaghan Toohey
for Scriptum, Rome

Layout
Mirco Ameglio

Production Coordinator
Antonio Micelli

Editorial Assistant
Ondina Granato

Photo Editor
Alessandra Olivari, Silvia Sala

Press Office
Lidia Masolini, press@silvanaeditoriale.it

ISBN 9788836638598

Silvana Editoriale S.p.A.
via dei Lavoratori, 78
20092 Cinisello Balsamo, Milano
tel. 02 453 951 01
fax 02 453 951 51
www.silvanaeditoriale.it

Reproductions, printing and binding
in Italy
Printed March 2019

Available through ARTBOOK | D.A.P.
155 Sixth Avenue, 2nd Floor, New York, N.Y. 10013
Tel: (212) 627-1999 Fax: (212) 627-9484

All the photo credits are in the captions accompanying the photos except

cover
Angénieux Optimo Ultra 12x at Roger Waters' concert in Amsterdam at the Ziggodome, June 19, 2018. Us & Them Productions / Grasshopper Films
©Yke Erkens from Cam-a-lot (rental house in Netherlands)

p. 14 Class of 1928–29 at the École supérieure d'optique (École supérieure d'optique collection)

p. 17 Portrait of Pierre Angénieux (Angénieux archives)

p. 28 Bell & Howell BH70 camera, turret with three Angénieux lenses (Angénieux archives, photo by Marc Salomon)

p. 31 6x9 folding camera with a
« P. Angénieux Ing. Paris Anastigmat 1 :4,5F=105 n° 11 1938 » zoom
(Patrice-Hervé Pont collection, photo by Eric Perrin)

p. 56 The first 16mm zoom (Angénieux archives)

p. 59 Pierre Angénieux with the Saint-Héand factory in the backdrop (Angénieux archives)

p. 102 *Apollo 11* mission, July 20, 1969 © NASA

p. 105 August 28, 1963, the 12-120 zoom (standing camera) films the famous speech "I have a dream" delivered by Martin Luther King during the March on Washington for Jobs and Freedom. (photo by AFP / Getty Images)

p. 114 Television studio (Angénieux archives)

p. 170 Lens polishing (Angénieux archives, photo by Nicolas Grognet)

p. 173 Focal engraving (Angénieux archives)

p. 192 Cinematographer Renan Oztruck and the Optimo Style 16-40 (photo by Robert Thoren)

p. 195 Optimo Style 16-40 in *Great Wide Open* directed by Jared Leto. *Great Wide Open*, a Paradox/Budweiser production, is a series of five short movies celebrating America's National Parks and the incredible adventures people can experience.
(DP Renan Ozturk, photo by Matt Irving)

p. 224 Scientific and Engineering Award, 2009 (Angénieux archives, photo AMPAS)

p. 227 Cannes 2013: tribute to Philippe Rousselot (photo by Jon Fauer)

p. 235 Cinec Award 2018 (photo by Albrecht GmbH)

p. 237 Photo by Pauline Maillet

p. 248 and p. 251 Angénieux Optimo Ultra 12x at Roger Waters' concert in Amsterdam at the Ziggodome, June 19, 2018. Us & Them Productions / Grasshopper Films
©Yke Erkens from Cam-a-lot (rental house in Netherlands)